Models in Ecosystem Science

Models in Ecosystem Science

EDITORS

Charles D. Canham

Jonathan J. Cole

William K. Lauenroth

PRINCETON UNIVERSITY PRESS

PRINCETON AND OXFORD

Copyright © 2003 by Princeton University Press

Published by Princeton University Press, 41 William Street, Princeton,
New Jersey 08540

In the United Kingdom: Princeton University Press, 3 Market Place,
Woodstock, Oxfordshire OX20 1SY

All Rights Reserved

Library of Congress Cataloging-in-Publication Data
Models in ecosystem science / Charles D. Canham, Jonathan J. Cole, William
    K. Lauenroth, editors.
        p. cm.
    Includes bibliographical references and index.
    ISBN: 0-691-09288-5 (cl. : alk. paper)—ISBN 0-691-09289-3 (pbk. : alk.
paper)
        1. Ecology—Mathematical models. I. Canham, Charles Draper William,
    1954– II. Cole, Jonathan J. III. Lauenroth, William K.

    QH541.15.M3M74 2003
    577'.01'5118—dc21   2003043323

The publishers would like to acknowledge the editors of this volume for
providing the camera-ready copy from which the book was printed.

British Library Cataloging-in-Publication Data is available

The book has been composed in Times New Roman

Printed on acid-free paper.∞

www.pupress.princeton.edu

Printed in the United States of America

1 3 5 7 9 10 8 6 4 2

Contents

Part III. The Role of Models in Environmental Policy and Management

Acknowledgments

This book is a product of the ninth Cary Conference, held in May 2001 at the Institute of Ecosystem Studies, Millbrook, N.Y. The biennial Cary Conference series originated in 1985 and is designed to bring together a diverse group of participants to discuss topics of broad interest to the field of ecosystem science. Cary Conference IX, which explores the role of quantitative models in ecosystem science, is thematically related to two prior Cary Conferences, which explored the roles of long-term studies, and comparative approaches (CC II and CC III, respectively) in ecology. Although the topic of quantitative models has come up in many of the past conferences, we felt that it was the right time to devote an entire conference to this rich area. The full list of past Cary Conferences and the books that came from them can be seen at the website of the Institute of Ecosystem Studies (www.ecostudies.org).

Many people contributed to the planning of the conference. We would like to thank our steering committee—John Aber, Ingrid Burke, Stephen Carpenter, Carlos Duarte, Gene Likens, Pamela Matson, José Paruelo, Hank Shugart, and Doris Soto—for their help throughout the process. Our local IES advisory committee of Nina Caraco, Peter Groffman, Kathleen Hogan, Gary Lovett, Michael Pace, David Strayer, and Kathleen Weathers generously shared their advice and experiences from planning previous conferences.

The conference was supported by grants from the National Science Foundation, the U.S. Department of Energy, the U.S. Environmental Protection Agency, and the USDA Forest Service. We thank all of them for their financial support, their feedback on conference planning, and in many cases, for their participation in the conference. Both financial and logistical support were provided by the Institute of Ecosystem Studies. We thank the Director, Gene Likens, for his vision in creating the Cary Conference series and for his strong support for the conferences over the years. We also thank American Airlines and Delta Airlines for providing special reduced airfares for conference attendees.

Bringing together over a hundred people for three days of presentations and discussions is an enormous task. Claudia Rosen did a superb job as conference coordinator, handling everything from complex travel arrangements to ensuring that attendees were well housed and well fed. Keeping over a hundred academics to a demanding conference schedule is no mean feat, but she pulled it off with style and wit. She was ably assisted by a team of graduate students and IES staff, including Darren Bade, Emily Bernhardt, Feike Dijkstra, Jennifer Funk, Chloe Keefer, Kate Macneale, Michael Papaik, Pamela Templer, and Justin Wright. Tanya Rios did her usual superb job preparing proposals and conference materials.

We would particularly like to thank the authors for their insights and diligence in translating the conference into a form that can be shared with the rest of the scientific community. Tanya Rios deserves special thanks for handling all of the manuscripts, from first drafts through to camera-ready copy. We would also like to thank Sam Elworthy at Princeton University Press for his encouragement and patience.

Charles Canham
Jon Cole
Bill Lauenroth

Contributors and Participants

John D. Aber, Natural Resources CSRC, 445 Morse Hall, University of New Hampshire, Durham, NH 03824, USA

Edward A. Ames, Mary Flagler Cary Charitable Trust, New York, NY 10168, USA

Jeffrey S. Amthor, U.S. Department of Energy, SC-74, 19901 Germantown Road, Germantown, MD 20874-1290, USA

Robert A. Armstrong, Marine Sciences Research Center, 145 Discovery Hall, State University of New York, Stony Brook, NY 11794-5000, USA

Dominique Bachelet, Bioresource Engineering, Oregon State University, Corvallis, OR 97331, USA

Darren L. Bade, Center for Limnology, 680 North Park Street, University of Wisconsin, Madison, WI 53706, USA

A. John Bailer, Department of Mathematics, Miami University, Oxford, OH 45056, USA

Jill S. Baron, Natural Resource Ecology Lab, Colorado State University, Fort Collins, CO 80523-1499, USA

Alan R. Berkowitz, Institute of Ecosystem Studies, Box R, Millbrook, NY 12545, USA

Emily S. Bernhardt, Department of Ecology and Evolutionary Biology, Corson Hall, Cornell University, Ithaca, NY 14853 and Institute of Ecosystem Studies, Box AB, Millbrook, NY 12545, USA

Joyce K. Berry, 101 Natural Resources, Colorado State University, Fort Collins, CO 80523, USA

Seth W. Bigelow, Institute of Ecosystem Studies, Box AB, Millbrook, NY 12545, USA

Suzanne P. Bird, College of Natural Resources, Colorado State University, Fort Collins, CO 80523, USA

Harald K.M. Bugmann, Mountain Forest Ecology, Department of Forest Sciences, Swiss Federal Institute of Technology, CH-8029 Zurich, Switzerland

Ingrid C. Burke, Forest Sciences, 206 Natural Resources, Colorado State University, Fort Collins, CO 80523, USA

Mary L. Cadenasso, Institute of Ecosystem Studies, Box AB, Millbrook, NY 12545, USA

Charles D. Canham, Institute of Ecosystem Studies, Box AB, Millbrook, NY 12545, USA

Nina F. Caraco, Institute of Ecosystem Studies, Box AB, Millbrook, NY 12545, USA

Zoe G. Cardon, Department of Ecology and Evolutionary Biology, University of Connecticut, Storrs, CT 06269, USA

Stephen R. Carpenter, Center for Limnology, 680 North Park Street, University of Wisconsin, Madison, WI 53706, USA

Jonathan J. Cole, Institute of Ecosystem Studies, Box AB, Millbrook, NY 12545, USA

Jeffrey D. Corbin, Department of Integrative Biology, University of California, Berkeley, CA 94720, USA

Kathryn L. Cottingham, Department of Biological Sciences, Dartmouth College, Hanover, NH 03755, USA

William S. Currie, University of Maryland Center for Environmental Sciences, Appalachian Laboratory, Frostburg, MD 21532-2307, USA

Hélène Cyr, Department of Zoology, Ramsay Wright Zoological Labs, University of Toronto, 25 Harbord Street, Toronto, ON M5S 3G5, Canada

Carla M. D'Antonio, Department of Integrative Biology, University of California, Berkeley, CA 94720, USA

Donald L. DeAngelis, U.S. Geological Survey, Biological Resources Division, and University of Miami, Department of Biology, P.O. Box 249118, Coral Gables, FL 33124, USA

C. Lisa Dent, Center for Limnology, 680 North Park Street, University of Wisconsin, Madison, WI 53706, USA

Feike A. Dijkstra, Wageningen University, Netherlands, and Institute of Ecosystem Studies, Box AB, Millbrook, NY 12545, USA

Carlos M. Duarte, Instituto Mediterráneo de Estudios Avanzados, CS IC-UiB, c/Miquel Marqués, 21, 07190 Esporles, Mallorca, Spain

Bridgett A. Emmett, Centre for Ecology and Hydrology, Bangor Orton Building, Deiniol Road, University of Wales, Bangor Gwynedd LL57 2UP, UK

Howard E. Epstein, Environmental Services Department, University of Virginia, Charlottesville, VA 22901-4123, USA

Holly A. Ewing, Institute of Ecosystem Studies, Box AB, Millbrook, NY 12545, USA

Stuart E.G. Findlay, Institute of Ecosystem Studies, Box AB, Millbrook, NY 12545, USA

Ross D. Fitzhugh, Institute of Ecosystem Studies, Box AB, Millbrook, NY 12545, USA

Marie-Josée Fortin, School of Resource and Environmental Management, Simon Fraser University, Burnaby, BC V5A 1S6, Canada

Jennifer Funk, State University of New York - Stony Brook and Institute of Ecosystem Studies, Box AB, Millbrook, NY 12545, USA

Robert H. Gardner, University of Maryland Center for Environmental Science, Appalachian Laboratory, Frostburg, MD 21531, USA

Sarah E. Gergel, National Center for Ecological Analysis and Synthesis, Santa Barbara, CA 93101-3351, USA

Anne E. Giblin, The Ecosystems Center, Marine Biological Laboratory, Woods Hole, MA 02543-0105, USA

Brett J. Goodwin, Institute of Ecosystem Studies, Box AB, Millbrook, NY 12545, USA

Peter M. Groffman, Institute of Ecosystem Studies, Box AB, Millbrook, NY 12545, USA

Louis J. Gross, Ecology and Evolutionary Biology, 569 Dabney/Buehler, 639 Science and Engineering Research Facility, University of Tennessee, Knoxville, TN 37996, USA

Lars Håkanson, Department of Earth Sciences, Uppsala University, Villav., 16, S-752-36, Uppsala, Sweden

Sonia A. Hall, College of Natural Resources, Colorado State University, Fort Collin, CO 80523, USA

Graham P. Harris, SIRO Land and Water, GPO Box 1666, Canberra, ACT 2601, Australia

Kathleen Hogan, Institute of Ecosystem Studies, Box R, Millbrook, NY 12545, USA

Lorraine L. Janus, New York City Department of Environmental Protection, 465 Columbus Avenue, Valhalla, NY 10595, USA

Jennifer C. Jenkins, USDA Forest Service, Northern Global Change Program, Burlington, VT 05402, USA

Clive G. Jones, Institute of Ecosystem Studies, Box AB, Millbrook, NY 12545, USA

K. Bruce Jones, National Exposure Research Lab/ORD, Environmental Protection Agency, Las Vegas, NV 89193-3478, USA

Linda A. Joyce, USDA Forest Service Rocky Mountain Research Station, 240 West Prospect, Fort Collins, CO 80521, USA

Jason P. Kaye, Department of Forest Sciences, Colorado State University, Fort Collins, CO 80523, USA

Chloe Keefer, Institute of Ecosystem Studies, Box AB, Millbrook, NY 12545, USA

Felicia Keesing, Department of Biology, Box 5000, Bard College, Annandale-on-Hudson, NY 12504, USA

Ann P. Kinzig, Department of Biology, PO Box 871501, Arizona State University, Tempe, AZ 85287, USA

James F. Kitchell, Limnology Laboratory, 680 North Park Street, University of Wisconsin, Madison, WI 53706, USA

William K. Lauenroth, Department Forest, Rangeland, and Watershed Stewardship, Colorado State University, Fort Collins, CO 80523, USA

Gene E. Likens, Institute of Ecosystem Studies, Box AB, Millbrook, NY 12545, USA

Kathleen M. LoGiudice, Institute of Ecosystem Studies, Box AB, Millbrook, NY 12545, USA

Orie L. Loucks, Department of Zoology, Miami University, Oxford, OH 45056, USA

Gary M. Lovett, Institute of Ecosystem Studies, Box AB, Millbrook, NY 12545, USA

Kate Macneale, Department of Ecology and Evolutionary Biology, Corson
    Hall, Cornell University, Ithaca, NY 14853, and Institute of Ecosystem
    Studies, Box AB, Millbrook, NY 12545, USA
Roxane J. Maranger, Institute of Ecosystem Studies, Box AB, Millbrook, NY
    12545, USA
Neo D. Martinez, Romberg Tiburon Center, Department of Biology, San
    Francisco State University, Tiburon, CA 94920, USA
Rebecca L. McCulley, College of Natural Resources, Colorado State University,
    Fort Collins, CO 80523, USA
Cheryl L. McCutchan, Institute of Ecosystem Studies, Millbrook, NY 12545,
    USA.
James H. McCutchan Jr., Institute of Ecosystem Studies, Box AB, Millbrook,
    NY 12545, USA
Wolf M. Mooij, Netherlands Institute of Ecology, Centre for Limnology,
    Rijksstraatweg 6, 3631 AC Nieuwersluis, The Netherlands
Ransom A. Myers, Department of Biology, Dalhousie University, Halifax, NS
    B3H 4J1, Canada
Naomi Oreskes, History-0104, 9500 Gilman Drive, University of California at
    San Diego, LaJolla, CA 92093-0104, USA
James T. Oris, Department of Zoology, Miami University, Oxford, OH 45056,
    USA
Richard S. Ostfeld, Institute of Ecosystem Studies, Box AB, Millbrook, NY
    12545, USA
Michael L. Pace, Institute of Ecosystem Studies, Box AB, Millbrook, NY
    12545, USA
Michael J. Papaik, University of Massachusetts and Institute of Ecosystem
    Studies, Box AB, Millbrook, NY 12545, USA
William J. Parton, Natural Resource Ecology Lab, B233 Natural and
    Environmental Sciences, Colorado State University, Fort Collins, CO
    80523-1499, USA
José M. Paruelo, Depto. Ecología and IFEVA, Facultad de Agronomia,
    Universidad de Buenos Aires, Buenos Aires, Argentina
John Pastor, Department of Biology and the Natural Resources Research
    Institute, Center for Water and Environment, University of Minnesota,
    Duluth, MN 55811, USA
Debra P.C. Peters (nee Coffin), USDA-ARS, Jornada Experimental Range,
    New Mexico State University, Las Cruces, NM 88003-0003, USA
Steward T.A. Pickett, Institute of Ecosystem Studies, Box AB, Millbrook, NY
    12545, USA
Roger A. Pielke Jr., Environmental and Societal Impacts Group, National
    Center for Atmospheric Research, Boulder, CO 80301, and University of
    Colorado/CIRES, Center for Science and Technology Policy Research, 133
    Grand Avenue, Campus Box 488, Boulder, CO 80309-0488, USA
Edward B. Rastetter, The Ecosystems Center, Marine Biological Laboratory,
    Woods Hole, MA 02543-0105, USA

Kenneth H. Reckhow, Box 90328, Duke University, Durham, NC 27708, USA

Brian Reynolds, Centre for Ecology and Hydrology, Bangor Orton Building, Deiniol Road, University of Wales, Bangor Gwynedd LL57 2UP, UK

Paul G. Risser, Office of the President, Oregon State University, Corvallis, OR 97331-2126, USA

William Robertson, Andrew W. Mellon Foundation, 140 East 62nd Street, New York, NY 10021, USA

Steven W. Running, Numerical Terradynamic Simulation Group, School of Forestry, University of Montana, Missoula, MT 59812, USA

Don Scavia, National Centers for Coastal Ocean Science, National Oceanic and Atmospheric Administration, Room 13508, 1205 East West Highway, Silver Spring, MD 20910, USA

Kenneth A. Schmidt, Department of Biology, Williams College, Williamstown, MA 01267, USA

Gericke L. Sommerville, College of Natural Resources, Colorado State University, Fort Collins, CO 80523, USA

Doris Soto, Facultad de Pesquerias y Oceanografia, Laboratoria de Ecologia Acuatica, Universidad Austral de Chile, Casilla 1327, Puerto Montt, Chile

Robert S. Stelzer, Institute of Ecosystem Studies, Box AB, Millbrook, NY 12545, USA

David L. Strayer, Institute of Ecosystem Studies, Box AB, Millbrook, NY 12545, USA

Lee M. Talbot, Program on Environmental Science and Policy, 656 Chilton Court, George Mason University, McLean, VA 22101-4422, USA

Pamela H. Templer, Department of Ecology and Evolutionary Biology, Corson Hall, Cornell University, Ithaca, NY 14853, and Institute of Ecosystem Studies, Box AB, Millbrook, NY 12545, USA

Monica G. Turner, Department of Zoology, University of Wisconsin, Madison, WI 53706, USA

Dean L. Urban, Landscape Ecology Laboratory, Nicholas School of the Environment, School of the Environment and Earth Sciences, Duke University, Durham, NC 27708-0328, USA

Joseph J. Vallino, The Ecosystems Center, Marine Biological Laboratory, Woods Hole, MA 02543, USA

Rodney T. Venterea, Institute of Ecosystem Studies, Box AB, Millbrook, NY 12545, USA

Kathleen C. Weathers, Institute of Ecosystem Studies, Box AB, Millbrook, NY 12545, USA

Thorsten Wiegand, Center for Environmental Research, Department of Ecological Modeling, UFZ Centre for Environmental Research Leipzig-Halle, D-04318 Leipzig, Germany

Justin P. Wright, Department of Ecology and Evolutionary Biology, Corson Hall, Cornell University, Ithaca, NY 14853, and Institute of Ecosystem Studies, Box AB, Millbrook, NY 12545, USA

Models in Ecosystem Science

1

Models in Ecosystem Science

Charles D. Canham, Jonathan J. Cole, and
William K. Lauenroth

The Role of Modeling in Ecosystem Science

Quantitative models play an important role in all of the sciences. Models can range from simple regression equations and analytical models to complex numerical simulations. Their roles can vary from exploration and problem formulation to sophisticated predictions upon which management decisions are based. In the most basic sense, models express the logical consequences of a set of hypotheses and generate predictions (in the strictest sense) that can be compared with observations in the quest to falsify those hypotheses. Beyond this, the definitions and utility of models become controversial, and further discussion of models usually sparks an often intense debate over a host of both practical and philosophical issues. The ninth Cary Conference, held May 1–3, 2001, at the Institute of Ecosystem Studies, was designed to explore those debates, and to evaluate the current status and role of modeling in ecosystem science.

Beyond their fundamental use in testing hypotheses, models serve a number of functions in our quest to understand ecosystems. Quantitative models allow the investigator to *observe* patterns embedded in the data, to *synthesize* data on disparate components into an integrated view of ecosystem function, and ultimately to *predict* the future behavior of some aspects of the ecosystem under given scenarios of future external drivers (Figure 1.1). While the participants of Cary Conference IX found broad consensus for these uses of quantitative models, the conference also revealed strongly held preferences for different approaches to modeling. One of the major axes of contention, for example, was the tension between favoring simple or parsimonious models (Chapters 4 and 8) versus models that were more mechanistically rich (Chapter 5). Under the surface of this usually jovial disagreement between modelers of different schools lie deep philosophical differences about the nature of scientific understanding itself. In Chapter 2, Oreskes, the lone philosopher at the conference has articulated some of the relationships between science, philosophy, and modeling.

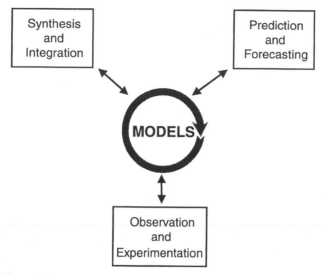

Figure 1.1

For the purposes of the Conference, we highlighted the roles of models in three distinct components of ecosystem science: observation and experimentation; synthesis and integration; and prediction and forecasting (Figure 1.1).

Observation and Experimentation

There are many examples in which models have provided the motivation for intensive empirical research. The most noteworthy is undoubtedly the "missing carbon sink" in the global carbon balance, although the carbon is only "missing" in the context of our models and/or our measurements. The mass-balance principles that led to the focus on the missing sink represent an important and useful constraint on ecosystem models. Pastor (Chapter 15) provides a powerful example of the use of mass-balance principles to suggest new field experiments for the study of plant competition.

Nonetheless, it is relatively rare to see tight integration between development of models in ecosystem science and the field research needed to generate the parameter estimates for the models. Modeling texts are replete with examples of failures when modeling is brought in as an afterthought (e.g., Starfield et al. 1990). There are many reasons for the lack of integration, including the generally weak modeling skills of many ecosystem scientists (Chapters 3 and 23). Ecosystem experiments are frequently designed to test fairly narrow hypotheses that do not require a formal model. In general, classical experiments answer qualitative hypotheses (Does treatment X have any effect on Y?). Quantitative models, on the other hand, require parameterization of the functional relationship between X and Y. This is often better accomplished through comparative studies and regression than through the much more widely taught (and respected) analysis of variance and formal experimental design. Thus, experiments are particularly valuable in deciding what to include in a model but poorly suited to generate the functional relationships needed by the model.

Synthesis and Integration

There is a strong tradition of holism in ecosystem science, but it could be argued that much of the current research in the field focuses on new mechanisms and processes (boxes and arrows) as elaborations on traditional and simpler models. Ecologists can be justly accused of reveling in the complexity of nature, and ecology is frequently touted as the science of connections, but it is obvious that not all connections are equally important in governing any particular ecosystem state or process. Quantitative models can play an important role in helping us determine the most important processes and components for any given question (Chapter 6). Sensitivity analyses and the related techniques of path analyses and structural equation modeling (e.g., Grace and Pugesek 1997) can be used in the traditional sense of identifying parameter values that deserve rigorous attention in order to reduce model uncertainty, but they also serve a much broader role in helping us understand the strength of individual processes within the complex web of potential interactions that occur in ecosystems (Gardner et al. 1981; Fennel et al. 2001).

In a survey of attitudes about modeling, members of the Ecological Society of America (Chapter 3) were asked the most important reason for their use of models. The two most frequent responses were (1) to clarify conceptualization of system structure and (2) to clarify quantitative relationships between and among system components (> 40% of respondents, combined). The use of models as an important tool for prediction was the third most common response, given by only 15% of respondents. While it is easy to focus on the details of the quantitative output of models, and many of the chapters of this volume address the quantitative evaluation of models (e.g., Chapters 8 and 13), many of them stress the more critical role of models in synthesizing our understanding of ecosystems (e.g., Chapters 4, 5, and 6) and in the teaching of ecology (e.g., Chapters 22 and 23).

Prediction and Forecasting

There was a great deal of discussion at the conference of the nature and limitations of prediction in ecosystem science. Beyond the narrow, scientific use of models to test hypotheses, ecosystem models are frequently used in public policy and natural resource management (Clark et al. 2001). There is considerable debate over our success at predicting the temporal dynamics of ecosystems (Chapter 2), and even over the philosophical validity of such predictions

(Oreskes et al. 1994). Within the narrow confines of science, all model output can be defined as a prediction (see below), but as Bugmann points out in Chapter 21, there is real danger that the public perceives far more certainty in model predictions than is warranted. Pielke (Chapter 7) argues that conventional approaches to modeling are poorly suited to simultaneously meet scientific and decision-making needs.

There is considerable variation in the terminology used to describe model output, both within the chapters in this volume and within the field as a whole. As Pielke (Chapter 7) and Bugmann (Chapter 21) point out, the problem is compounded by differences between the narrow, scientific use of a term and the range of meaning imparted by the same term in a public arena. We considered but rejected the idea of trying to standardize terminology either before the conference or in this volume and, instead, present an attempt at a lexicon of modeling terminology later in this chapter.

The Status of Modeling in Ecosystem Science

Some divisions remain, but there appears to be broad acceptance of the important role of models in ecosystem science (Chapter 3). In contrast, a relatively small proportion of the papers in premier ecological journals have employed quantitative models (Chapter 3). Duarte et al. (Chapter 24) outline a number of impediments to both the development and the achievements of models. The impediments to model development are more prosaic, and technological advances constantly lessen the barriers through development of modeling software and availability of computing power. The impediments to the achievements of models are more fundamental and include limits to both prediction and understanding (Chapters 24 and 26).

Despite widespread acceptance of the value of models, modeling skills remain elusive. "Lack of training" was the most often cited limitation on the use of modeling by the respondents of the survey conducted by Lauenroth et al. (Chapter 3). One of the discussion groups at the conference focused on strategies to increase modeling skills among ecologists and identified a number of specific types of modeling skills that need to be developed, as well as specific suggestions for addressing those needs (Chapter 23). A second group considered the role of modeling in undergraduate education (Chapter 22). Their framework for improving the use of models in undergraduate ecology education is based on the premise that undergraduates at all levels would benefit from more explicit training in modeling.

A number of the chapters address another major limitation on ecosystem models: the availability of critical data for both model parameterization and model testing (e.g., Chapters 3, 12, 13, and 27). This is, in part, a reflection of insufficient integration of modeling and empirical research (i.e., a disconnect between the needs of models and the objectives of field researchers). It also reflects the time and expense of collecting the necessary data, particularly for models that span large space and time scales.

Simplicity versus Complexity in Ecosystem Models

Models should be made as simple as possible, but not simpler.—adapted from a quote about theories attributed to Albert Einstein

All ecological models are, by definition, simplifications of nature (Chapter 2). Oreskes et al. (1994) argue that there is little empirical evidence that the world is actually simple or that simple accounts are more likely than complex ones to be true. They suggest that predilections for simplicity are largely an inheritance of seventeenth century theology. While Ockham's razor was originally sharpened for theological arguments and may not necessarily be the path to a full understanding of nature, there are many compelling reasons to keep models as simple as possible (e.g., Chapters 2 and 5).

Ecologists appear to differ widely in their predilection for simplicity and abstraction. These differences are apparent in the chapters in this volume (e.g., Chapters 4 and 5). We feel that the differences are healthy and that a diversity of approaches to modeling is as important as a diversity of approaches to science. How simple a model should be is part art form, part personal preference, but it is always determined by the nature of the question (Chapter 6). Moreover, our standards for what constitutes a "simple" model are likely to evolve as both our modeling abilities and our detailed understanding of natural phenomena evolve. Armstrong (Chapter 14) presents the arguments for the need to incorporate the size structure of organisms in ocean ecosystem models, particularly in the context of global carbon questions. Pastor (Chapter 15) provides an example in which very traditional competition models that ignore ecosystem science can be recast in the light of simple mass-balance principles. DeAngelis and Mooij (Chapter 5) argue for the benefits of "mechanistically rich" models. One of the benefits of such models is the rich array of outputs generated by the model. This variety allows comparison of diverse outputs against empirical data, providing more means to evaluate to model and develop confidence in the model (Chapter 24). A related limitation of such models is that associated with this rich array of outputs is a large amount of uncertainty (Chapter 8).

A Selective Lexicon for Evaluating Ecosystem Models

There are two ways of constructing a model: One way is to make it so simple that there are obviously no deficiencies, and the other way is to make it so complicated that there are no obvious deficiencies. The first method is far

more difficult.—adapted from a quote by computer scientist C.A.R. Hoare on the subject of software design

A significant portion of the conference was devoted to the issue of evaluating ecosystem models. It became clear early in the conference that there was considerable difference of opinion not only over approaches to model evaluation, but also over the terminology used in this important effort. Conscious efforts to standardize terminology are almost always futile in science. In lieu of that, we present here a selective lexicon of the major terms and common usages expressed at the conference. We focus on two areas: model testing and the nature of model output.

Model Testing

Validation. As Gardner and Urban point out in Chapter 10, the process of model validation has "been surrounded with an inordinate degree of confusion." The on-line *Merriam-Webster Dictionary* (www.m-w.com) defines "validate" as "to support or corroborate on a sound or authoritative basis." Oreskes et al. (1994, 642) echoed this definition by arguing that validation of models "does not necessarily denote an establishment of truth.... Rather, it denotes establishment of legitimacy." Thus, "a model that does not contain known or detectable flaws and is internally consistent can be said to be valid" (Oreskes et al. 1994, 642). As they point out, the term is commonly used in a much broader (and to their minds, inappropriate) sense as a general determination that the model provides an accurate representation of nature. Hilborn and Mangel (1997), in a monograph on "confronting models with data," don't include the term "validation" in their index, although there is some discussion of the issue in a section on distinguishing between models and hypotheses. As they point out, there is a common view that models should be "validated" through comparisons between model predictions and data. However, all models will disagree with some of the data. Thus, "models are not validated; alternate models are options with different degrees of belief" (Hilborn and Mangel 1997, 31; see under "Confirmation," below). Burnham and Anderson (1998) provided a detailed summary of the statistical methods for evaluation of alternate models, using the principles of maximum likelihood and information theory.

    The usage advocated by Oreskes et al. (1994) (and implicitly by Hilborn and Mangel 1997) focuses on model structure rather than on model output. This is a subtle but important distinction. As Rastetter points out in Chapter 12, evaluation of alternate model structures can present much greater challenges than determination of the goodness of fit of any particular model structure to a set of data (Chapter 8). While alternate model formulations may not differ significantly in their fit to a particular data set, they may invoke vastly different mechanisms, with important consequences when the model is used in novel conditions (Chapters 12 and 13). Burke et al. (Chapter 13) provide an example of this through an analysis of the implications of seemingly minor differences in the equations used to characterize the temperature dependence of decomposition in biogeochemical models. The process of evaluating model structure is clearly critical enough to warrant a specific term, and "validation" appears to be the best candidate.

Calibration. As Oreskes et al. (1994) pointed out, we frequently have better data on ecosystem responses (the dependent variables) than on the processes that drive those responses (the independent variables). They define "calibration" as "the manipulation of the independent variables to obtain a match between the observed and simulated" dependent variables. Aber et al. (Chapter 11) note that most large ecosystem simulation models are calibrated, in the sense that free parameters (unconstrained by actual measurements) have been adjusted to make the model output match the observed data (or to simply produce reasonable patterns). Aber et al. considered this a weakness of those models.

Regression is a form of calibration in which rigorous statistical procedures can be used to determine the values of parameters that optimize the fit between observed data and the predictions of the regression model. The principles of maximum likelihood (Hilborn and Mangel 1997), information theory (Burnham and Anderson 1998), and Bayesian statistics (Chapter 9) extend the familiar concepts of regression to provide a very powerful framework for rigorous parameter estimation and testing of alternate models. These principles are often used in the development of component submodels within large ecosystem simulation models, but they are also eminently suitable for the simpler statistical models presented by Håkanson in Chapter 8.

Confirmation. There is a natural temptation to claim that a match between observed data and predicted model results "confirms" the model. As Oreskes et al. (1994) pointed out, this is a logical fallacy ("affirming the consequent"). The concordance could be a result of chance rather than of the verity of the model. In contrast, if the match is poor, the model can logically be called flawed in some way. On the other hand, scientists consider hypotheses that are not refuted by repeated comparisons to data to gradually gain "confirmation." The bottom line is that we can never truly verify a model, just as we can never fully "prove" a hypothesis. We can, however, develop various degrees of confidence in models.

Adequacy and Reliability. Gardner and Urban (Chapter 10) suggest replacing the term "validation" with more specific terms that measure the utility and explanatory power of a model. "Adequacy" is the degree to which a model "explains" the observed set of ecological dynamics. "Reliability" is the degree to which model behaviors or predictions are within the observed set of ecosystem behaviors. They present a formal method (the receiver-operator (ROC) curve) based on signal detection literature for quantifying these terms.

Predictive Power and Goodness of Fit. Presumably all ecosystem scientists are familiar with the concept of goodness of fit. At least in the case of model predictions that come in the form of continuous variables, there are well-developed and intuitive statistical measures of the goodness of fit of a model to a dataset. These include consideration of the related concept of bias. Håkanson

(Chapter 8) explores the concept of goodness of fit in considerable detail and presents the concept of "predictive power" when the goodness of fit of a model can be tested a number of times (i.e., in different systems or settings).

Model Output

We follow Harald Bugmann's lead in Chapter 21 and consider four characterizations of model output. When applied to statements of the future states of ecosystems, the four terms are generally interpreted to imply differing degrees of certainty.

Prediction. Merriam-Webster's on-line dictionary defines "prediction" as "foretell on the basis of observation, experience, or scientific reason." There are at least two distinctly different usages in ecosystem modeling. Modelers adhere to the common definition when they refer to temporal predictions from dynamic models (i.e., statements about the future state of an ecosystem based on model output). Modelers commonly depart from the standard usage when they refer to any qualitative or quantitative output of a model as a "prediction," regardless of whether the model is static or dynamic. For example, a regression model "predicts" the primary productivity of a lake (without explicit reference to time) as a function of phosphorus loading (Chapter 8).

As Bugmann points out in Chapter 21, the common definition is often interpreted by the public (and resource managers) to imply a high degree of certainty. Scientists don't necessarily make this assumption and instead rely on a variety of measurements of the goodness of fit of the model predictions to observed data (Chapter 8). Oreskes (Chapter 2) argues that the predictive ability of ecosystem models is fundamentally limited because ecosystems are not "closed" systems and because important physical and biological forcing functions are necessarily treated as externalities. This clearly imposes limitations on goodness of fit of temporal predictions.

Forecast. Merriam-Webster's on-line dictionary defines "forecast" as "to calculate or predict some future event or condition, usually as a result of study and analysis of available pertinent data; to indicate as likely to occur." In the lexicon of modeling, the critical distinction between a prediction of a future event and a forecast lies in the assessment of the likelihood of the occurrence of the event (Chapter 21). As Clark et al. (2001) define it, an ecosystem forecast is "[a prediction of the] state of ecosystems…with fully specified uncertainties." They do not completely resolve the question of what would constitute "full specification" of uncertainty.

Projection. Merriam-Webster's on-line dictionary defines "projection" as "an estimate of future possibilities based on a current trend." The common usage seems to imply less certainty than either a prediction or a forecast (Chapter 21). In technical usage, it would appear to be most appropriately applied to the results of the broad range of techniques for extrapolating to future ecosystem states from past data, based on a statistical model.

Scenario. Defined as "an account or synopsis of a possible course of action or events" (Merriam-Webster's on-line dictionary), the term "scenario" appears to be most commonly used by ecosystem modelers in the sense of an "if/then" statement referring to the hypothetical predictions of a model under a specified set of parameter values, initial conditions, and external forcing functions (Chapter 21). While "scenarios" may not contain any statement of the likelihood of an actual, future state of a real ecosystem, the use of quantitative models to explore the logical consequences of alternative scenarios (i.e., given the structure of the model and its associated assumptions) is a powerful motivation for modeling in both basic and applied science (Chapter 7).

Use of Models in Ecosystem Management

*For every problem there is a model that is simple, clean and wrong.—*adapted from a quote by H.L. Mencken on solutions to societal problems

If all models are simplifications of nature (Chapter 2) and therefore never fully capture the range of behavior of real ecosystems (Chapter 10), how "wrong" can a model be and still be useful in a management context? Håkanson (Chapter 8) defines a set of quantitative indices of how "wrong" models are in the context of errors of prediction. More generally, Pielke argues strongly for better communication of the limitations of both our basic understanding and the inherent predictability of ecosystems (Chapter 7). As he points out, management decisions are, and always will be, made in the face of imperfect knowledge. Very little is served (and real damage can be done) by the failure of scientists to clearly communicate the nature and uncertainty of model predictions.

Models serve a number of purposes in ecosystem management other than prediction and forecasting. These include providing decision-support systems for focusing consideration of diverse issues and providing an explicit framework for adaptive management (Chapter 16). As Harris et al. point out in Chapter 16, the use of models in ecosystem management is now ubiquitous. It can be argued that the most innovative work on the development of ecosystem modeling as a tool in science is being done in the context of resource management.

References

Burnham, K.P., and D.R. Anderson. 1998. *Model Selection and Inference: A Practical Information-Theoretic Approach.* New York: Springer-Verlag.

Clark, J.S., S.R. Carpenter, M. Barber, S. Collins, A. Dobson, J.A. Foley, D.M. Lodge, M. Pascual, R. Pielke Jr., W. Pizer, C. Pringle, W.V. Reid, K.A. Rose, O. Sala, W.H. Schlesinger, D.H. Wall, and D. Wear. 2001. Ecological forecasts: An emerging imperative. *Science* 293: 657–660.

Fennel, K., M. Losch, J. Schroter, and M. Wenzel. 2001. Testing a marine ecosystem model: Sensitivity analysis and parameter optimization. *Journal of Marine Systems* 28: 45–63.

Gardner, R.H., R.V. O'Neill, J.B. Mankin, and J.H. Carney. 1981. A comparison of sensitivity analysis and error analysis based on a stream ecosystem model. *Ecological Modelling* 12: 177–194.

Grace, J.B., and B.H. Pugesek. 1997. A structural equation model of plant species richness and its application to a coastal wetland. *American Naturalist* 149: 436–460.

Hilborn, R., and M. Mangel. 1997. *The Ecological Detective: Confronting Models with Data.* Princeton, NJ: Princeton University Press.

Oreskes, N., K. Shrader-Frechette, and K. Belitz. 1994. Verification, validation, and confirmation of numerical models in the earth sciences. *Science* 263: 641–646.

Starfield, A.M., K.A. Smith, and A.L. Bleloch. 1990. *How to Model It: Problem Solving for the Computer Age.* New York: McGraw-Hill.

Part I

The Status and Role of Modeling in Ecosystem Science

2

The Role of Quantitative Models in Science

Naomi Oreskes

Summary

Models in science may be used for various purposes: organizing data, synthesizing information, and making predictions. However, the value of model predictions is undermined by their uncertainty, which arises primarily from the fact that our models of complex natural systems are always open. Models can never fully specify the systems that they describe, and therefore their predictions are always subject to uncertainties that we cannot fully specify. Moreover, the attempt to make models capture the complexities of natural systems leads to a paradox: the more we strive for realism by incorporating as many as possible of the different processes and parameters that we believe to be operating in the system, the more difficult it is for us to know if our tests of the model are meaningful. A complex model may be more realistic, yet, ironically, as we add more factors to a model, the certainty of its predictions may decrease even as our intuitive faith in the model increases. For this and other reasons, model output should not be viewed as an accurate prediction of the future state of the system. Short timeframe model output can and should be used to evaluate models and suggest avenues for future study. Model output can also generate "what if" scenarios that can help to evaluate alternative courses of action (or inaction), including worst-case and best-case outcomes. But scientists should eschew long-range deterministic predictions, which are likely to be errone-ous and may damage the credibility of the communities that gener-ate them.

The Role of Quantitative Models in Science

What is the purpose of models in science? This general question underlies the specific theme of this volume: What should be the role of quantitative models in ecosystem science? Ultimately the purpose of modeling in science must be congruent with the purpose of science itself: to gain understanding of the natural world. This means understanding both processes and products, the things in the world and the ways in which they interact. Historically, scientists have sought understanding for many reasons: to advance utilization of earth resources, foster industrialization, improve instruments and techniques of warfare, prevent or treat disease, generate origins stories, reflect on the glory and beneficence of the world's creator, and satisfy human curiosity. None of these goals has proved itself superior to any of the others; science has advanced under all of these motivations.

Until the twentieth century, the word "model" in science typically referred to a physical model—such as the seventeenth-century orreries built to illustrate the motions of the planets. Physical models made abstract ideas concrete and rendered complex systems visualizable, enabling scientists to think clearly and creatively about complex systems. Models provided analogies: in the early twentieth century, the orbital structure of the solar system provided a cogent analogy for the orbital structure of the atom. Models also served as heuristic devices, such as the nineteenth-century mechanical models that used springs and coils to interrogate the mechanism of light transmission through the ether, or the twentieth-century wooden models of sticks and balls used to explore possible arrangements of atoms within crystal structures. Perhaps most ambitiously, physical models were used as arguments for the plausibility of proposed causal agents, such as the nineteenth-century scale models of mountains, in which various forms of compression were applied in hopes of simulating the fold structures of mountain belts to demonstrate their origins in lateral compression of the earth.

In our time, the word "model" has come to refer primarily to a computer model, typically a numerical simulation of a highly parameterized complex system. The editors of this volume suggest that quantitative models in ecosystem science have three main functions: synthesis and integration of data; guiding observation and experiment; and predicting or forecasting the future. They suggest that scientists are well aware of the value of models in integrating data and generating predictions, but are less well informed about the heuristic value of models in guiding observation and experiment. In their words,

> Quantitative models provide a means to test our understanding of ecosystems by allowing us to explore the interactions among observations, synthesis, and prediction. The utility of models for synthesis and prediction is obvious. The role of quantitative models in informing and guiding observation and experimentation is perhaps less often appreciated, but equally valuable. (Canham ct al. 2001)

There is, however, a generous literature on the heuristic value of models and their use in guiding observation and experiments, particularly in the physical sciences, which need not be reiterated here (e.g., Cartwright 1983; Tsang 1991, 1992; Konikow 1992; Konikow and Bredehoeft 1992; Beven 1993, 2000, 2001, 1999; Oreskes et al. 1994, Rastetter 1996; Narisimhan 1998; Oreskes 1998; Morgan and Morrison 1999). The focus of this essay is therefore on challenging the "obvious"—that is, challenging the utility of models for prediction.

To be sure, many scientists have built models that can be run forward in time, generating model output that describes the future state of the model system. Quantitative model output has also been put forward as a basis for decision making in socially contested issues such as global climate change and radioactive waste disposal. But it is open to question whether such models generate reliable information about the future, and therefore in what sense they could reasonably inform public policy (Oreskes et al. 1994; Pilkey 1994; Shackley et al. 1998; Evans 1999; Morgan 1999; Sarewitz and Pielke 2000, Sarewitz et al. 2000; Oreskes 2000a; Oreskes and Belitz 2001). Moreover, it is not even clear that time-forward model output necessarily contributes to basic scientific understanding. If our goal is to understand the natural world, then using models to predict the future does not necessarily aid that goal. If our goal is to contribute usefully to society, using models to predict the future may not do that, either.

Why should we think that the role of models in prediction is obvious? Simply because people do something does not make its value obvious; humans do many worthless and even damaging things. To answer the question of the utility of models for prediction, it may help to step back and think about the role of prediction in science in general. When we do so, we find that our conventional understanding of prediction in science doesn't work for quantitative models of complex natural systems precisely because they are complex. The very factors that lead us to modeling—the desire to integrate and synthesize large amounts of data in order to understand the interplay of various influences in a system— mitigate against accurate quantitative prediction.

Moreover, successful prediction in science is much less common than most of us think. It has generally been limited to short-duration, repetitive systems, characterized by small numbers of measurable variables. Even then, success has typically been achieved only after adjustments were made based on earlier failed predictions. Predictive success in science, as in other areas of life, usually ends up being a matter of learning from past mistakes.

Models Are Open Systems

The conventional understanding of scientific prediction is based on the hypothetico-deductive model. Philosophers call it the deductive-nomological model, to convey the idea that, for our predictions to be correct, they must derive from stable scientific laws. Whatever one prefers to call it, this model assumes that the principal task of science is to generate hypotheses, theories, or laws and compare their logical consequences with experience and observations in the natural world. If our predictions match the facts of the world, then we can say that the hypothesis has been confirmed, and we can feel good about what we have done so far. We are on the right track. If the observations and experiences don't match our predictions, then we say that the hypothesis has been refuted and we need to go back and make some adjustments.

The problem with the hypothetico-deductive model, as many scientists and philosophers have realized, is that it works reliably only if we are dealing with closed systems. The hypothetico-deductive model is a logical structure of the form "$p \rightarrow q$," where our proposition, p, does in fact entail q, if and only if $p \rightarrow q$ is a complete description of the system. That is, if and only if the system is closed. But natural systems are never closed: they always involve externalities and contingencies that may not be fully specified, or even fully known. When we attempt to test a hypothesis in the real world, we must invoke auxiliary assumptions about these other factors. This means all the additional assumptions that have to be made to make a theory work in the world: frictionless surfaces, ideal predators, purely rational humans operating in an unfettered free market. When we test a theory by its consequences, other potentially influential factors have to be held constant or assumed not to matter. This is why controlled experiments play such a pivotal role in the scientific imagination: in the laboratory we have the means to control external conditions in ways that are not available in ordinary life. Yet even in the laboratory, we still must assume—or assert—that our controlled factors are in fact fully controlled and that the factors we consider negligible are in fact so. If a theory fails its test, we cannot be certain whether the fault lies in the theory itself or in one of our other assumptions.

Another way to understand this is to compare a model of a natural system with an artificial system. For example, in our commonly used arithmetic, we can be confident that if $2 + 2 = 4$, then $4 - 2 = 2$, because we have *defined* the terms this way and because no other factors are relevant to the problem. But consider the question: Is a straight line the shortest distance between two points? Most of us would say yes, but in doing so we would have invoked the auxiliary assumption that we are referring to a planar surface. In the abstract world of Euclidian geometry, or on a high-school math test, that would be a reasonable assumption. In high school, we'd probably be classified as a smart aleck if we wrote a long treatise on alternative geometrical systems. But if we are referring to natural systems, then we need additional information. The system, as specified, is open, and therefore our confident assertion may be wrong. The shortest distance between two points can be a great circle.

Furthermore, in order to make observations in the natural world, we invariably use some form of equipment and instrumentation. Over the course of history, the kinds of equipment and instruments that have been used in science have tended to become progressively more sophisticated and complex. This means that our tests have become progressively more complex, and apparent failures of theory may well be failures of equipment, or failures on our part to understand the limitations of our equipment.

The most famous example of this in the history of science is the problem of stellar parallax in the establishment of the heliocentric model of planetary motions (Kuhn 1957). When Nicolaus Copernicus proposed that model in the early sixteenth century, it was widely recognized that this idea would have an observable consequence: stellar parallax—the apparent changing position of a star in the heavens as the earth moved through its orbit. If the earth stood still, its position relative to the stars would be constant, but if it moved, then the position of the stars would seem to change from winter to summer and back again. One could test the Copernican model by searching for stellar parallax. This test was performed and no parallax was found, so many astronomers rejected the theory. It had failed its experimental test.

Four hundred years later, we can look back and see the obvious flaw in this test: it involved a faulty auxiliary hypothesis—namely, that the universe is small. The test assumes that the diameter of the earth's orbit around the sun is large relative to the distance to the star and that the stellar parallax is a large angle. Today, we would say this is a conceptual flaw: the stars are almost infinitely far away and the parallax is therefore negligibly small.

The experimental test of stellar parallax also involved an assumption about equipment and instrumentation: namely, that the available telescopes were accurate enough to perform the test. Today, astronomers can detect stellar parallax, which is measurable with the instruments of the twenty-first century, but sixteenth-century telescopes were simply inadequate to detect the change. Of course, our equipment and instrumentation are far more sophisticated today, but the same kinds of assumptions of instrumental adequacy are built into our tests as were built into theirs.

This brings us to the kind of models that most of us work with today. The word "model" can be problematic because it is used to refer to a number of different things, but this discussion will assume that we are referring to a mathematical model, typically a numerical simulation, realized on a digital computer. However, the points may apply to other kinds of models as well.

All models are open systems. That is to say, their conclusions are not true by virtue of the definition of our terms, like "2" and "+," but only insofar as they encompass the systems that they represent. Alas, no model completely encompasses any natural system. By definition, a model is a simplification—an idealization—of the natural world. We simplify problems to make them tractable, and the same process of idealization that makes problems tractable also makes our models of them open. This point requires elaboration.

There are many different ways in which models are open, but there are at least three general categories into which this openness falls. First, our models are open with respect to their conceptualization (how we frame the problem). When we create a model, we abstract from the natural world certain elements that we believe to be salient to the problem we wish to understand, and we omit everything else. Indeed, good science *requires* us to omit irrelevancies. For example, consider a model of predator-prey relations in the Serengeti Plain. We can be fairly confident that the color of my bedroom is irrelevant to this model.

There is no known reason why it should matter to animals in Africa; indeed, there is no way (so far as we know) that the animals could even be aware of it. But a little imagination reveals that there could be other factors that we consider irrelevant but that might in the future be shown to matter. (The history of science is full of connections and correlations that were previously unsuspected but later demonstrated. It is also full of examples of correlations that were later rejected.) Moreover, there may be factors that we know or suspect *do* matter, but which we leave out for various reasons—we lack time, computational power, or other resources to incorporate them; we lack data or analytical methods to represent them; or we lack confidence about their significance (Ascher 1993; Oreskes and Belitz 2001). Or we may simply frame the problem incorrectly (for a more detailed discussion, with specific examples of this, see Oreskes and Belitz 2001). At every level there remains the question whether the model conceptualization is adequate.

Our models are also open with respect to the empirical adequacy of the governing equations (how well our mathematical representations map onto natural processes). We often call these equations *laws*, but as philosopher Nancy Cartwright (1983) has cogently shown, this usage is really rather misleading. Scientific laws are idealizations that map onto the natural world to a greater or lesser degree depending on the circumstances. Moreover, while the use of the term "law" was borrowed from the political realm, laws of nature are different from laws of the state.

Political laws do not attempt to describe an actual state of affairs, but rather the opposite: they announce how we want things to be. Political laws are declarations of intent, and we adjudicate them based on rather formal procedures. Laws of nature are not declarations. They are our best approximations of what we think is really going on, and there is no formal standard by which we judge them. In a numerical model, we assume, based on prior experience, that the equations used are adequate, but we have no logical way to demonstrate that this assumption is correct. Therefore the model remains open.

Third, models are open with the respect to the input parameterization (how well numerical variables represent elements of the system). In some systems, like planets orbiting around the sun, we can count the number of objects involved in the problem—nine planets[1] and the sun—and make our best measurements of their size, shape, distance, etc. But in many other problems, it is an open question as to exactly what the relevant variables are. Even if we know what they are—for example, permeability in a hydrological model—we still have to make choices about how best to represent and parameterize it.

It is important to underscore that calling a model open is not the same as calling it "bad." The more we don't know about a system, the more open our model will be, even if the parts we do know about are very well constrained. A card game is often taken as a good example of a closed system because we know how many cards are in the deck and precisely what those cards have to be, assuming no one is cheating. But there's the rub. To be confident of our odds, we have to make the assumption of honest dealing. In fact, this paradigmatically closed system is, in real life, open. Poker is a good game, but in real life it's an open system. Similarly, it is possible to imagine a model in which the available empirical data are well measured and in which the governing equations have stood the test of time, yet in which important relevant parameters have not yet been recognized. Such a model might be "good" in terms of the standards of current scientific practice yet remain highly open and therefore fail to make reliable predictions.

The Complexity Paradox

The openness of a model is a function of the relationship between the complexity of the system being modeled and the model itself. The more complex the natural system is, the more different components the model will need to mimic that system. Therefore we might think that by adding components we can make the model less open. But for every parameter we add to a model, we can raise a set of questions about it: How well does it represent the object or process it purports to map onto? How well constrained is our parameterization of that feature? How accurate are our measurements of its specific values? How well have we characterized its interrelations with other parameters in the model? Even as we increase our specifications, the model still remains an open system.

This might suggest that simpler models are better—and in some cases no doubt they are—but in ecosystems modeling we do not want to abandon complexity because we believe that the systems we are modeling are in fact complex. If we can demonstrate that certain parameters in the model are insignificant, then we can omit them, but in most cases that would be assuming the thing we wish to discover: What role does this parameter play? How does it interact with other parameters in the system? Indeed, in many cases it is the very complexity of the systems that has inspired us to model them in the first place—to try to understand the ways in which the numerous parts of the system interact.

Moreover, complexity can improve accuracy by minimizing the impact of errors in any one variable. In an analysis of ecological models of radionuclide kinetics in ecosystems, O'Neill (1973) showed, as one might expect, that systematic bias resulting from individual variables decreased as the number of variables in a model increased. However, uncertainty *increased* as the measurement errors on individual parameters accumulated. Each added variable added uncertainty to the model, which, when promulgated in a Monte Carlo simulation, contributed to the uncertainty of the model prediction.

These considerations may be summarized as the "complexity paradox." The more we strive for realism by incorporating as many as possible of the different processes and parameters that we believe to be operating in the system, the more difficult it is for us to know if our tests of the model are meaningful. Put another way, the closer a model comes to capturing the full range of processes and parameters in the system being modeled, the more difficult it is to ascertain whether or not the model faithfully represents that system. A complex model may be more realistic yet at the same time more uncertain. This leads to the ironic situation that as we add more factors to a model, the certainty of its predictions may decrease even as our intuitive faith in the model increases. Because of the complexities inherent in natural systems, it may never be possible to say that a given model configuration is factually correct and, therefore, that its predictions will come true. In short, the "truer" the model, the more difficult it is to show that it is "true."

Successful Prediction in Science

At this point some readers will be thinking, "But surely there are many cases in which scientists have made successful predictions, many areas of science where we do a good job and our predictions have come true." This intuition may be recast as a question: Where *have* scientists developed a track record of successful prediction? What do we learn when we examine the nature of those predictions and how they have fared? Viewed this way, we find a surprising result: successful prediction in science is less common than most of us think, and it has developed as much through a process of trial and error as through the rigorous application of scientific law. Consider three areas of science that have a large literature on their predictive activities: weather, astronomy, and classical mechanics.

Example 1: Meteorology and Weather Prediction

Meteorology is the science most closely associated with prediction in the public mind and the only science that regularly earns time on the evening news. What do we know about weather prediction? First, that it is non-deterministic. Weather forecasts are not presented in the form, "It will rain two inches tomorrow beginning at 3:00 o'clock and lasting until 4:30." They are presented in the form, "There is a 20% chance of rain tomorrow afternoon to early evening." Moreover, if rain is expected, forecasters typically offer a range of plausible values, such as 1–2 inches. In a sense, we could say that meteorologists hedge their bets, and there has been considerable debate within the meteorological community over just how weather forecasts should be presented (e.g., Murphy 1978). The use of probabilistic forecasting is partly a response to experience: history has demonstrated just how difficult specific, quantitative prediction of a complex system is.

Second, weather prediction involves spatial ambiguity. If my local forecast calls for an 80% chance of rain tomorrow in San Diego County, that forecast will be deemed accurate if it does in fact rain, even though some parts of the county may remain dry. Some of us have seen the phenomenon where it is raining at our house and dry across the street. Weather can be very local; forecasts are spatially averaged; and typically they become more accurate as they become more general and less accurate as they attempt greater specificity.

Third, and perhaps most important, accurate weather prediction is restricted to the near term. In meteorology, a "long-range" forecast is 3–5 days (or lately perhaps a bit longer.) The great successes in the history of meteorology, like the celebrated D-Day forecasts, are a case in point (Petterssen 2001). If you need an accurate forecast for the weather on January 27 next year, you simply cannot get it. (Anyone who has tried to plan an outdoor wedding is familiar with this problem. In fact there are now folks who *will* sell you a forecast for next June 10, and there are also folks who still sell snake oil.)

Partly in response to this problem, atmospheric scientists have developed a distinct terminology to deal with long-term change—they speak of (general) climate rather than (specific) weather. Meteorologists can accurately predict the average temperature for the month of January because the dominant control on monthly weather is the annual journey of Earth around the Sun, coupled with the tilt of Earth's axes, factors that are themselves quite predictable. Yet these planetary motions are not the only relevant variables: natural climate variation also depends on solar output, gases and dust from volcanoes, ocean circulation, perhaps even weathering rates controlled by tectonic activity. These latter processes are less regular than the planetary motions and therefore less predictable. Hence our predictive capacity is constrained to the relatively near future: we can confidently predict the likely average temperature of for a specific month within the next couple of years, but not so the average temperature of that month in 2153 (claims to the contrary not withstanding). The more extended the period, the more difficult the forecasting task becomes.

We know something about why long-range weather forecasting is so difficult: weather patterns depend upon external forcing functions, such as the input of solar radiation, small fluctuations in which can produce large fluctuations in the system. Weather systems are also famously chaotic, being highly sensitive to initial conditions. Many systems of interest to ecologists are similar to climate: they are strongly affected by exogenous variables. We know that these factors are at play, and we may even understand the reasons why they vary, but we cannot predict how their variations will alter our systems in the years to come. This is why a model can be useful to guide observation and experiment yet be unable to make accurate predictions: we can use a model to test which of several factors is most powerful in affecting the state of a system and use that information to motivate its further study, but we cannot predict which of these various factors will actually change and therefore what the actual condition of the system will be.

There is an additional point to be made about weather prediction. The reason we can make accurate predictions at all is because our models are highly calibrated. They are based on enormous amounts of data collected over an extended time period. In the United States and Europe, weather records go back more than a century, and high quality standardized records exist for at least four decades. In the United States, there are 10 million weather forecasts produced by the National Weather Service each year (Hooke and Pielke 2000)! Predictive weather models have been repeatedly adjusted in response to previous bad predictions and previous failures. Compare this situation with the kinds of models currently being built in aid of public policy: general circulation models, advective transport models, forest growth models, etc. None of these has been subject to the kind of trial and error—the kind of learning from mistakes—that our weather models have been.

We can readily learn from mistakes in weather modeling because weather happens every day and because of the enormous, worldwide infrastructure that has been created to observe and forecast weather. A similar argument can be made about another area of successful prediction in science: celestial mechanics.

Example 2: Celestial Mechanics and the Prediction of Planetary Motions

Celestial mechanics is an area in which scientists make very successful predictions. The motion of the planets, the timing of comets and eclipses, the position and dates of occultations—scientists routinely predict these events with a high degree of accuracy and precision. Unlike rain, we can predict to the minute when a solar eclipse will occur and precisely where the path of totality will be. Given this success, we can ask ourselves: What are the characteristics of these systems in which we have been able to make such successful predictions?

The answer is, first, that they involve a small number of measurable parameters. In the case of a solar eclipse, we have three bodies—Earth, its moon, and the Sun—and we need to know their diameters and orbital positions. This is a relatively small number of parameters, and each has a fixed value. Put another way, the variables in the system do not actually vary. The diameter of Earth is not, for all intents and purposes, changing (or at least not over the time frame relevant to this prediction). Second, the systems involved are highly repetitive. Although eclipses don't happen every day, they happen a lot, and we can track the positions of the planets on a daily basis. Moreover, planets return to their orbital positions at regular intervals. When we make an astronomical prediction, we can compare it with observations in the natural world, generally without waiting too long. If our predictions fail because we have made a mistake, we can find this out fairly quickly and make appropriate adjustments to the model. Third, as in the case of weather, humans have been making and recording observations of planetary motions for millennia. We have an enormous database with which to work.

However, precisely because of its successful track record, the development of accurate planetary prediction raises one of the most serious concerns for modelers: that faulty models may make accurate predictions. Returning to Copernicus, the failure of the prediction of stellar parallax was a serious problem for the heliocentric model, but there was another problem more pressing still: the Ptolemaic system, which the heliocentric model aspired to replace, made highly accurate predictions. In fact, it made *more* accurate predictions than the Copernican system until many decades later when Johannes Kepler introduced the innovation of elliptical orbits. Astronomers did not reject the Copernican view because they were stubborn or blinded by religion. They rejected it because the model they were already working with was better, as judged by the accuracy of its predictions. Yet in retrospect we believe that it was wrong at its very foundations.

This should give any modeler pause, for it shows that accurate predictions are not proof of a model's conceptual veracity: false models can make true predictions. In this case, repeated accurate predictions serve to bolster what in hindsight will be viewed as misplaced confidence. Conversely, a model that is conceptually well grounded may still fail in its predictions if key elements of the model are incorrect. Therefore, building a model that makes accurate predictions may not necessarily aid the goal of improved basic understanding. Indeed, it may actually impede it.

Many modelers achieve successful prediction through a process of model calibration: they obtain observations from the natural world and adjust model parameters until the model successfully reproduces the observational data. But this kind of calibration ensures that the model cannot fail—effectively, it makes the model refutation-proof (Oreskes et al. 1994; see also Anderson and Bates 2001). If there is a conceptual flaw in the model, unconstrained calibration will hide the flaw from view. This is what the Ptolemaic astronomers did: they added epicycles upon epicycles to "save the phenomenon"—to make the model fit the data—all the while preserving a system that was conceptually flawed at its root.

Example 3: Classical Mechanics

A third example may be drawn from classical mechanics. This is the area of science that most closely matches the hypothetico-deductive ideal of deterministic laws that generate specific quantitative predictions. We all learned these laws in high school physics: $f = ma$, $s = \frac{1}{2} at^2$, $p = mv$, etc. What do we really know about these laws, which most of us think of as the best example of deterministic laws that make accurate predictions? None is literally true. To generate accurate predictions, they all require *ceteris paribus* clauses: that there is no friction, no air resistance, or that an ideal ball is rolling down a perfectly frictionless plane. In real life, the laws do not work as stated, because in real life these kinds of ideal systems do not exist. Philosopher Nancy Cartwright calls this how the laws of physics "lie"—they posit an imaginary world, not the world that we live in (Cartwright 1983; see also Cartwright 1999). To make the laws of physics work in practice requires adjustments and modifications that are *not* based upon deterministic laws but rather on past experience and earlier failed attempts. So what starts out looking like a successful case of clean, deterministic prediction becomes a great deal messier as you get down into the nitty-gritty details.

Where does this leave us? First, we can draw the conclusion that accurate prediction in science is a special case (Cartwright 1999). It tends to be restricted to systems characterized by small numbers of measurable parameters or to systems in which the events at issue are naturally repetitive, like the orbits of the planets, or easily repeated, like balls rolling down hills. Second, accurate predictions have usually been achieved only after adjustments to the model in response to earlier failed predictions. This is why the interplay between modeling and observation is so important. It also is why repetitive systems are more likely to be predictable even if they are chaotic. Whether a system is intrinsically deterministic or intrinsically chaotic, we are more likely to be able to predict its behavior successfully if we have observed it many times before. Third, even when we achieve a match between quantitative prediction and empirical observation, it does not guarantee an accurate conceptualization of a system. Faulty conceptual models can make accurate predictions. Predictive capacity by itself is a weak basis for confidence.

Model Testing, Forecasting, and Scenario Development

Our focus here has been on branches of science that have a track record of prediction—meteorology, astronomy, and physics. The natural sciences that deal with complex nonrepetitive systems—geology, biology, and ecology—have no historic track record of predictive success at all. Indeed, until very recently, no one ever expected scientists in these disciplines to make predictions (Oreskes 2000b). If our science has no track record of predictive success, then it behooves us to ask, what makes us think we are different?

To answer this question, it may help to distinguish between short-term and long-term prediction, or what might be better referred to as model testing versus forecasting. Clearly, short-term predictions that can be used as a means to evaluate the model are extremely important, because they can provide a test of our understanding. The Copernican example shows that agreement between prediction and observation is not proof that a model is right, but disagreement can be evidence that something is wrong. Moreover, short-term predictions can be used to compare alternative models: to give us a handle on which of a number of possible conceptualizations or parameterizations does a better job of describing a given system, or to see what range of outcomes is suggested by current understandings. The latter in particular may be highly relevant for policy decisions, for example, in developing worst-case scenarios of climate change. Understood this way, prediction—that is to say, short-term prediction—becomes one of the heuristic functions of modeling.

However, long-range predictions—or *forecasts*—cannot be tested, and therefore do nothing to improve scientific understanding. A concrete example may help to clarify this point. Recently, the U.S. Department of Energy has issued forecasts of radionuclide releases at the proposed nuclear waste repository at

Yucca Mountain ten thousand years from now. These projections are interpreted to be required by law to ensure that the repository will satisfy regulatory standards. If the law demands such forecasts, then they will be made, and they are not necessarily useless: if the values are considered too high they may inspire modification to the repository design, or perhaps even stop the repository from going forward. In this way, such forecasts can be socially and politically useful (or harmful, depending upon your point of view). But from a scientific point of view, such long-range forecasts offer very little value, for they cannot be compared with empirical results and therefore cannot be used to improve scientific knowledge. If we understand our purpose as fundamentally a scientific one—to improve our understanding of natural systems—then long-range forecasting cannot aid this goal. Only when predictions are on a short enough time frame that we can compare them to events in the natural world can they play a role in improving our comprehension of nature.

This brings us to the domain of models and public policy. The desideratum of policy-relevant scientific information is frequently cited as a major motivation for modeling in ecology, atmospheric science, geochemistry, hydrology, and many other areas of natural and social science (Oreskes et al. 1994, Oreskes 1998). Several papers in this volume state that one of the important reasons for building the model under discussion is its relevance to public policy.

Some questions of public policy involve short time frames, and these questions may be usefully informed by model results. However, questions about short-duration events are typically questions about particular localities whose details are unique (Beven 2000). Given the uncertainties of the details of most site-specific models, predictive modeling is unlikely to be a substitute for monitoring. The problem here, as recently emphasized by Anderson and Bates (2001), is that data collection and monitoring are frequently viewed as "mundane," whereas modeling is viewed as "cutting-edge." Perhaps for this reason, data-collection programs have proved difficult to sustain. Scientists should ponder why this is so and consider whether the public interest is being served by our emphasis on models. A better balance between modeling and data collection may be called for.

Better data collection for model testing and monitoring may improve the usefulness of models used for short-term policy decisions, but the policy issues that models are commonly held to illuminate often involve long-term change in natural systems. Global climate change is the most obvious, but many, perhaps most, ecological models involve questions whose import will be realized over time frames of at least years, if not decades. Yet all of the available evidence suggests that long-term forecasts are likely to be wrong and may very well misinform public policy.

Given how many models have been built in the past decades, it is remarkable how few have been evaluated after the fact to determine whether their forecasts came true. Where such post hoc evaluations have been done, the results are extremely disconcerting: the vast majority of the predictions failed, often they failed rapidly, and in many cases the actual results were not only outside the error bars of the model, but they were in an entirely different direction (Ascher 1978, 1981, 1989, 1993; Ascher and Overholt 1983; Konikow 1986; Konikow and Patten 1985; Konikow and Person 1985; Konikow and Swain 1990; Leonard et al. 1990; Pilkey 1990, 1994, 2000; Nigg 2000; see also Sarewitz et al. 2000 and Oreskes and Belitz 2001).

The most detailed work on the question of forecasting accuracy has been done by political scientist William Ascher, working on economic and social science models (Ascher 1978, 1981, 1989, 1993; Ascher and Overholt 1983). For example, in a study of the rate of return on development projects, he found that in most cases the return was not only lower than predicted, but it was below the designated cut-off criterion for the funding agency. Indeed, it was so low that had it been accurately predicted, the project would never have been funded in the first place (Asher 1993). Another example is the celebrated "World Model" of the Club of Rome, published in the 1970s best-seller, *The Limits to Growth* (Meadows et al. 1972). This model predicted major shortages of natural resources by the end of the twentieth century that would cause commodity prices to skyrocket. In fact, not only did prices fail to increase at the predicted rate, they did not increase at all. At the end of the twentieth century, the prices for nearly all natural resources were lower than at the time the model was built (Moore 1995).

Natural scientists may be inclined to discount results from the social sciences because we have been trained to think of the systems they deal with as poorly constrained and subject to the intrinsic unpredictability of human behavior. Yet if one considers why social sciences models are messy—particularly the difficulty of specifying and quantifying system variables—we find much in common with models of complex natural systems. Moreover, if part of the motivation for our model-building is policy-relevance, then by definition we are looking at systems which in some way either affect or have been affected by humans, in which case the unpredictability of human behavior may well be relevant. For example, in his post audit of hydrological models, Leonard Konikow found that models typically failed because of unanticipated changes in the forcing functions of the system, and these forcing functions commonly involved human activities such as groundwater pumping, irrigation, and urbanization (Konikow 1986).

Faulty Forecasts Undermine Scientific Credibility

If the forecasts of a model are wrong, then sooner or later they *will* be refuted. For complex systems that are poorly constrained, or for which there are limited historical data, experience suggests that sooner is the more likely timeframe. If the forecasts are presented as facts—and they typically are—then this ultimately undermines the credibility of the community that generated them. A good example of this comes from the 1997 Grand Forks flood, an example of prediction gone wrong that has been analyzed by Pielke (1999) and Changnon (2000).

In the spring of 1997, the Red River of the North, which flows past Grand Forks, North Dakota, crested at the historically unprecedented level of 54.11 feet, causing massive flooding and $1–2 billion in property damage. Several weeks before, the National Weather Service had issued two flood "outlooks," one based on a scenario of average temperature and no additional precipitation, the other based on average temperature with additional precipitation. The two outlooks were 47.5 and 49 feet, respectively. Because spring flooding in this region is largely controlled by snowmelt, one might expect such forecasts to be reliable—after all, the snowpack can be measured. So the town of Grand Forks prepared. Residents added a couple of feet "just in case" and prepared for flood crests of up to 51 feet (Pielke personal comm. 2002). When the waters rose to 54 feet, massive damage ensued.

Although the scientists involved surely understood the uncertainties in their forecasts, and the Weather Service appended qualitative disclaimers, these uncertainties were not "received" by local officials. Rather they interpreted the outlooks as "facts"—either that the river would crest between 47.5 and 49 feet or that it would crest no higher than 49 feet. As these numbers were repeated by local officials, the media, and ordinary citizens, their status as facts hardened. When these "facts" proved false, blame for the disaster was laid at the feet of the Weather Service. While one might argue that local officials were at least in part responsible—for misunderstanding the scientific data and not preparing for worst-case scenarios—or that this was simply a tragic but uncontrollable natural disaster—an act of God or other unforeseeable event—this is not how the disaster was interpreted by those involved. Rather, it was interpreted as a failure on the part of the Weather Service. As Pielke (1999) quotes the mayor of Grand Forks, "they blew it big."

Examples like these illustrate how scientists can and will be blamed for faulty predictions, even if disclaimers or error bars are included and even if scientists have done their best in the face of an unprecedented natural event. Given this, it is hard to see how it can be in the long-term interest of the scientific community to make confident assertions that will be soon refuted. At best, the result is embarrassment; at worst, the risk is a loss of confidence in the significance and meaning of scientific information and ultimately a loss of public support for our endeavors.

The alternative is to shift our focus away from specific quantitative predictions of the future and toward policy-relevant statements of scientific understanding (see Sarewitz and Pielke 2000, and this volume chapter 7, for related views). Rather than attempt a specific prediction that is likely to fail, we can develop "what if" scenarios that highlight, for policy-makers and other interested parties, what the most likely consequences will be of alternative possible courses of action. To say that something is policy-relevant is to say that there is some possible course of human action (or inaction) that bears on the future state of affairs and that our information might plausibly affect which course of action humans will take.

We cannot say what will happen in the future, but we can give an informed appraisal of the possible outcomes of our choices. In the case of Grand Forks, Pielke argues that the National Weather Service intended to convey—and thought it had conveyed—the message that the communities involved should prepare for unprecedented flooding. Imagine then, for a moment, if instead of issuing their quantitative outlooks, the Weather Service had said just this: "All the available evidence suggests that we need to prepare for unprecedented flooding." The outcome, both for the reputation of the Weather Service and, more importantly, for the people of Grand Forks, might have been very different.

Acknowledgments. I am grateful to the organizers of Cary Conference IX for inviting me to present my ideas there and in this volume, and to Roger Pielke Jr. and an anonymous reviewer for very helpful reviews. This paper represents a summary and amalgamation of ideas presented and published elsewhere over the past several years; I am indebted to Kenneth Belitz, John Bredehoeft, Nancy Cartwright, Leonard Konikow, Dale Jamieson, Roger Pielke Jr., and Daniel Sarewitz for ongoing conversations about the meanings and purposes of modeling, and to Dennis Bird, who first inspired me to think hard about experience and observation in the natural world. I completed the revisions to this paper while visiting in the Department of History of Science, Harvard University, where I was generously provided with an ideal environment for thinking and writing about science in all its dimensions.

Note

1. Or maybe eight, depending upon your opinion of the status of Pluto.

References

Anderson, M.G., and P.D. Bates. 2001. Hydrological science: Model credibility and scientific integrity. Pp. 1–10 in M.G. Anderson and P.D. Bates, editors. *Model Validation: Perspectives in Hydrological Science.* Chichester: John Wiley and Sons.

Ascher, W. 1978. *Forecasting: An Appraisal for Policy-Makers and Planners.* Baltimore: Johns Hopkins University Press.

———. 1981. The forecasting potential of complex models. *Policy Sciences* 13: 247–267.

———. 1989. Beyond accuracy. *International Journal of Forecasting* 5: 469–484.

————. 1993. The ambiguous nature of forecasts in project evaluation: Diagnosing the over-optimism of rate-of-return analysis. *International Journal of Forecasting* 9: 109–115.

Ascher, W., and W.H. Overholt. 1983. *Strategic Planning and Forecasting: Political Risk and Economic Opportunity*. New York: John Wiley and Sons.

Beven, K. 1993. Prophecy, reality, and uncertainty in distributed hydrological modeling. *Advances in Water Resources* 16: 41–51.

————. 2000. Uniqueness of place and process representations in hydrological modeling. *Hydrology and Earth Systems Science* 4: 203–213.

————. 2001. Calibration, validation and equifinality in hydrological modeling: A continuing discussion. Pp. 43–55 in M.G. Anderson and P.D. Bates, editors. *Model Validation: Perspectives in Hydrological Science*. Chichester: John Wiley and Sons.

Canham, C.D., J.J. Cole, and W.K Lauenroth. 2001. Understanding ecosystems: The role of quantitative models in observation, synthesis, and prediction. Paper presented at Cary Conference IX, May 1–3, 2001, Institute of Ecosystems Studies, Millbrook, NY.

Cartwright, N. 1983. *How the Laws of Physics Lie*. Oxford: Clarendon Press.

————. 1999. *The Dappled World: A Study of the Boundaries of Science*. Cambridge: Cambridge University Press.

Changnon, S.A. 2000. Flood prediction: Immersed in the quagmire of national flood mitigation strategy. Pp. 85–106 in D. Sarewitz, R.A. Pielke Jr, and R. Byerly, editors. *Prediction: Decision Making and the Future of Nature*. Washington, DC: Island Press.

Evans, R. 1999. Economic models and policy advice: Theory choice or moral choice? *Science in Context* 12: 351–376.

Hooke, W.H., and R.A. Pielke Jr. 2000. Short-term weather prediction: an orchestra in search of a conductor. Pp. 61–84 in D. Sarewitz, R.A. Pielke Jr, and R. Byerly, editors. *Prediction: Decision Making and the Future of Nature*. Island Press: Washington, DC.

Konikow, L. 1986. Predictive accuracy of a ground-water model: Lessons from a post-audit. *Ground Water* 24: 173–184.

————. 1992. Discussion of 'The modeling process and model validation' by Chin-Fu Tsang. *Ground Water* 30: 622–623.

Konikow, L., and J.D. Bredehoeft. 1992. Groundwater models cannot be validated. *Advances in Water Resources* 15: 75–83.

Konikow, L., and E.P. Patten Jr. 1985. Groundwater forecasting. Pp. 221–270 in M.G. Anderson and T.P. Burt, editors. *Hydrological Forecasting*. Chichester: John Wiley and Sons.

Konikow, L., and M. Person. 1985. Assessment of long-term salinity changes in an irrigated stream-aquifer system. *Water Resources Research* 21: 1611–1624.

Konikow, L. and L.A. Swain. 1990. Assessment of predictive accuracy of a model of artificial recharge effects in the upper Cochella Valley, California. Pp. 433–439 in E.S. Simpson and J.M. Sharp, editors. *Selected Papers on*

Hydrology from the 28th International Geological Congress (1989). Vol. 1. Hannover: Heinz Heise.

Kuhn, T.S. 1957. *The Copernican Revolution*. Cambridge, MA: Harvard University Press.

Leonard, L., T. Clayton, and O.H. Pilkey. 1990. An analysis of replenished beach design parameters on U.S. east coast barrier islands. *Journal of Coastal Research* 6: 15–36.

Meadows, D.H., D.L. Meadows, and J. Randers. 1972. *The Limits to Growth: A Report for the Club of Rome's Project on the Predicament of Mankind*. New York: Universe Books.

Moore, S. 1995. The coming age of abundance. Pp. 110–139 in R. Bailey, editor. *The True State of the Planet*. New York: Free Press.

Morgan, M.S. 1999. Learning from models. Pp. 347–388 in M.S. Morgan and M. Morrison, editors. *Models as Mediators*. Cambridge: Cambridge University Press.

Morgan, M.S., and M. Morrison, editors. 1999. *Models as Mediators*. Cambridge: Cambridge University Press.

Murphy, A.H. 1978. Hedging and the mode of expression of weather forecasts. *Bulletin of the American Meteorological Society* 78: 371–373.

Narisimhan, T.N. 1998. Quantification and groundwater hydrology. *Ground Water* 36: 1.

Nigg, J. 2000. The issuance of earthquake 'predictions': Scientific approaches and strategies. Pp. 135–156 in D. Sarewitz, R.A. Pielke Jr, and R. Byerly, editors. *Prediction: Decision Making and the Future of Nature*. Washington, DC: Island Press.

O'Neill, R.V. 1973. Error analysis of ecological models. Pp. 898–908 in D.J. Nelson, editor, *Proceedings of the Third National Symposium on Radioecology*, No. 547. Environmental Sciences Division, Oak Ridge, TN: Oak Ridge National Laboratory.

Oreskes, N. 1998. Evaluation (not validation) of quantitative models. *Environmental Health Perspectives* 106: 1453–1460.

———. 2000a. Why believe a computer? Models, measures and meaning in the natural world. Pp. 70–82 in J.S. Schneiderman, editor. *The Earth around Us: Maintaining a Livable Planet*. New York: W.H. Freeman.

———. 2000b. Why predict? Historical perspectives on prediction in earth science. Pp. 23-40 in D. Sarewitz, R.A. Pielke Jr, and R. Byerly, editors. *Prediction: Decision Making and the Future of Nature*. Washington, DC: Island Press.

Oreskes, N., and K. Belitz. 2001. Philosophical issues in model assessment. Pp. 23–41 in M.G. Anderson and P.D. Bates, editors. *Model Validation: Perspectives in Hydrological Science*. Chichester: John Wiley and Sons.

Oreskes, N., K. Shrader-Frechette, and K. Belitz. 1994. Verification, validation, and confirmation of numerical models in the earth sciences. *Science* 263: 641–646.

Petterssen, S. 2001. *Weathering the Storm: Sverre Petterssen, the D-Day Forecast, and the Rise of Modern Meteorology*, J.R. Fleming, editor. Boston: American Meteorological Society.

Pielke, R.A. Jr. 1999. Who decides? Forecasts and responsibilities in the 1997 Red River Flood. *Applied Behavioral Science Review* 7: 83–101.

Pilkey, O.H. Jr. 1990. A time to look back at beach nourishment (editorial). *Journal of Coastal Research* 6: iii–vii.

———. 1994. Mathematical modeling of beach behaviour doesn't work. *Journal of Geological Education* 42: 358–361.

———. 2000. Predicting the behavior of nourished beaches. Pp. 159–184 in D. Sarewitz, R.A. Pielke Jr, and R. Byerly, editors. *Prediction: Decision Making and the Future of Nature*. Washington, DC: Island Press.

Rastetter, E.B. 1996. Validating models of ecosystem response to global change. *Bioscience* 46: 190–198.

Sarewitz, D., and R.A. Pielke Jr. 2000. *Breaking the Global-Warming Gridlock, Atlantic Monthly July 2000*. http://www/theatlantic.com/issues/2000/07/sarewitz.htm.

Sarewitz, D., R.A. Pielke Jr, and R. Byerly, editors. 2000. *Prediction: Decision Making and the Future of Nature*. Island Press: Washington, DC.

Shackley, S., P. Young, S. Parkinson, and B. Wynne. 1998. Uncertainty, complexity, and concepts of good science in climate change modeling: Are GCMs the best tools? *Climatic Change* 38: 159–205.

Tsang, Chin-Fu. 1991. The modeling process and model validation. *Ground Water* 29: 825–831.

———. 1992. Reply to the preceding discussion of "The modeling process and model validation." *Ground Water* 30: 622–624.

3

The Status of Dynamic Quantitative Modeling in Ecology

William K. Lauenroth, Ingrid C. Burke,
and Joyce K. Berry

Summary

Dynamic quantitative modeling in the broadest sense has been a part of
ecology since early in the twentieth century. It began in the form of
mathematical population theory and was expanded in midcentury by
the addition of systems analysis and ecosystem modeling. Because
modeling is an important and productive activity for a scientific discipline, the status of modeling can be used as an indicator of the state of
the discipline. Our objective was to evaluate the current state of ecology with respect to dynamic quantitative modeling and to speculate
about profitable future directions for incorporation of dynamic quantitative modeling. We evaluated the state of modeling in ecology in two
ways. First, we assessed the attitudes of ecologists about quantitative
modeling, and second, we evaluated the degree to which quantitative
modeling is being used as a research tool in ecology.

We assessed the attitudes of ecologists toward quantitative modeling by conducting a survey of a random 20% sample of the membership of the Ecological Society of America. We sent 1,535 e-mail
invitations to participate in the survey, and 350 ecologists responded.
As a separate analysis we evaluated the degree to which quantitative
modeling is currently being used as a research tool in ecology by assessing the frequency and kind of modeling that was published in the
journals *Ecology* and *Ecological Applications* in 1996, 1998, and 2000.

We found that ecologists are overwhelmingly supportive of an important role for dynamic quantitative modeling in both past discoveries
and future advancement of knowledge. This was true for population,
ecophysiological, community, and ecosystem ecologists. By contrast,
we discovered that only 17% of the papers published in what are arguably the premier journals in the field employed dynamic quantitative
modeling. This difference, representing what individuals say and what
is represented in our journals, may be explained by limited availability

of training in quantitative modeling. While most ecologists recognize the potential value, relatively few have appropriate training to allow them to make effective use of dynamic quantitative modeling. Our results suggest that an effective way to address this problem is to establish opportunities for relationships between undergraduate and graduate students and ecologists who are using modeling in their research. Our conclusion, conditioned on the limitation of our sample, is that ecologists are ready to make much more effective use of dynamic quantitative modeling than they have in the past. Our major challenge is to decide how to most effectively make the necessary skills available to ecological researchers.

Introduction

Dynamic quantitative modeling in the broadest sense has been a part of ecology from early in the twentieth century. It began in the form of mathematical population theory and was expanded in midcentury by the addition of systems analysis and ecosystem modeling. Population modeling had peaks of activity in the 1920s and the 1950s. In the 1920s, Lotka, Volterra, Pearl, Verhulst, and others were providing the foundation for mathematical population ecology, much of which survives to the present (Kingsland 1985; MacIntosh 1985). While the others were primarily interested in populations, A.J. Lotka had much broader interests and his writings, particularly *Elements of Physical Biology*, contained the seeds of systems analysis. Those seeds were to lie dormant until the 1950s, when Bertalanffy and others developed general systems theory and what was to become modern systems analysis (Kingsland 1985). The 1950s and 1960s saw a resurgence of interest in theoretical population ecology as the result of the enthusiasm injected into the field by Robert MacArthur and of his interactions with most of the other major players in population ecology (Kingsland 1985; MacIntosh 1985).

Systems analysis began as a topic in engineering but was rapidly transferred to ecology by H.T. Odum, J.S. Olson, D. Garfinkel, and others in the late 1950s and early 1960s (Shugart and O'Neill 1979). Systems analysis in ecology became ecosystem modeling with the advent of U.S. participation in the International Biological Program (IBP). The key figures during this period were J.S. Olson, G.M. VanDyne, and B.C. Patten (MacIntosh 1985; Golley 1993). VanDyne, as director of the IBP Grassland Biome program, was a driving force for systems analysis in the entire U.S. IBP effort (MacIntosh 1985; Golley 1993). Although interest in ecosystem modeling peaked in the 1970s with the IBP, it is to some degree having a rebirth with the recent interest in predicting ecosystem effects of global change (but see Oreskes this volume and Pielke this volume).

Modeling is an important and productive activity for a scientific discipline (Morgan and Morrison 1999), and therefore the status of modeling can be used as one important indicator of the state of the discipline. Our objective was to evaluate the current state of ecology with respect to dynamic quantitative modeling and to speculate about what role it could and should play in the future. We evaluated the state of quantitative modeling in ecology in two ways. First, we assessed the attitudes of ecologists about quantitative modeling, and second, we evaluated the degree to which quantitative modeling is being used as a research tool in ecology. Our definition of a dynamic quantitative model followed standard conventions of an equation or set of equations that describes the change in state of a system with time (Forrester 1961; Luenberger 1979; Percival and Richards 1982). We extended this definition to include models that describe the change in system state over space.

Assessing Attitudes of Ecologists

We assessed the attitudes of ecologists by conducting a survey consisting of 32 questions (Appendix I, http://www.ecostudies.org/cary9/appendices.html). The questions were designed to address three components of the attitudes of ecologists toward quantitative modeling. The first consisted of a series of questions designed to query attitudes about modeling, and the second included questions about the respondents training in modeling techniques. We also included standard questions about demography to help us detect trends among age, gender, or subfield groups and to provide benchmarks against which we could ask whether our survey was representative.

WebSurveyor Corporation implemented the survey on the Web and potential respondents were invited to participate via e-mail. The survey was pre-tested on a group of 110 ecologists associated with the Graduate Degree Program in Ecology at Colorado State University. This enabled us to identify weak questions and programming problems. For the final survey, e-mail invitations were sent to 1,535 members of the Ecological Society of America (ESA), representing a 20% random sample of those members with e-mail addresses. We chose the ESA for both conceptual and practical reasons. Conceptually, we thought that the ESA represents one of a small number of associations of ecologists whose members demonstrably define the forefront of ecological knowledge. Practically, the ESA is technologically advanced, and after furnishing appropriate justification and assurances, we were able to obtain a sample of the e-mail list of the membership. That list contained 7,677 addresses, and the membership of the society in 2000 was 7,807. Two percent of the membership of the society in 2000 did not have an e-mail address and were therefore excluded from the survey.

Of the 1,535 invitations we sent, 70 of the invitations were undeliverable because of errors in the addresses, and 350 members (24%) responded by completing the survey. The only information we were able to obtain to assess the representativeness of our sample was the gender split of the membership. The society cautioned us that the information they requested from members was voluntary and may not be definitive in characterizing the membership. Of the 5,979 members reporting gender, 28% were females and 72% males. In our survey results, respondents were 29% female and 71% male. For our remaining demographic data we have no benchmarks for comparison. All of the quantitative data from the survey are reported in Appendix I (http://www.ecostudies.org/cary9/appendices.html).

Evaluating Publications

We assessed the status of dynamic quantitative models in ecological publications by looking at every regular article printed in three volumes of *Ecology* and *Ecological Applications*. ESA publishes both journals, and we chose these journals for the same reasons that we chose to survey the membership of ESA. They are arguably two of the most important outlets for ecological research; they both have high impact factors, and they are among the top journals in the ecology group (ISI 1999). Of the 90 journals in the Institute for Scientific Information's ecology group, *Ecology* is ranked number 8 and *Ecological Applications* is number 15 in terms of impact factor.

We recognize that the ESA journals, especially *Ecology*, are not often thought of as prime outlets for modeling papers. Some might argue that at least past editorial policy has implicitly or explicitly made it very difficult to get modeling papers published. We do not deny these concerns but argue that the employment of dynamic quantitative modeling as an important part of the theoretical underpinning and/or as an important part of the analysis of empirical results has always been a part of what gets published in these journals. We further argue that if quantitative modeling is important in advancing the state of knowledge in ecology, it should be an integral part of the manuscripts that get published in the flagship journals of one of the leading societies of ecologists. What often does not get published in the ESA journals are papers that focus entirely on the description and analysis of a quantitative model. Other journals such as *Ecological Modelling* are primary outlets for such papers.

For the two journals, we looked at every issue published in 1996, 1998, and 2000 and determined whether each regular article contained the use of dynamic quantitative modeling. For those papers that used dynamic models, we determined the subfield they best represented (population, ecophysiology, community, or ecosystem) and whether they used an analytical or simulation model. Each paper was allowed only a single model type–subfield category (i.e. analytical population, community simulation, etc.). In cases in which a portion of the analysis was completed analytically and the remainder via simulation, the paper was categorized as simulation.

We used two criteria to identify papers containing dynamic quantitative modeling. We first required the paper to contain results from manipulating an analytical model or solving a simulation model through time or over space. We also required that the model be used as a conceptual framework or theory for designing the analysis and/or interpreting the results. The model needed to be both a tool for analysis and a structure for placing the results into a larger framework. We used this definition in the introduction to our survey to convey to the respondents what we meant by quantitative models (Appendix I). Statistical models, time series analyses, and geostatistical analyses were not considered to be dynamic quantitative models for the purposes of this paper.

What Ecologists Say about Quantitative Modeling

The initial presentation of our survey results will focus on the following four questions.

1. Do ecologists consider modeling important?
2. Is there variability to the answer to question 1 depending upon subfield?
3. What kinds of models do ecologists use?
4. Is there variability to the answer to question 3 depending upon subfield?

The answer to our first question was overwhelmingly affirmative (Figure 3.11a). Ninety-eight percent of the ecologists surveyed agreed that modeling was an important tool, and 52% strongly supported the importance of modeling. Separating the results by subfield suggested that there are differences compared to the average response across subfields (Figure 3.1). The distribution of answers (strongly agree, agree, tend to agree, tend to disagree, disagree, strongly disagree) for population ecologists ($\chi^2 = 15.94$; $df = 4$) and for community ecologists ($\chi^2 = 12.69$; $df = 5$) was significantly different ($P \leq 0.05$) from the answers for all ecologists combined. The key difference for population ecologists was that more of them strongly agreed about the importance of modeling than all ecologists combined. Conversely, the key difference for community ecologists was that fewer of them strongly agreed about the importance of modeling than all ecologists combined. Lumping the data into two categories, "agreed" or "disagreed," suggested that only physiological ecologists differed from the average for all ecologists ($\chi^2 = 6.10$; $df = 1$) and they all fell into the "agreed" category.

The respondents also overwhelmingly supported the idea that quantitative modeling had contributed to advancement of knowledge in the past and would continue to contribute in the future (Figure 3.2). Ninety-five percent of respondents thought that modeling had contributed substantially in the past, and 96% thought that it would continue to contribute in the future.

There was a relatively small difference in the use of dynamic models between simulation and analytical with 45% favoring simulation and 39% favoring analytical. Nine percent of the respondents indicated that they do not use models, and 6% use other kinds of models, largely statistical. The choice of model type differed among subfields. Both population ecologists ($\chi^2 = 21.88$; $df = 2$) and ecophysiologists ($\chi^2 = 7.68$; $df = 2$) used simulation more frequently than the average.

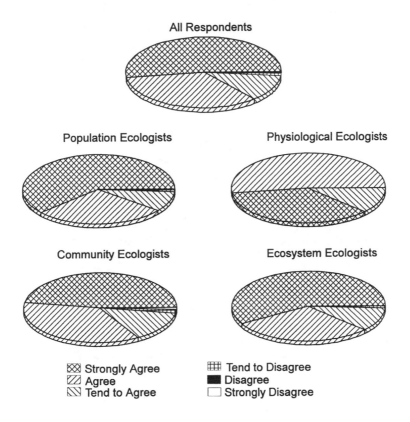

All Respondents

Population Ecologists

Physiological Ecologists

Community Ecologists

Ecosystem Ecologists

Strongly Agree   Tend to Disagree
Agree   Disagree
Tend to Agree   Strongly Disagree

Figure 3.1. The distribution of responses to the statement "Quantitative modeling is an important tool for ecologists" for all respondents and for respondents separated into subfield categories.

a.   b.

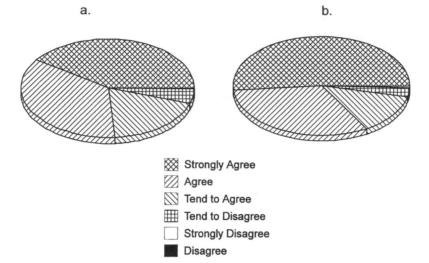

☒ Strongly Agree
▨ Agree
◫ Tend to Agree
▦ Tend to Disagree
☐ Strongly Disagree
■ Disagree

Figure 3.2. The distribution of answers to the statements (a) "Quantitative modeling has contributed substantially to the advancement of knowledge in ecology in the past"; and (b) "Quantitative modeling will be important to the advancement of knowledge in ecology in the future."

What Ecologists Publish

We sampled 627 papers in *Ecology* and 234 in *Ecological Applications* in 1996, 1998, and 2000. Our analysis of the journal data will focus on four questions that were designed to be parallel to those for the survey data. The questions are

1. Do ecologists use models in the research?
2. Is there variability to the answer to question 1 depending upon subfield?
3. What kinds of models do ecologists use?
4. Is there variability to the answer to question 3 depending upon subfield?

We found that 15% of the *Ecology* papers (91) and 23% of the *Ecological Applications* papers (54) contained some use of dynamic quantitative modeling. While it is possible that we categorized a paper as not containing quantitative modeling when it in fact did, we are quite certain that the opposite was not possible. Therefore our 15% for *Ecology* and 23% for *Ecological Applications* can be interpreted as absolute minima. Importantly, the bulk of the papers in both journals, 536 in Ecology and 180 in *Ecological Applications*, did not contain quantitative modeling. Our findings were fairly consistent for all three years. In *Ecology*, the number of modeling papers ranged from 29 in 1996 to 32 in 2000. The modeling papers in *Ecological Applications* were more variable and ranged from 13 in

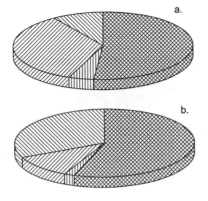

a.

b.

Figure 3.3. The total distribution of modeling papers by subfield published in (a) *Ecology* and (b) *Ecological Applications* in 1996, 1998, and 2000.

▨ Ecosystem   ▨ Population
▥ EcoPhys   ▧ Community

2000 to 21 in 1998. There was no clear increasing or decreasing trend in the use of models over time in either journal although three years of data are likely too few to detect a trend.

Population models were used in greater than 50% of the modeling papers in both journals (Figure 3.3); ecophysiological models were the least-used category. There was a large difference between the two journals in the representation of community and ecosystem models. In *Ecological Applications*, 31% of the modeling papers referred to ecosystems and 11% to community models (Figure 3.3b), while in *Ecology*, 9% were ecosystem models and 34% were community models (Figure 3.3a). There was substantial variability among years in each of the subfield categories, with the most poorly represented category, ecophysiological models, being the most variable (table 3.1).

Table 3.1. Categorization of modeling papers from *Ecology* and *Ecological Applications* by subfield.

	2000		1998		1996	
	Number	%	Number	%	Number	%
Ecology						
Population	6	50	18	60	12	41
Ecophysiology	3	9	1	3	1	3
Community	11	34	7	23	14	48
Ecosystem	2	6	4	13	2	7
Ecological Appl.						
Population	4	31	13	62	13	65
Ecophysiology	1	8	0	0	0	0
Community	3	23	2	10	1	5
Ecosystem	5	38	6	29	6	30

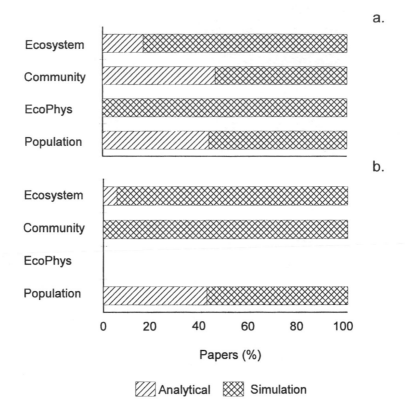

Figure 3.4. The total distribution of the kinds of models used by subfield for papers published in (a) *Ecology* and (b) *Ecological Applications* in 1996, 1998, and 2000.

Simulation models were by far the most common model type, representing 63% of the modeling papers in *Ecology* and 76% of the modeling papers in *Ecological Applications*. Since simulation was used as the priority category in classifying the papers, some of the papers in this group used analytical solutions as well. Investigation of the variability among subfields in their use of the two model types revealed largely the expected answers (Figure 3.4). The subfield that most commonly used analytical models was population ecology, with 40% of the population models in both journals categorized as analytical. We were somewhat surprised that 40% of the community models published in *Ecology* were also analytical models; these were all age/stage class matrix models. The majority of the ecosystem models were simulation models in both journals.

The Current Status of Dynamic Quantitative Modeling in Ecology

We suggested in the introduction that modeling is an important and productive activity for any field of science (Morgan and Morrison 1999). Our survey results indicate strong and widespread support for the importance of modeling by ecologists. Ninety-eight percent of ecologists surveyed agreed that quantitative modeling was an important tool, and similar percentages agreed that it had contributed substantially to the field in the past and would continue to into the future. While we found variability among subfields, none of it contradicted this general positive attitude toward modeling. The respondents endorsed the idea that modeling is important and productive for ecology.

The literature survey produced different results. Combining the data from *Ecology* and *Ecological Applications* and putting the total forward as an indication of how knowledge is generated in the field of ecology, 83% of the published material did not employ the use of quantitative modeling and 17% did. Does this mean that while ecologists agree that modeling is important, they do not often use dynamic models to answer ecological questions, or is there another explanation? Before we can make statements about the status of dynamic quantitative modeling in ecology, we must attempt to reconcile this issue. To accomplish this we will explore the answers to two related questions: (1) Is this a contradiction? And (2) Does it matter that few ecologists employ dynamic quantitative modeling in their research?

Should we be surprised that when asked, a majority of respondents expressed strong support for dynamic quantitative modeling, yet when we evaluated the literature, a minority of papers included the use of such models? One could argue that the two issues are independent. Agreement by most ecologists with statements about the importance of modeling does not provide any information about the frequency with which we should expect to encounter dynamic quantitative modeling in the literature. Alternatively, one could argue that dynamic modeling represents such a powerful tool that we should expect a relationship similar to the one we would expect for statistics: expression of strong support for its importance should translate into a high frequency of encounter in the literature. Regardless of which interpretation is appropriate for ecology, the low frequency with which we recorded dynamic quantitative modeling in the literature raises questions about why this should be so. We have additional data from our survey that may be useful in explaining the apparent differences between the attitude of ecologists, as reflected in our survey, and the published ecological literature. We asked how often the respondents used quantitative modeling in their research and discovered that 76% used quantitative modeling at least occasionally and 24% almost never or never. While this helps explain part of the difference, it does not take care of all of it. When we asked the respondents if they were modelers as opposed to users of models, 35% answered that they were users of models, and only 19% indicated that they were modelers (Figure 3.5). We asked the respondents to choose the five most important reasons that limit their use of models from a list and found that lack of

Figure 3.5. The distribution of responses to the statement "Would you describe your-self as…"

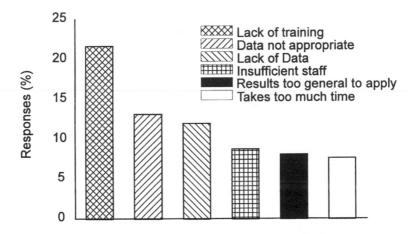

Figure 3.6. Responses to the question "What are the most important reasons limiting your use of quantitative models? (Indicate top 5 reasons)."

training was the most frequently cited reason (Figure 3.6). If we include "insuffi-cient staff" as another component of lack of expertise, the two categories account for 31% of the responses. These additional data explain part of the difference be-tween the survey results and the journal analysis. While an overwhelming number of ecologists think modeling is important and has a high potential to advance knowledge, a much smaller number are currently able to use quantitative modeling in their research, largely because of a lack of expertise.

If we are interested in increasing the number of ecologists who are trained in quantitative modeling, it will be useful to understand where and at what stage in their careers the current contingent of modelers and perhaps model users got their training (Figure 3.7a). Surprisingly, only 25% of modelers obtained their training in modeling classes. The majority is self-taught or received their training working with experienced modelers. Most were undergraduate or graduate students when

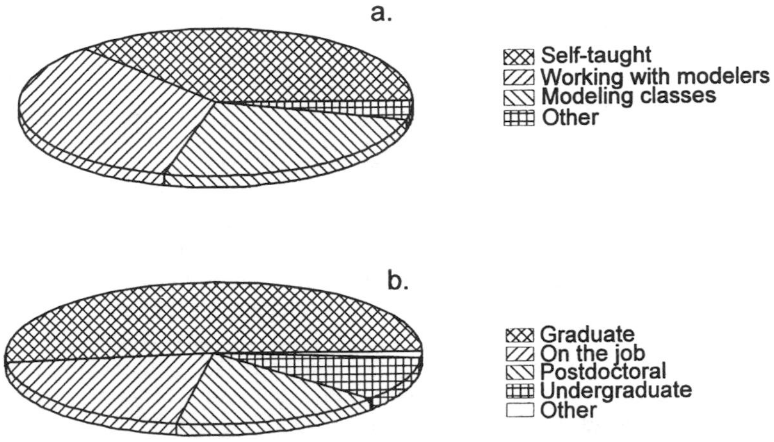
a.
Self-taught
Working with modelers
Modeling classes
Other
b.
Graduate
On the job
Postdoctoral
Undergraduate
Other

Figure 3.7. Answers to questions about how and when respondents obtained their modeling skills: (a) "If you are modeler, how did you obtain your modeling skills?" and (b) "At what stage in your career did you obtain your modeling skills?"

they received their training (Figure 3.7b). Postdoctoral experience accounted for a small fraction of the training. How might we increase the number of ecologists with training in modeling? Our results suggest that programs that connect undergraduate and graduate students to scientists who are actively using quantitative modeling in their research have a high probability of being effective. Such exposure could both introduce young ecologists to the value of modeling and to the necessary skills.

The Future of Modeling in Ecology

Does it matter that few ecologists employ dynamic quantitative modeling in their research? Modeling represents an important way in which we use theory in ecology. Dynamic quantitative models have the potential to satisfy the key criteria by which theories are evaluated (Barbour 1966): agreement with data, internal logical consistency, and generality. The simple alternative to a field with a strong theoretical component is unconstrained empiricism and domination by case studies. Viewed through this lens, the picture of ecology that emerges from our assessment of the literature is of an intensely empirical field, in which the participants have widespread appreciation for the potential benefits of quantitative theory but employ it very little as a tool in their research. Eighty-three percent of the papers published in *Ecology* and *Ecological Applications* in 1996, 1998, and 2000 were empirical and 17% had a quantitative theoretical component. Is this too much empiricism or too little theory development? We don't know of a simple basis for answering this question, but we can speculate that the current approach will lead to an enormous quantity of unconnected data about ecological systems (i.e., case studies). This was amply borne out by our experience of looking at every published paper in six journal-years. Ecologists have an enormous amount of knowledge about individual ecological objects of study repeated over different locations but have few general frameworks that provide information about how these objects are or can be related to one another.

Another of the responses from our survey points to data limitations on the use of quantitative modeling (Figure 3.6). Combining the two categories "data not appropriate" and "lack of data" suggests that data limitations account for 25% of the responses to the question "What are the most important reasons limiting your use of quantitative modeling?" Perhaps it is the presence of what we call "the data paradox" in ecology, in which we have both too much and too little data. We have too much data of the type that Lauenroth et al. (1998) referred to as empirical fragments and too little that have been collected within the framework of a quantitative theory or dynamic model. This is not a recommendation for fewer data and more model development. This is a plea for more model development and more theory-guided data collection.

The criticism that ecology is awash in data is not new and is in many ways a restatement of a portion of the criticism that was leveled at ecology by an aspiring theoretical community in the 1950s and 1960s (MacIntosh 1985). Two things are noteworthy today. First, it is almost fifty years later and ecology still seems vulnerable to the criticism. Second, we are currently not in the midst of two revolutions in the field. In the 1950s and 1960s, theoretical population ecologists were locked in a struggle with anti-theorists and the stakes were assumed to be the future of ecology (MacIntosh 1985). Simultaneously, another group was splitting the field in a different direction by promoting what they called "systems ecology" with its daunting collection of mathematical, statistical, and computational tools borrowed from systems engineering (Shugart and O'Neill 1979; MacIntosh 1985). Additionally, the two revolutionary groups had little respect for each other and the fighting within each group was at times fierce (Kingsland 1985).

At the turn of the twenty-first century we have a much more sophisticated understanding of quantitative modeling and continuing needs for theory development under the broadest possible definition of theory. Presumably we are beyond the idea of a single unifying theory for ecology, but the need for contingent theories is everywhere evident. While theory remains a controversial term in ecology, what seems unquestionable is the need in ecology for the kind of organization of empirical fragments that "theory" has brought to other fields. We should acknowledge here that we recognize the existence and value of ecological theories that do not meet our definition of quantitative theory. Whether we call them theories, conceptual frameworks, or operational frameworks, dynamic quantitative models have an important role to play in fulfilling the need for organization of our enormous depth

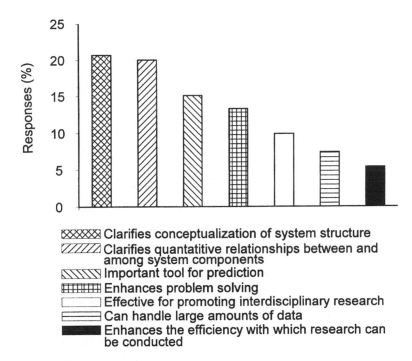

Figure 3.8. Answers to the question "What are the most important reasons why you use quantitative models?"

of knowledge about ecological systems and for guiding future knowledge generation (Lauenroth et al. 1998).

Additional results from our survey of attitudes may provide some insight into the issue of the utility of quantitative modeling. From a list of reasons why they use quantitative models, the most frequently selected reasons were that they clarify conceptualization of system structure and clarify quantitative relationships between and among system components (Figure 3.8). On a related topic, when asked to what degree the use of quantitative models has stimulated new ideas or directions in their research, 81% of the respondents answered "very often," "often," or "occasionally" (Figure 3.9a). Finally, 56% of the respondents indicated that they thought models were most useful as heuristic tools (Figure 3.9b). Sixteen percent of the respondents checked the "other" category on this question, and the majority of their reasons were closely related to the value of models for placing data in a larger context. While this information does not directly support our contention that models are needed to help ecologists deal with the huge amount of data the field has accumulated, we think it is related to this issue.

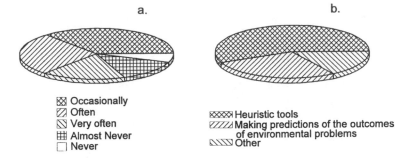

a.   b.

☒ Occasionally
☑ Often
☒ Very often   ☒ Heuristic tools
⊞ Almost Never   ☒ Making predictions of the outcomes
☐ Never   of environmental problems
                                         ☒ Other

Figure 3.9. Answers to the question "To what degree, in your research, has quantitative modeling stimulated new ideas or directions?" (a) and the statement "In my opinion, models are most useful as…" (b).

Lauenroth et al. (1998) dealt in some detail with this issue and here we elaborate one of their examples of such a use for a quantitative model. The CENTURY model is arguably the best theoretical framework and synthesis of the biogeochemistry of grasslands in existence (Parton et al. 1987). CENTURY represents the key elements of the biogeochemical structure of grasslands with three compartments into which the carbon, nitrogen, phosphorus, sulfur, etc. are divided (Jenkinson and Rayner 1977; Anderson 1979). These are conceptual rather than physical pools, and each has a characteristic turnover time. The passive pool represents the most recalcitrant material and has a turnover time of hundreds to thousands of years. The slow pool is intermediate material and has a turnover of decades to half century. The active pool represents the most labile material and turns over in months to several years. This structure has proven to be very fruitful in terms of stimulating thinking about the biogeochemistry of grasslands and in terms of stimulating research. Substantial effort has been and continues to be expended in search of physical analogs of these pools (Cambardella and Elliot 1994; Kelly et al. 1996). In addition to a structure that has been found to be remarkably robust under considerable testing, CENTURY also includes representations of processes that are simultaneously consistent with available data and have been influential in terms of shaping thinking about controls on processes. Decomposition is an excellent example. While there is much controversy about whether temperature or water availability controls decomposition (see Burke et al. this volume), CENTURY uses a synthetic parameter that combines information about temperature and water availability to control decomposition and that has proven to be quite robust in simulating decomposition across ranges of ecosystems and litter quality (Gholz et al. 2000). Because of the importance of the representations of structure and processes in CENTURY, it has become a de facto standard against which grassland biogeochemical data sets are compared. While there are many well-known limitations to CENTURY, it is an enormously valuable tool for researchers in grassland biogeochemistry.

We think that tools such as CENTURY are not only possible but also absolutely necessary for other topic areas in ecology in order for us to progress beyond the empirical phase in which we currently find ourselves. We are not suggesting that every topic area needs to have a single dynamic quantitative model that captures the key issues, although there may be value in that. Rather, what is needed is an active quantitative modeling component. Based upon the strong support of dynamic quantitative modeling among the membership of the Ecological Society of America, ecology seems ready to make dynamic quantitative modeling a much larger part of its future than it has been of its past.

References

Anderson, D.W. 1979. Processes of humus formation and transformation in soils of the Canadian Great Plains. *Journal of Soil Science* 30: 77–84.

Barbour, I.G. 1966. *Issues in Science and Religion.* New York: Harper and Row.

Cambardella, C.A., and E.T. Elliott. 1994. Carbon and nitrogen dynamics of soil organic-matter fractions from cultivated grassland soils. *Soil Science Society of America Journal* 58: 123–130.

Forrester, J.W. 1961. *Industrial Dynamics.* Cambridge: M.I.T. Press.

Gholz, H.L., D.A. Wedin, S.M. Smitherman, M.E. Harmon, and W.J. Parton. 2000. Long-term dynamics of pine and hardwood litter in contrasting environments: Toward a global model of decomposition. *Global Change Biology* 6: 751–765.

Golley, F.B. 1993. *A History of the Ecosystem Concept in Ecology.* New Haven: Yale University Press.

ISI. 1999. Journal citation reports. Compact disc. Philadelphia: ISI (Institute for Scientific Information).

Jenkinson, D.S., and J.H. Rayner. 1977. The turnover of soil organic matter in some of the Rothamsted classical experiments. *Soil Science* 123: 298–305.

Kelly, R.H., I.C. Burke, and W.K. Lauenroth. 1996. Soil organic matter and nutrient availability responses to reduced plant inputs in shortgrass steppe. *Ecology* 77: 2516–2527.

Kingsland, S.E. 1985. *Modeling Nature.* Chicago: University of Chicago Press.

Lauenroth, W.K., C.D. Canham, A.P. Kinzig, K.A. Poiani, W.M. Kemp, and S.W. Running. 1998. Simulation modeling in ecosystem science. Pp. 404–415 in M.L. Pace and P.M. Groffman, editors. *Successes, Limitations, and Frontiers in Ecosystem Science.* New York: Springer-Verlag.

Luenberger, D.G. 1979. *Introduction to Dynamic Systems: Theory, Models, and Applications.* New York: John Wiley and Sons.

MacIntosh, R.P. 1985. *The Background of Ecology.* Cambridge: Cambridge University Press.

Morgan, M.S., and M. Morrison. 1999. *Models as Mediators.* Cambridge: Cambridge University Press.

Parton, W.J., D.S. Schimel, C.V. Cole, and D.S. Ojima. 1987. Analysis of factors controlling soil organic matter levels in Great Plains grasslands. *Soil Science Society of America Journal* 51: 1173–1179.

Percival, I., and D. Richards. 1982. *Introduction to Dynamics.* Cambridge: Cambridge University Press.

Shugart, H.H., and R.V. O'Neill. 1979. *Systems Ecology.* Stroudsburg, PA: Dowden, Hutchington and Ross.

4

The Utility of Simple Models in Ecosystem Science

Michael L. Pace

Summary

Ecosystem models are often envisioned as complex, multicompartment simulations. This vision, however, misses a large class of models that are "simple" in the sense that they consist of only one or a few equations. Three uses of simple models in ecosystem research are comparisons of empirical relationships, evaluation of alternate states, and prediction. Comparisons of ecosystems often reveal empirical patterns. For example, general patterns of the sinking flux of organic matter as a function of primary production differ between lakes and ocean systems. An evaluation of the underlying relationships reveals a distinction between oligotrophic systems that suggests fundamental differences between lake and ocean primary producers. Simple dynamic models represent a second type of modeling approach that is particularly useful for exploring conditions that promote shifts in ecosystem states. An example of these alternate states is the shift from turbid to clear-water conditions in shallow lakes. Simple models also provide predictions and may outperform more complex models in forecasting. A recent evaluation of models that forecast the dynamics of the 1997–1998 El Niño reveals that well-developed dynamic and statistical models were no better, and in some cases much worse, than a simple empirical model based on historical data.

Simple models have a wide utility in ecosystem science. They are useful for exploring general questions and depicting ideas as well as for evaluating specific research problems. Simple models are pathways to a better understanding of the patterns and underlying mechanisms that structure ecosystems, and they can aid communication with other scientists, managers, and the public.

Introduction

Imagine asking a group of ecologists to envision a model of an ecosystem. What would they see? My guess is that regardless of whether this model was of a grassland, forest, ocean, or stream, most would think of an ecosystem in a box-and-arrow format consisting of numerous compartments, flows, and feed-backs. These types of modest to highly complex models have been the domi-nant approach to the analysis of ecosystems since the initial studies of the Inter-national Biosphere Program (McIntosh 1985; Lauenroth et al. 1998). Current models of this genre include well-known terrestrial ecosystem models such as CENTURY (Parton et al. 1987), PnET (Aber and Federer 1992), and GEM (Rastetter et al. 1991). A common feature of these models is that they are rela-tively complex: they are based on more than a few equations and require that a number of parameters be estimated.

This vision of ecosystem modeling, however, is incomplete. It misses a large class of models that are "simple" in the sense that they consist of only one or a few equations. Simple models most often describe some process or aspect of ecosystems. This mathematical description might be strictly empirical (correla-tive) or more mechanistically based, for example, on equations that conserve mass flow. The purpose and goals of simple models are highly diverse, encom-passing many aspects of ecosystem science. These models are ubiquitous tools and, hence, deserve consideration in any discussion of ecosystem modeling.

This chapter considers simple models by discussing their utility in ecosystem science, but rather than providing a comprehensive review, it instead empha-sizes a few general points. First, simple empirical models are useful for inte-grating the results of comparative studies and can lead to important, novel in-sights about ecosystems. Second, simple dynamic models aid the understanding of alternate conditions in ecosystems and the possible mechanisms that promote or inhibit shifts between states. Third, simple models make useful predictions, and these predictions may be better than those derived from more complex models. Each point is illustrated by a specific example, but the specifics serve only as a means toward a larger goal of assessing the uses, advantages, and con-tributions of simple models.

There is no specific definition of simple models, but here I will consider them to be models with only a few equations. This discussion of the benefits of simple models should not be interpreted as a criticism of more complex models based on many equations. The advantages and shortcomings of large ecosystem models is a separate subject, and the advocacy for the use of simple models made here does not imply a lesser role for more complex models.

Simple Empirical Models and Comparisons of Ecosystems

Simple empirical models such as linear regressions are often used to integrate data from studies of numerous ecosystems. For example, many relationships between soil respiration and temperature often are established from a variety of ecosystems with different soils and vegetation (e.g., Raich and Schlesinger 1992, Lloyd and Taylor 1994). These relationships set a standard for evaluating new data from different systems, serve as equations in models that require estimates of soil respiration, and summarize the overall pattern of how respiration changes with temperature.

In addition to summarizing major patterns, comparative studies can also lead to unexpected insight about similarities and differences of ecosystems. The comparison by Baines et al. (1994) of the sinking flux of organic carbon in ocean and lake ecosystems provides an example. Oceanographers have long been interested in the losses of carbon (C) from the photic, surface layer to the deep ocean, because this flux of C is important in the global carbon cycle. The transfer of organic carbon from surface to deep waters is known as the "biological pump" (Volk and Hoffert 1985). Inorganic carbon is taken up by phytoplankton. The resulting organic carbon has three possible fates that are mediated by a combination of nutrient, phytoplankton, and food web interactions. Most of the phytoplankton carbon is eventually respired, some is converted to dissolved organic carbon (DOC), and the balance sinks as particulate organic matter (POC). The sinking of POC, either passively as dead organic matter or actively by organism migration, and net losses of DOC during deep mixing are the two means whereby the biological pump moves carbon from surface to deep water.

The efficiency of upper-ocean systems is often assessed by comparing the sinking flux of POC (measured in sediment traps) out of the photic zone relative to primary production (Wassman 1998). The sinking flux ratio (SFR) tends to increase with primary production in oceanic ecosystems—a pattern observed in a number of different studies (Figure 4.1a). However, measurements of SFR in lakes reveal a very different pattern (Baines and Pace 1994): SFR tends to decrease with primary production (Figure 4.1a).

Why do these patterns differ so strikingly? One way to answer the question is to decompose SFR into its components and ask if there are fundamental differences between oceanic systems and lakes. The sinking flux ratio (SFR) is a function of three variables, sinking flux, SF (mg C m^{-2} d^{-1}); volumetric primary production, PP (mg C m^{-3} d^{-1}); and depth, D (m).

$$SFR = SF \div (PP * D) \tag{4.1}$$

All of these variables relate to the concentration of phytoplankton biomass as measured by chlorophyll *a*. Baines et al. (1994) derived independent, empirical equations for lakes and oceans relating chlorophyll to SF, PP, and D. The relationships between chlorophyll and SF as well as chlorophyll and D were sufficiently similar for lakes and oceans that these could not explain the difference. There was, however, a strong difference between the relationships of chlorophyll and PP (Figure 4.1b). There is much greater production per unit biomass in the ocean relative to lakes at low concentrations of chlorophyll ($<$ 1

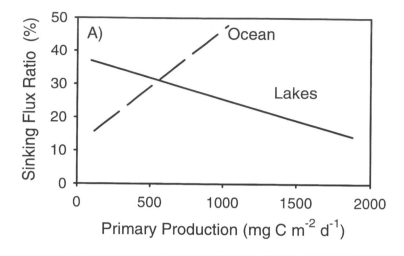

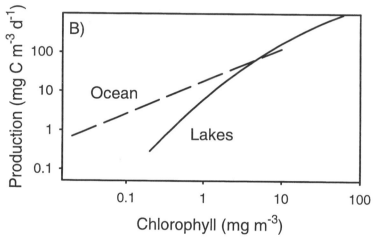

Figure 4.1. (A) The relationship between primary production and percent sinking flux of particulate organic matter. Solid line is for lakes and represents a linear regression fit to data assembled by Baines et al. (1994). Dashed line is an approximation of several ocean models; see Baines et al. 1994 for details. (B) Relationships for chlorophyll versus primary production for lake (solid line) and ocean (dashed line) systems. For both panels A and B, confidence intervals have been excluded to emphasize the patterns. Details on the uncertainties of the relationships are provided in the original study. (Used with permission from Baines et al. 1994, 214 and 219.)

mg m^{-3}). There are many possible reasons for this, one of which is the very small size of oceanic primary producers in the low biomass systems of the most oligotrophic regions of the ocean. These areas are dominated by phytoplankton such as *Prochlorococcus* (Chisholm et al. 1988), and this group has not been observed in lakes.

Whatever the reason, the differences of the relationships for chlorophyll and PP are sufficient to account for the distinctions between lakes and oceans in SFR across productivity gradients. Substituting the individual empirical relations based on chlorophyll (Chl) into equation 4.1 results in the following equations for oceans (4.2) and lakes (4.3):

$$\log \text{SFR} = -0.67 + 0.30 \log \text{Chl} \tag{4.2}$$

$$\log \text{SFR} = -0.22 - 0.41 \log \text{Chl} + 0.27 \log \text{Chl}^2 \tag{4.3}$$

The positive slope in the ocean equation results in an increase in SFR with increases in phytoplankton biomass, while in freshwater systems SFR initially declines and eventually flattens with increases in phytoplankton biomass across the range of observations. These equations for SFR as a function of chlorophyll make distinct predictions that conform to data (Baines et al. 1994).

This study illustrates how empirical models can be used to establish and evaluate ecosystem patterns and to make unexpected discoveries about differences among ecosystems. The insights gained are about ecosystems in general and not about a specific system. The evaluation of simple statistical models not only leads to significant generalizations but also raises important questions about the reasons for the difference in how lakes and oceans function as revealed by the Baines et al. (1994) study. While much of ecological science necessarily focuses on method, measurement, and mechanism, synthesis of existing data into simple models as in this example is an equally significant activity that often helps identify critical conceptual questions.

Dynamic Models and Ecosystem States

Another class of models that are more explanatory and are based on dynamic differential equations is useful for analyzing and characterizing states and state transitions in ecosystems. These models help determine if changes observed in ecosystems are consistent with a relatively small set of mechanisms represented by only a few equations with a small number of parameters. Alternate states and state transitions arise from certain combinations of parameters, which may represent relatively common or relatively rare conditions. For example, Carpenter and Pace (1997) used a simple model of lakes to address whether the conditions of dystrophy (brown water due to high organic carbon loading) and eutrophy (green water due to high nutrient loading) constituted alternate stable states. Analysis of the model revealed that less than 10% of the parameter sets

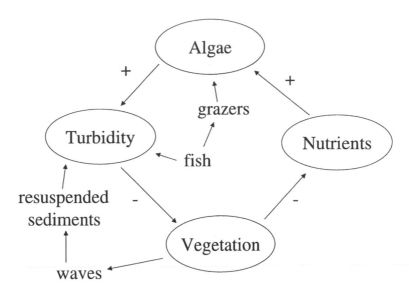

Figure 4.2. Depiction of interactions that lead to alternate states in shallow lakes. Figure is redrawn from Scheffer 1998 and used with permission.

produced model solutions that were consistent with alternate stable states, suggesting that internal processes do not normally cause lakes to change rapidly between green and brown water conditions. Instead, the model indicates that these lake states are ends along a continuum that is determined by relative rates of nutrient and organic matter loading.

Better evidence for alternate states comes from studies of shallow lakes. Some shallow lakes shift between a vegetation-dominated state and a turbid state where algae and/or suspended sediments dominate the water column and inhibit the growth of vegetation (Scheffer et al. 1993). Lake Zwenlust in the Netherlands provides an example (Van Donk and Gulati 1995; Scheffer 1998). Prior to 1988, algae dominated the lake with massive sustained blooms occurring each year. The lake was manipulated to remove most of the fish, and this allowed grazers to flourish, holding the phytoplankton blooms in check. With increased transparency, the favorable light conditions allowed submersed vegetation to colonize and grow on the bottom. The lake shifted to a clear-water state with little phytoplankton and a dominance of submerged, rooted vegetation that was maintained through much of the 1990s. Subsequently, the lake has shifted back to an algae-dominated state. Changes of this type have been observed in a number of different shallow lakes and appear to be driven by several mechanisms, including changes in nutrient loading, storms, changes in water level, and shifts in food web structure (Scheffer 1998).

Several switches appear to cause the shift from an algae-dominated state to a clear-water, submersed vegetation state (Figure 4.2). One set of switches is due to the food web. Shallow, eutrophic lakes often support large populations of planktivorous fish that severely limit grazer populations. If these fish are removed, grazers can exert some control over algae and promote a shift to vegetation dominance. Another switch is related to resuspension of sediments and turbidity due to algae that collectively determine light availability for submersed plants. Once some vegetation is established, there is an inhibition of wave formation and resuspension of sediments allowing greater light penetration with further development of vegetation. These interactions can promote a switch from algae to vegetation dominance. Fish can also have differential net effects based on the relative importance of benthic- versus pelagic-feeding forms. Through their feeding activities, benthic fish help to sustain turbid-water states and inhibit shifts toward a vegetated state.

The effects of turbidity and vegetation can be modeled by two equations that take into account turbidity and the fraction of the lake covered by vegetation together with feedback effects of each variable on the other (Scheffer 1998). In absence of vegetation, turbidity depends on the concentration of algae, which is a function of nutrient loading. This maximum turbidity is modified by vegetation cover, with turbidity declining as vegetation increases. The fraction of the lake covered by vegetation is similarly influences by turbidity.

When run with standard parameters for a shallow Dutch lake, this type of model leads to two stable equilibria (Figure 4.3). The lake exists either in a turbid state or in a vegetated state; intermediate conditions are unstable. Importantly, the behavior of this type of model is consistent with numerous observations. In addition to establishing that this conceptualization of the functioning of shallow lakes is consistent with alternate stable states, the model makes a number of qualitative predictions about how these systems change. It indicates that there are discontinuous responses at thresholds, shifts among states exhibiting hysteresis, and dramatic responses to large perturbations. These predictions agree with various observations, experiments, and long-term studies of shallow lakes, and the general concept of alternate states driven by switches is now being used in lake restoration and management (Moss 1996).

In a discussion of modeling shallow lakes, Scheffer (1998) suggests that simple models are, "best used to study the properties of isolated mechanisms... not to obtain quantitative insights into the relative importance of different mechanisms." Yet, ecosystems are complex and subject to multiple mechanisms that interact and operate simultaneously or in sequence. So are simple dynamic models simply too simplistic to be worthwhile? The answer depends on the scientific question. My sense is that many of the more interesting questions in ecosystem science are about key processes that determine the similarities and differences within and among systems. For this purpose, simple models can be illuminating, helping to isolate features of ecological systems and assessing whether specific mechanisms are consistent with dynamics. This perspective suggests that simple models help unfold the complexity of ecological

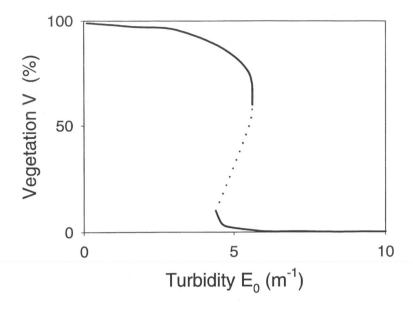

Figure 4.3. Results of model for vegetation and turbidity given parameters typical for a shallow Dutch lake. Solid and dotted lines indicate stable and unstable equilibrium solutions respectively. Figure is redrawn from Scheffer 1998 and used with permission.

systems and provide a pathway for further investigation of critical features (e.g., see Scheffer et al. 2001).

Prediction

Prediction is an important goal for ecology and for modeling in particular (Clark et al. 2001; Pace 2000). Predictions are formally defined as derivations from the logical structure of theory that serve as a means of testing and evaluating theory (Pickett et al. 1994). The means for deriving theoretical predictions is frequently via mathematical models such as the turbidity-vegetation model of shallow lakes discussed above. Predictions are also needed for the application of ecological knowledge to specific situations and to inform the public and policy makers of the likely consequences of current actions and social trends (but see Chapters 2 and 7 for pitfalls and cautions in this context). Uncertainty accompanies most ecological predictions, and the explicit description of uncertainty is another reason that ecological research benefits from a predictive orientation (Clarke et al. 2001; Pace 2001).

The term "prediction" is used informally to describe a number of activities that are associated closely with modeling. These include forecasting and scenario building. At the Cary Conference, there was considerable discussion of what types of predictions could be expected from ecosystem modeling especially with regard to global environmental change. In this context, predictions will be required in a number of different forms, including the consequences of projecting current trends into the future (e.g. Sala et al. 2000; Tilman et al. 2001). Predictions of this nature, however, may tell us little about the real future and may not reduce uncertainty for decision making (see Chapter 7). When entering this arena, ecologists must be clear about the shortcomings and limitations of their models and explicit that changes in human activity will alter the trajectories of many forecasts, such as those related to climate change effects, loss of biodiversity, alterations of habitat, cutting of forests, accumulations of toxins, and degradation of water resources. Forecasting the future precisely is not going to be possible, but evaluating consequences of current trajectories and casting alternate possibilities via scenarios remains a need.

There are many examples where simple models provide useful predictions and clearly demonstrate the uncertainty associated with them; one such example is a study of Lake Mendota, Wisconsin. Because of heavy loading of nutrients from urban and agricultural areas in the watershed, Lake Mendota has long experienced nuisance blooms of cyanobacteria. To assess what level of nutrient loading resulted in blooms, Stow et al. (1997) analyzed a 19-year data set on cyanobacterial dynamics and phosphorus. While the data are noisy, a reasonably good fit was obtained between spring total phosphorus concentrations and summer cyanobacterial biomass with a logistic regression model that assumes error in both variables. Having an established model, its implications can be recast in terms of probabilities of a bloom at specific spring total phosphorus (TP) concentration. Defining a nuisance bloom as exceeding a specific cyanobacterial biomass, the model indicates that there is less than a 5% chance of a bloom at TPs below 0.08 mg L^{-1}, but at TPs of 0.12 mg L^{-1}, the chance of a bloom is greater than 60%. Management to limit blooms might effectively target a spring TP concentration of 0.08 mg L^{-1} or less.

This example illustrates how a management-relevant prediction can be derived from a simple empirical model. While the model is not mechanistic, it is strongly based on underlying principles about nutrient limitation of phytoplankton in lakes and extensive experience in lake management. The uncertainties of the model are clear and their basis transparent. Moreover, this simple model had an important influence on efforts to study the sources of nutrients deriving from the watershed of Lake Mendota and on management plans to reduce these loadings (S. Carpenter, pers. comm.).

A second problem related to models and prediction concerns how well simple models compare to models that are more complex in their ability to forecast. There are relatively few examples of forecasting in the ecological literature, let alone comparisons of forecasts made by different models (Chapter 2). There are examples from other disciplines, such as a recent model comparison for forecasting El Niño–Southern Oscillation (ENSO) events. ENSO events are large-scale ocean-atmosphere interactions that result in the signature condition of a warming of the eastern tropical Pacific Ocean.

Because of the widespread impact of ENSO on global climate, numerous simulation and statistical models forecast this phenomenon. A major event occurred during 1997–1998 for which there were prior model forecasts. Landsea and Knaff (2000) compared these forecasts to assess the "skill" of the models relative to a base model that they viewed as having no skill since it was simply a set of multiple regressions derived from past data. El Niños vary in the rapidity with which they become established, their magnitude and duration, and the rate of decline during the decay phase. The goal of the model comparison for the 1997–1998 event was to determine how well the model forecasts fit the entire dynamic including the onset, duration, and decay of the El Niño.

Landsea and Knaff (2000) compared the models for a number of different forecast lead times. One of their results compared predictions versus observations using a linear correlation coefficient. Models were compared for forecasts that were two seasons ahead (i.e., 6 months). The more complex models were never as good as the base statistical model in this comparison, and some of the forecasts were quite poor (Figure 4.4). Landsea and Knaff made a number of comparisons with various metrics for model comparison (e.g., residual mean square errors). Their primary findings were that all models underestimated warming by at least 50% in 6- and 12-month forecasts, that most models assessed the timing of onset and/or decay of El Niño poorly, and that the complex dynamic simulation models as well as a variety of sophisticated statistical models were no better than a base statistical model derived from historical data. Landsea and Knaff concluded that the best performing model for forecasting the entirety of the very strong 1997–98 El Niño was the base statistical model. Thus, the use of more complex, physically realistic dynamical models does not automatically provide more reliable forecasts.

There is a significant lesson here for ecosystem modeling. The dynamic ocean-atmosphere models used to make El Niño forecasts are well developed and based in many aspects on established and well-understood physical interactions. They are analogous to the best ecosystem models in that they include multiple mechanisms and rich detail. Yet, their predictive ability, at least by the criteria used, was not great. It seems unlikely that in the near term most ecosystem studies will have the data, the resources, or the depth of knowledge about the modeled system that exists for ENSO. Thus, the ability of complex ecosystem models to make predictions is called into question by this example. When prediction is a critical goal, more condensed models of the type I have discussed might often be particularly useful. There are, however, many sources of uncertainty and limits to ecological prediction even for simplest empirical models (see Chapter 8). Nevertheless, there is a developing capacity and clear need for prediction and ecological forecasting, and this will require continued efforts to improve ecosystem models and foster better interaction and communication with decision makers (Clark et al. 2001; Chapter 7).

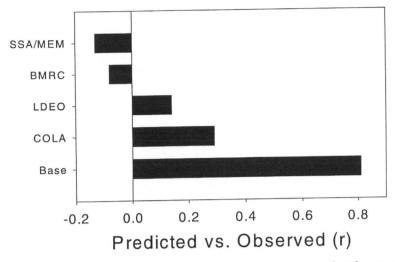

Figure 4.4. Linear correlation coefficients of predictions versus observations for several model forecasts of ENSO compared to a base model. Models are the singular spectrum analysis/maximum entropy statistical model (SSA/MEM), the Bureau of Meteorology Research Centre simplified dynamic model (BMRC), the Lamont-Doherty simplified model (LDEO), the Center for Ocean-Land-Atmosphere Studies dynamic model (COLA), and a "Base" empirical model. Forecasts are for two seasons ahead. Data are from Table 2 of Landsea and Knaff 2000 for the "Niño 3" region. The reader should consult the original source for details and citations for the models and comparisons abstracted in this figure.

Domain of Simple Models

Simple models are not a tool for all purposes but are useful for the many purposes illustrated in Figure 4.5. The two-way arrows indicate an interaction between models and uses. Perhaps the most important use of simple models is to ask general questions and depict ideas. In this context, models serve as a means of evaluation or testing questions or of exploring and illustrating new or different ideas. One of the great advantages of simple models is that they are easily discarded and replaced by something better (Chapter 25). In many ways, the science of ecology is the choice among models, and alternate simple models can be readily compared. The discussion and examples in the earlier part of this chapter have emphasized the utility of simple models in the areas of pattern and mechanism evaluation as well as prediction. Simple models can also play a role in study design by assessing, for example, an appropriate temporal scale of sampling or the number of ecosystems needed to assess a pattern given an expected trend and error.

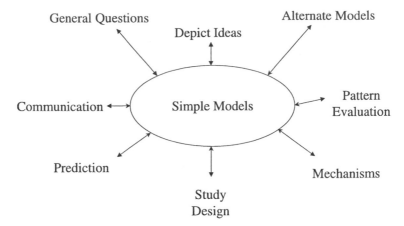

Figure 4.5. The utility of simple models in ecosystem science. The connections suggest that simple models can be effective tools toward progress in the various areas of research depicted.

Finally, simple models play a critical role in communication. Ecosystem science has many audiences, including scientists in other disciplines, managers, and the public. Simple models can facilitate communication of scientific results and implications to all these audiences. In this regard, simple models of ecosystems and ecosystem processes are critical in communicating the problems of ecosystem research to other scientists, including ecologists who do not focus on ecosystems. The patterns that simple models attempt either to describe (i.e., statistical models) or explain often represent important problem areas. Have ecosystem scientists been effective in broadcasting both their understanding and the remaining uncertainties for such general problems? Consider basic ecology texts, where the familiar general patterns of latitudinal gradients in species diversity, allometric relationships, and other hallmarks of population and community ecology are well represented, but some of the greatest successes of ecosystem studies, such as the generalized responses of aquatic systems to nutrient loading (Smith 1997), are often left out. Ecosystem scientists need to be more proactive in demonstrating these patterns and associated models. Simple models play a critical role in presenting ecosystem science to a wide audience.

Acknowledgments. I thank the Cary Conference organizers—Charles Canham, Jonathan Cole, and William Lauenroth—for encouragement. Helpful comments on a draft version of the paper were provided by an anonymous reviewer.

References

Aber, J.D., and C.A. Federer. 1992. A generalized, lumped-parameter model of photosynthesis, evapotranspiration and net primary production in temperate and boreal forest ecosystems. *Oecologia* 92: 463–474.

Baines, S.B., and M.L. Pace. 1994. Sinking fluxes along a trophic gradient: Patterns and their implications for the fate of primary production. *Canadian Journal of Fisheries and Aquatic Sciences* 51: 25–36.

Baines, S.B., M.L. Pace, D.M. Karl. 1994. Why does the relationship between sinking flux and planktonic primary production differ between lakes and the ocean? *Limnology and Oceanography* 39: 213–226.

Carpenter, S.R., and M.L. Pace. 1997. Dystrophy and eutrophy in lake ecosystems: implications of fluctuating inputs. *Oikos* 78: 3–14.

Chisholm, S.W., R.J. Olson, E.R. Zettler, R. Goericke, J. Waterbury and N. Welschmeyer. 1988. A novel free-living prochlorophyte abundant in the oceanic euphotic zone. *Nature* 334: 340–343.

Clark, J.S., S.R. Carpenter, M. Barber, S. Collins, A. Dobson, J.A. Foley, D.M. Lodge, M. Pascual, R. Pielke Jr., W. Pizer, C. Pringle, W.V. Reid, K.A. Rose, O. Sala, W.H. Schlesinger, D.H. Wall, and D. Wear. 2001. Ecological forecasts: An emerging imperative. 2001. *Science* 293: 657–660.

Landsea, C.W., and J.A. Knaff. 2000. How much skill was there in forecasting the very strong 1997–98 El Niño? *Bulletin of the American Meteorological Society* 81: 2107–2119.

Lauenroth, W.K., C.D. Canham, A.P. Kinzig, K.A. Poiani, W.M. Kemp, and S.W. Running. 1998. Simulation modeling in ecosystem science. Pp. 404–415 in M.L. Pace and P.M. Groffman, editors. *Successes, Limitations, and Frontiers in Ecosystem Science*. New York: Springer-Verlag.

Lloyd, J., and J.A. Taylor. 1994. On the temperature dependence of soil respiration. *Functional Ecology* 8: 315–322.

McIntosh, R.P. 1985. *The Background of Ecology*. New York: Cambridge University Press.

Moss, B., J. Medgewick, and G. Phillips. 1996. *A Guide to the Restoration of Nutrient-Enriched Shallow Lakes*. United Kingdom: Norwich Broads Authority.

Pace, M.L. 2001. Prediction and the aquatic sciences. *Canadian Journal of Fisheries and Aquatic Sciences* 58: 63–72.

Parton, W.J., D.S. Schimel, C.V. Cole, and D.S. Ojima. 1987. Analysis of factors controlling soil organic matter levels in Great Plains grasslands. *Soil Science Society of America Journal* 51: 1173–1179.

Pickett, S.T.A., J. Kolasa, and C.G. Jones. 1994. *Ecological Understanding*. San Diego: Academic Press.

Raich, J.W., and W.H. Schlesinger. 1992. The global carbon dioxide flux in soil respiration and its relationship to vegetation and climate. *Tellus* 44B: 81–99.

Rastetter, E.B., M.G. Ryan, G.R. Shaver, J.M. Melillo, K.J. Nadelhoffer, J.E. Hobbie, and J.D. Aber. 1991. A general biogeochemical model describing the responses of the C and N cycles in terrestrial ecosystems to changes in CO_2, climate and N deposition. *Tree Physiology* 9: 101–126.

Sala, O.E., F.S. Chapin III, J.J. Armesto, E. Berlow, J. Bloomfield, R. Dirzo, E. Huber-Sanwald, L.F. Huenneke, R.B. Jackson, A. Kinzig, R. Leemans, D.M. Lodge, H.A. Mooney, M. Oesterheld, N.L. Poff, M.T. Sykes, B.H. Walker, M. Walker, and D.H. Wall. 2000. Global biodiversity scenarios for the year 2100. *Science* 287: 1770–1774.

Scheffer, M. 1998. *Ecology of Shallow Lakes.* New York: Chapman and Hall.

Scheffer M., S. Carpenter, J.A. Foley, C. Folke, and B. Walker. 2001. Catastrophic shifts in ecosystems. *Nature* 413: 591–591.

Scheffer, M., S.H. Hosper, M.L. Meijer, and B. Moss. 1993. Alternative equilibria in shallow lakes. *Trends in Ecology and Evolution* 8: 275–279.

Smith, V. 1997. Cultural eutrophication of inland, estuarine, and coastal waters. Pp. 7–49 in M.L. Pace and P.M. Groffman, editors. *Successes, Limitations, and Frontiers in Ecosystem Science.* New York: Springer-Verlag.

Stow, C.A., S.R. Carpenter, and R.C. Lathrop. 1997. A Bayesian observation error model to predict cyanobacterial biovolume from spring total phosphorus in Lake Mendota, Wisconsin. *Canadian Journal of Fisheries and Aquatic Sciences* 54: 464–473.

Tilman, D., J. Fargione, B. Wolff, C. D'Antonio, A. Dobson, R. Howarth, D. Schindler, W.H. Schlesinger, D. Simberloff, and D. Swackhamer. 2001. Forecasting agriculturally driven global environmental change. *Science* 292: 281–284.

Van Donk, E., and R.D. Gulati. 1995. Transition of a lake to a turbid state six years after biomanipulation: Mechanisms and pathways. *Water Science and Technology* 32: 197–206.

Volk, T., and M.I. Hoffert. 1985. Ocean carbon pumps: Analysis of relative strengths and efficiencies in ocean-driven atmospheric CO_2 changes. Pp. 99–110 in E.T. Sundquist and W.S. Broecker, editors. *The Carbon Cycle and Atmospheric CO_2: Natural Variations Archaeon to Present.* Washington, DC: American Geophysical Union.

Wassmann, P. 1998. Retention versus export food chains: Processes controlling sinking loss from marine pelagic systems. *Hydrobiologia* 363: 29–57.

5

In Praise of Mechanistically Rich Models

Donald L. DeAngelis and Wolf M. Mooij

Summary

Simplicity is widely regarded as a virtue in ecological models. We suggest that the quest for the simplest models necessary to capture the relevant details of an ecosystem will not necessarily lead to models that are simple in terms of containing few variables, parameters, and mechanisms. In particular, we argue that "mechanistically rich" models, in which conceptual models incorporating complex causal networks are translated into a quantitative framework, offer distinct advantages over much simpler models in areas ranging from hypothesis testing to model evaluation. Mechanistically rich models can, in fact, be transparent, readily analyzed with current computing power, and less subject to error propagation than simple models in which many different mechanisms are typically aggregated into phenomenological variables that can not be easily related to field data. Mechanistically rich models are particularly suited to the specific "tactical" questions of primary concern in many areas of applied ecology.

Theory and Models

Ecological theory and many areas of applied ecology are dominated by approaches based on simple statistical or analytic models. It is not difficult to find explicit claims for the superiority of models that are analytic, or that at least contain few variables and parameters, over more complex simulation models of ecological systems. Doak and Mills (1994, 623) advised that "modelers of real species or communities should seek to formulate the simplest models necessary to capture the relevant biological details of their system," and Beissinger and Westphal (1998) recommended the use of the simplest models that data can support. Likewise, much research is devoted to simplification of models through deriving mean field (Levin and Pacala 1997) and spatially implicit formulations of spatial models (Slatkin and Anderson 1984; Roughgarden 1997). We agree that such simplification, where possible, is eminently sensible. We also agree that it is advantageous, where possible, to use models that have well-known mathematical properties, such as matrix models, over "crude simulation exercises" (Caswell 2000).

Yet, we do not believe that the quest for "the simplest models necessary to capture the relevant biological details of a system" will or necessarily should lead to models that are simple in terms of containing low numbers of variables, parameters, and mechanisms. Complex models already do much of the heavy lifting in the environmental sciences and their importance will probably increase. Following the fate of pollutants in an ecosystem (Montoya et al. 2000), describing processes such as carbon dynamics in the soil (Fu et al. 2000), or representing the coupling of physical and biological processes in a lake (Riley and Stefan 1988) are complex problems. These problems require detailed simulation models with numerous mechanisms and spatial heterogeneity. We believe that not only these obvious examples but also many other problems of applied ecology, and perhaps even theoretical issues such as community structure, will require complex models. We will refer to such models as "mechanistically rich," because they may include a wide array of mechanisms, including abiotic, trophic, physiological, and behavioral processes. Mechanistically rich models contrast with those based on simple sets of equations (often only a pair of equations), which we will refer to as "simple" models.

Our advocacy of mechanistically rich models stems from our belief in their explanatory power and their ability, in some cases, to make useful projections of the future. Of course, complex simulation models have a long heritage in ecology and were an important component of the International Biological Program of the late 1960s and early 1970s. Admittedly, some of the optimism engendered at that time for complex models in ecology was not fulfilled. However, we believe that conceptual and technical progress have accelerated over the past decade such that mechanistically rich modeling is now a far more incisive tool in understanding ecological systems. This progress includes the development of individual-based models beginning in the early 1970s (e.g., Botkin et al. 1972) and their use to study, among other things, the effects of individual plant species on primary production, nutrient cycling, and other ecosystem processes (e.g., Pastor and Post 1986) and the effects of behaviors of animals on community structure and vegetative patterns (e.g., Schmitz 2000). There is also increasing use of detailed simulation of the abiotic processes that underlie population dynamics (e.g., complex hydrology in the models of Possingham and Roughgarden 1990, and Wolff 1994).

Our objective here is to illustrate these trends by highlighting several examples. At the same time we address some persistent criticisms of complex models in ecology; that is, mechanistically rich models are often dismissed as being "data hungry," nontransparent, prone to error multiplication, and not amenable to rigorous testing.

Conceptual Models and Mechanistically Rich Quantitative Models

From the above it is clear that an important trend has been the recognition that understanding ecological phenomena may involve integrating a variety of mechanisms, such as abiotic processes, trophic interactions, movement in space, and behavior and physiology of the individual. Ecologists increasingly recognize this in their conceptual models of ecological systems, which is a first step in formulating a quantitative model. Consider, as an example, the problem of forecasting how a population (e.g., of birds) may respond to a change in environmental conditions. A researcher starts with a body of general theory about birds and how they interact with their environment. This is supplemented with information on the particular species and the particular local environment. From these sources of theory and information, the researcher formulates a conceptual model containing the key components and processes. Typically, the details or importance of some processes are not well known, leaving room for a variety of hypotheses; thus there will be alternative conceptual models. To test and compare these alternative conceptual models, predictions must be made, which usually means that the conceptual models must first be translated into quantitative models. The models are parameterized with relevant data and then used to make predictions of how the population will behave under the changed environmental conditions. Model output, predicting the response of the population, is compared with empirical data (Figure 5.1).

Each step in the process illustrated in Figure 5.1 is important, particularly the formulation of a conceptual model. At first, the conceptual model may be very simple. Suppose, in this example, that an important environmental variable is rainfall. A researcher may simply hypothesize that annual fluctuations in a bird population are related to rainfall variations. The null hypothesis that there is no relationship may be tested using statistical techniques to find a significant relationship between annual rainfall and changes in population size over a number of years.

If there is a relationship, it is likely that the researcher will next want to know much more about the causal details of the relationship. At this point, she might construct, based on field studies, a number of alternative conceptual models of possible causal relationships between the rainfall and the bird population dynamics, incorporating different hypotheses by which variation in rainfall pattern affects the birds. To compare these alternative conceptual models in terms of their abilities to account for the observational data, quantitative models are developed to produce testable predictions of each of the alternatives. Often, no hypothesis is satisfactory, and new hypotheses may be needed prompting new field studies and new models (Figure 5.1).

Conceptual models arising from this process often are complex, so that the corresponding quantitative models are likely to be mechanistically rich, as will

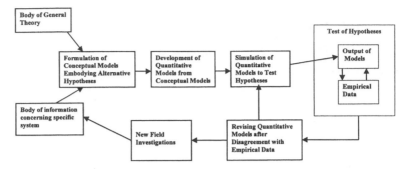

Figure 5.1. A simple diagram showing the stages in developing models and using them to test hypotheses. General theory and empirical data are used to help formulate conceptual models, incorporating alternative hypotheses. The conceptual models are translated into corresponding quantitative models, which are used to make predictions. The predictions, which are model outputs, are then compared with empirical data, using statistical approaches. If no hypotheses agree satisfactorily with data, the model may need revision or new field investigations may by needed, as shown in the bottom feedback loop.

be seen in examples below. Here we simply state two main points summarizing our views, which will be illustrated with examples later.

> Point 1. Simple models are often effective in ecology in helping to discern whether or not a particular type of effect is occurring in a system. However, ecological systems are complex and such models may not help elucidate the causal chains by which the effect operates.

> Point 2. To test between alternative hypotheses of the specific causes, it is often necessary to use quantitative models that can be described as "mechanistically rich" in congruence with the conceptual models of the system. A diagram of the possible stages in an ecological investigation using models is shown in Figure 5.2. The question at the top is addressed by the use of a simple statistical model to test for the existence of a suspected relationship. If there is a relationship, then models that are more complex are constructed to establish whether one hypothesized causal chain is more likely than another. Field observations or experiments are used to discriminate more carefully between the hypothesized relations.

Examples of Mechanistically Rich Models

We will briefly review a few genres of complex models. Some of these are focused on the dynamics of single populations, but we consider all to be ecosystem models in the sense that they contain rather detailed representations of the

**Is there a measurable effect of an environmental
variable on population variability?**

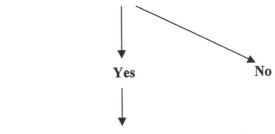

Yes   **No**

Construct models with different mechanisms

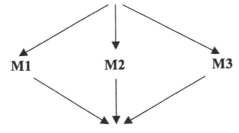

M1   **M2**   **M3**

Test model predictions against empirical data

Figure 5.2. Statistical models may be used to establish whether relationships exist between a population and environmental variables. If the models indicate a relationship,
models embodying alternative casual chains are developed and tested.

landscape and/or of ecosystem processes. We emphasize models that examine
the interface of individual organisms' physiology and behavior with the ecosystem and landscape.

Population models built on detailed modeling of ecosystem influences. Spatially explicit individual-based (SEIB) models of populations are becoming
commonplace. We cannot review the very large number of models of this type
here but will briefly discuss one of the growing number of models that are used
to simulate populations in realistic landscapes and that incorporate at least a
few ecosystem processes. This class includes models of ungulate grazers
(Turner et al. 1993), brown bears (Wiegand et al. 1998), Cape Sable seaside
sparrows (Nott 1998; Nott el al. 1998), European robins (Reuter and Breckling
1999), and snail kites (Mooij et al. 2002). Here we describe in detail the wood
stork model of Wolff (1994; see Table 5.1 for a listing of variables and mechanisms), which has the goal of testing hypotheses concerning the effects of hydrologic conditions on the reproductive success of a wood stork nesting colony.

Specifically, the model was used to test whether loss of short-hydroperiod marshes in the Everglades landscape could have led to a decline in wood storks. The conceptual model of the wood stork focuses on the indirect effects of changes in land and hydrology through the availability of fish prey. The crux of the model was to determine if, under particular hydrologic regimes, the wood stork parents would be able to bring enough fish to their offspring for them to survive to fledging. Therefore, the energy budgets for adults and nestlings were modeled. In addition, the time allocation of the adults, on 15-minute time intervals, was modeled to keep track of the time needed to find and bring prey back to the nest. The spatiotemporal pattern of water depths and fish concentrations around the nesting colony was also modeled. A large amount of input data was needed for the model, including the mean capture rate of fish biomass by a wood stork in particular water depths and fish concentrations; however, reasonable parameter estimates were available for all model processes. The model correctly predicts the delays in nesting observed in recent years. It has been used to support the hypothesis that the loss of short-hydroperiod wetlands can cause breeding colonies to be late in nest initiation and to fail because of the start of the wet season before fledging is completed (Fleming et al. 1994; Wolff 1994). The model also can be used to make predictions concerning the most effective foraging strategies (e.g., flocking vs. individual foraging) under different patterns of prey availability.

Tritrophic food chain. In many ecological systems, long-term data on the changes in population levels and other variables are lacking. An important question is whether short-term data, can be used in a simulation model to extrapolate behavior of the system over much longer time periods, such as several years. Schmitz (2000; see Table 5.1) showed that an SEIB model of a tritrophic food chain could make this extrapolation. The tritrophic food chain that was simulated had two plant types (grass and herbaceous plants), as well as grasshoppers and spiders. The growth and reproduction of the two plant types, as well as their exploitative competition for a nutrient pool, was simulated on the squares of a spatial grid. The grasshoppers were simulated as individuals that exhibited resource preference for grass over the less nutritious herb. The grasshoppers could grow and reproduce and, like the plants, exhibited population dynamics. The spiders were modeled as individuals but their population was fixed. They could affect the grasshoppers directly, by predation, and indirectly, by causing grasshoppers in their vicinity to move from grass to the safer herb sites. Seasonality was built into the model to simulate a New England climate.

Schmitz's model showed that behavioral details, especially risk avoidance by the grasshopper, did much to structure the trophic web. The presence of spiders, even when they did not prey on the grasshoppers but merely changed their behaviors, led to an increase in grass biomass compared with simulations where the spiders were not present. What the model also showed was that for this system short-term dynamics within a growing season determined the long-term structure of the community. This occurred because the non-growing season basically reset the system, and there was little long-term accumulation of effects from one growing season to the next. The model showed that although long interannual time series of data from the simulation would not indicate density-dependent regulation, such regulation clearly did occur within individual years.

Lake Mendota fish community. The individual-based modeling approach was used by McDermott and Rose (2000; see Table 5.1) to model the Lake Mendota fish community over several generations. Populations of six fish species were modeled at the individual level, with the lake modeled as three compartments; epilimnion, hypolimnion, and littoral zone. Daily temperature, dissolved oxygen, and prey densities in each region were used as forcing functions. The day-to-day feeding interactions with prey, growth, thermoregulation, reproductive behaviors, predation, and other mortality factors were modeled through time using a Monte Carlo representation for foraging and bioenergetics equations. The purpose of the model was to explore the biomanipulation of Lake Mendota (Carpenter and Kitchell 1992), in which intensive stocking of walleye and northern pike was performed to create a trophic cascade. The individual-based approach was used because feeding relationships among these fish are size-based, which is easy to implement in an individual-based model. Other individual-based models of fish dynamics have also revealed rich and often counterintuitive behaviors when size-based trophic interactions were combined with population dynamics (e.g., DeAngelis et al. 1993; Scheffer et al. 1995). Individual-based models can conveniently simulate such systems, in which interactions are complex and change over time, weight and length vary somewhat independently, and stochasticity on a daily time scale is important. Other size-structure modeling approaches lack this flexibility.

Intertidal mussels. Robles and Descharnais (2002; see Table 5.1) designed a stochastic cellular automata (CA) model, keeping track of individual mussels, their life cycle attributes, and their locations on a spatial grid. The model differs from the SEIB models described above in that the focus of the model is more on the state of grid cells (whether a cell is occupied and by what size mussel) rather than on the individuals themselves but is otherwise similar. The model incorporates several types of mechanisms. The issue of mechanisms in this model is particularly cogent to the overall question of mechanisms in models in general because the history of intertidal theory, as reviewed by the authors, has been one of recognition of the need to include an increasing number of mechanisms to describe the dynamics of sessile organisms such as mussels. Theory in the early twentieth century stressed the importance of physical factors, such as wave action and exposure to air. Competition between species was emphasized as an organizational factor later, especially after the work of Connell (1961). Top-down effects of predation were demonstrated by the elegant experiments of Paine (1966) and others. Finally, supply-side aspects, in which environmental factors affected the rate of recruitment, gained recognition in the early 1980s. Each of these types of mechanism was included in the mussel model. The model has provided the basis for testing a number of hypotheses, particularly in the way that gradients of elevation and intensity of wave action would affect the

Table 5.1. Examples of mechanistically rich models.

Source	Main Variables	Main Mechanisms
Wading birds model (Wolff 1994)	Water level and fish density in 500 × 500 m cells, nestlings; nesting and nonnesting adults; spatial location of foraging adult; energy status of adults and nestlings	Daily changes in water level; five-day changes in fish density; bioenergetic models of adults and nestlings; interactions of adults with mates; initiation of nesting by adult pairs, asynchronous egg laying and hatching; foraging flights from nests; methods of searching for prey; flocking or non-flocking; efficiency of foraging; feeding of nestlings; nest desertion
Tritrophic level model (Schmitz 2000)	Two-dimensional gridded landscape; nutrient resources; individual grass and herb plants; grasshoppers and spiders; detection radiuses of individuals	Depletion of resources; seasonality; foraging; consumption of resources; growth; metabolism; nutrient uptake and competition of plants; movement of animals; reproduction; withering of plants; predator avoidance by grasshoppers
Lake Mendota fish community model (McDermott and Rose 2000)	Lake divided into a littoral zone, epilimnion, and hypolimnion; daily water temperatures and DO; age, stage, sex, location (among three compartments); total length; and wet weight; of individuals in six fish species	Feeding; growth; spawning (dependent on female size and temperature); mortality; thermoregulation by seeking preferred temperature; biomanipulation
Intertidal mussel model (Robles and Descharnais 2002)	Spatial cells that can be empty or occupied; size and positions of individual mussels	Duration of inundation of cells, exposure to waves, recruitment of larval mussels, effects of elevation, wave intensity, and local mussel density on recruitment to various locations; effects of elevation and

Table 5.1. *Continued*

Source	Main Variables	Main Mechanisms
		wave action on growth; predation by seastars; effects of elevation, wave action, mussel size, and mussel aggregation on predation
Savanna ecosystem model (Jeltsch et al. 1996, 1998)	Individual 5 × 5 m spatial cells making up 50 ha area, with differing soil characteristics and two soil layers; water availability through rain and modified by local soil structure and texture; small scale heterogeneities in water availability in cells; biomasses of three matrix types of vegetation in each cell: grass, herbaceous, and shrubs; individual trees; at most one tree per cell; condition classes for all plants; dependent on moisture status	Changing water availability in two soil layers; fire; grazing; disturbances such as deposition of herbivore dung; growth in biomass; seed dispersal and establishment; vegetative spread; mortality; exploitative competition for water; water-stress-induced changes in condition; extinction of vegetation in a cell

patterns (size-structure, biomass) of mussels. In addition, the model was able to demonstrate that there was no need to include an additional specific mechanism of spatial refugia from predation for a certain fraction of mussels. The combined effects of all the other mechanisms created an "effective" refugium, an emergent phenomenon.

Savanna ecosystem. Traditional ecosystem modeling is very much alive and being used to examine questions at the whole ecosystem level. SAVANNA, a model with hundreds of parameters, was developed by Coughenour (1993) for savannas, temperate grasslands, and woodland ecosystems. A variation of SAVANNA has recently been applied to testing hypotheses for the decline in key woody plants, such as willows, in Rocky Mountain National Park. The complexity of the model is diminished through modularization of the various model components, the well-tested CENTURY soil-dynamics model (Parton et al. 1987) being one such module. The usefulness of such models grows as they are tested and as a large number of researchers become familiar with them.

Here, however, we will discuss another approach to modeling the savanna ecosystem. Jeltsch et al. (1996; see Table 5.1) used a 20,000-cell (5 × 5 m) cellular automata model to investigate the coexistence of trees and other vegetation (grasses, herbs, and shrubs, which collectively form the "matrix") in the Kalahari savanna. Previous non-spatial models, which assumed that trees and matrix vegetation exploit different levels of soil, were able to show that such niche differentiation allowed coexistence. However, because empirical work suggests that roots actually overlap, Jeltsch et al. sought an explanation in terms of other mechanisms in a spatial setting: grass fires, herbivory, and competition for water. In their model, some combinations of these processes were able to maintain a savanna consisting of matrix with strongly clumped distributions of trees; however, scattered, not clumped, trees characterize the southern Kalahari. This finding gave rise to new field investigations aimed at identifying the missing processes. Results of these field investigations and new simulation experiments indicated that the patchy distribution of tree seeds in the dung of herbivores was the most important factor producing small-scale heterogeneities that affect seedling establishment (Jeltsch et al. 1998). The addition of these mechanisms could give rise to the pattern of scattered trees in the savanna.

Evaluation of Mechanistically Rich Models

The above examples are only a small sampling of recent mechanistically rich models in the ecological literature. Nevertheless, they serve the purpose of addressing some of the pros and cons of these models.

Data Demands

SEIB, CA, and other mechanistically rich models are often described as being "data hungry." In an important paper reviewing SEIB models, Levin et al. (1997, 335) state that these models "provide a point of departure, but the amount of detail in such models cannot be supported in terms of what we can measure or parameterize." If Levin et al. are saying that it is impossible for any model to provide a completely accurate description of reality, we agree with this statement. If they are saying that in general mechanism-rich models cannot be supported by available data, we disagree. The models described above are not intended to fully describe reality but rather to make purposeful representations from which qualitative predictions can be made, so that data needs are limited to what serves that objective. Of the models discussed above, the Lake Mendota community and wood stork models were perhaps the most demanding of data, but they have been reasonably parameterized. All models were tailored to the available data and ongoing field studies. The tritrophic chain SEIB model, in fact, was designed because data were available for it, whereas data for population-level models were lacking. Within the scope of the questions or applications being addressed by this and other models, there were sufficient data.

Some, such as the Cape Sable sparrow, wood stork, and European robin models, are focused on a limited question or time period (e.g., the reproductive season). Such models are sometimes called "tactical," as opposed to "strategic" models, which address either larger questions and/or longer time scales. All of the above models are based directly on the conceptual ideas and data collected in ecological studies, are developed at the spatial and temporal scales of those data, and are tested against typical monitoring data. The spatially explicit models complement these species-specific data with high-resolution data on landscape characteristics that remote sensing is able to provide and, where necessary, with detailed abiotic models, such as models of landscape hydrology. Combining these types of information, these models can make a rich array of detailed and testable predictions about populations on a landscape. We can summarize as follows:

> Point 3. Data availability for mechanistically rich models is often great, particularly if the models focus on tactical questions and if they are flexible enough to use a variety of types of data.

Complexity and Error Multiplication in Models

Because input data and parameter values can only be estimated, the output of all models is subject to uncertainty. Models with numerous mechanisms, for which parameters must be estimated, have many potential sources of uncertainty. The question in applying any model, then, is whether the input in parameter uncertainty is so high, or multiplies to such an extent in the model, that output uncertainty is too high for useful predictions. A classic elucidation of the way that uncertainty may multiply in a complex model is given in a paper by Yodzis (1988), who studied the effects of changes in one species on others in steady-state solutions of highly connected nonlinear food web models. Yodzis found that small uncertainty in parameter values propagated in a nontransparent way through millions of indirect effects, leading to such uncertainty in predictions that it might be difficult to forecast even the direction of change in a population.

Such explosive growth in uncertainty, however, should not be generalized to all types of models that are characterized by many variables and interactions. The variables in SEIB models typically describe many individuals within a single species, or perhaps a few species, rather than describing many species linked through strong nonlinear interactions as in the models that Yodzis (1988) analyzed. Therefore, uncertainty propagation has to be looked at in a different way. For one thing, SEIB models do not promise to predict what happens to any individual organism, any more than a kinetic gas equation attempts to predict the paths of individual gas molecules. However, the SEIB model does ensure that individuals behave reasonably, which is an important factor in ensuring that larger-scale population patterns also behave reasonably (Mooij and Boersma 1996). Further, rather than introduce multiplicative effects, the individual organisms in the SEIB models operate largely in parallel, not in series.

Individual-level uncertainties "even out," rather than multiply, over the whole population. Just as importantly, the many mechanisms and parameters in these models are usually constraints that put limits on individuals. Therefore, what may look like a great deal of complexity in a model may in fact include a large amount of solid information that keeps the models behavior within biologically reasonable bounds. To take the case of the wood stork model, there are limits on the magnitude of foraging movements, hours of feeding, prey caught per unit time, amount of prey that can be carried to the nest, time required to fly between nests and foraging sites, numbers of eggs produced, etc. Such empirically based constraints on individuals and their processes provide a strong check that great uncertainties are not occurring at the population level. This is not to say that there are not uncertainties in the parameters or that these do not lead to population-level uncertainty. However, because the SEIB modeling approach typically addresses tactical problems such as changes in population during a single season and because it incorporates biological constraints, it is less likely than other approaches to have uncertainty propagation problems.

The potential problems of error multiplication are best seen in many models that are usually classified as "simple." Consider the simple model of the snail kite population developed by Beissinger (1995). This model consists of three parameters, a_{high}, a_{low}, a_{lag}, for fractional population changes, calculated from long-term bird field studies. The three parameters refer to population growth under "high water," "drought," and "lag" years. The predicted population size ten years in the future, given the population X_t at time t, is

$$X_{t+10} = a_{t+10} \, a_{t+9} \, a_{t+8} \, a_{t+7} \, a_{t+6} \, a_{t+5} \, a_{t+4} \, a_{t+3} \, a_{t+2} \, a_{t+1} \, X_t, \qquad (5.1)$$

where each of the parameters "a_{t+i}" can take on any of the three values, a_{high}, a_{low}, a_{lag}. Although we regard Beissinger's model as a valuable one and do not mean to criticize it, this type of model is highly vulnerable to slight errors in any of the parameters. Because this model lumps highly diverse spatial patterns of water levels into just three types aggregated over a large habitat, many nuances of geographical variation in disturbances are ignored, as closer examination of the spatial pattern of water levels in the snail kite's range has shown (Dreitz 2000). Although Equation 5.1 could be fit to time series data reasonably well, inevitably errors in the values of a_{high}, a_{low}, a_{lag} would multiply if the model were projected into the future.

Many years of data may be needed to test a model such as Equation 5.1. It is an advantage of most SEIB models (e.g., the tritrophic chain model) that much shorter studies are sufficient to provide data for testing. For that reason many SEIB models are designed for tactical predictions (e.g., a season to a few years). They are suited for understanding processes in the short term and for predicting spatial and short-term temporal patterns, based on the natural history of the species involved. By providing a detailed understanding of processes in the short term, such tactical models can provide input to management decisions that should help the population over the long term.

The fact that SEIB models are used primarily for short-term tactical questions does not mean they cannot be extended into a more strategic, or longer-term, domain. Instead of ignoring the detailed processes and lumping them into a few parameters, mechanistically rich models, if accompanied by insightful population monitoring, allow us to obtain better and better information on the processes controlling year-to-year population changes. Of course, uncertainty amplification can occur in attempting to make long-term predictions because data-rich models combine many independently collected pieces of information, each subject to error. We do not expect that even a very good SEIB model, based purely on microscale data concerning the details of behavior, physiology, and life cycles, will make excellent long-term predictions, any more than simple models such as Equation 5.1 can, because population data will always have some level of uncertainty.

What this means is that a SEIB model may use some long-term data at the population level for calibration. Grimm et al. (1996) have developed an approach to calibration termed "pattern-oriented modeling," which was applied by Wiegand et al. (1998) to an SEIB model of a population of brown bears in northern Spain. The model simulated the processes of reproduction and mortality, both dependent on environmental conditions, as well as the breakup of family groups (females with cubs) and density-dependent regulation. These individual-level processes could be modeled using a rich database on the brown bear. There was still a problem that data on some demographic parameters were missing, which caused high uncertainty in the model predictions (e.g., the risk of extinction or the growth rate of the population). However, because the SEIB model itself yielded a huge number of predictions concerning spatial and temporal patterns, it was possible to make statistical comparisons with an existing 14-year time series on the temporal and spatial patterns of females with cubs. This allowed adjustment of the formerly unknown parameters and a major reduction in uncertainty. Traditional models, such as matrix population models, would not have provided the opportunity to use such information, and extensive telemetric studies and trapping of animals would have been needed to parameterize such models.

A similar approach has been applied to the snail kite model (Mooij et al. 2002). Empirical information shows reliably that the population has increased at a rate of about 5.0% per year over a 20-year period, whereas the a priori SEIB model hindcasted a 3.0% growth rate during this time period. In this case, original estimates of some of the microscale parameters in the model, especially the least certain ones, had to be modified. Only some of the existing long-term data were used to calibrate the model, so that the remaining data, on the changing spatial patterns through time, could be used for model validation. Thus, we argue that mechanistically rich models not only can make use of data at the individual level but also can use long time series data opportunistically to reduce uncertainty. In this way, they may be effective in long-term simulations and population viability analyses. Our main point here is the following:

Point 4. Uncertainty can multiply through time in either simple or mechanistically rich models. Mechanistically rich models can access more information and data than simple models, allowing realistic constraints to be put on model performance. Because of the many intermediate-level predictions of mechanistically rich models, the potential for effects of uncertainty propagation and multiplication is greatly reduced. The use of long-term data, even in small amounts, to scale some of the less certain microscale parameters is a method that can further reduce this problem.

Transparency of Models

In mechanistically rich models, a large number of processes take place simultaneously. This type of complexity is often assumed to imply nontransparency of the models, such that it is impossible to understand what is going on in the model. However, we believe the models described here meet the criterion of transparency. The first reason is that they are based on clear conceptual models. The second is that the nature of the model output, in terms of individual organism behaviors, is easier to relate to basic biology than the output of traditional population-level models. Casti (1997) noted that transparency is not necessarily a function of a model having few variables. He discussed TRANSIMS, a transportation model that follows the daily movements of about 400,000 individual travelers on about 30,000 road segments (in Albuquerque). Casti states that "on the criterion of simplicity, TRANSIMS rates a measure of something like 'medium-simple'" (176). Despite the huge number of variables, the model is relatively transparent. Using visual interface for TRANSIMS, a modeler can focus in on an individual traveler, an individual road, or an intersection and see exactly what is happening at the detailed level. This examination of the detailed picture and whether it makes sense leads to confidence in the whole model.

Similarly, the SEIB models constructed for ecological populations can be examined in detail at the level of individual organism or the level of a small piece of the landscape by experts on those species. The Cape Sable seaside sparrow, wood stork, and European robin models (and perhaps some of the others) produce detailed output maps of locations of individuals in space. These can be carefully compared with data and an ecological researcher's personal observations. Although simple rules at the organism level tend to lead to highly complex spatial patterns and usually also to important emergent properties, the ecologist can easily judge if the theoretical assumptions that go into the model are faithfully represented in its behavior. Humans are equipped to interpret and gauge the validity of actions at the level of individuals, whereas our intuitions at the population level are seldom as sharp. Thus, from observing how the individuals behave in a model, we can judge easily if something inconsistent is occurring. From our training as biologists, we know that individual physiology and behavior rest on the soundest scientific data that exist in biology. Therefore, SEIB models are by nature capable of being extremely transparent, both in the sense that their results can be easily comprehended and that the rules rest on a solid scientific basis. This awareness leads to the following points:

Point 5. SEIB and CA models can simulate many individuals in a complex landscape; however, this does not mean that the models are conceptually complex. The rules for individual behavior are generally relatively simple, and the variables in these models are usually observable entities.

Point 6. When populations are represented as individuals rather than as aggregated state variables, one gets far closer to being able to prescribe real laws based on the physiological limits of individuals and on what is known of animal behavior. This promotes conceptual clarity because the model is grounded in sound science.

Amenability of Mechanistically Rich Models to Analysis

It is often said that a disadvantage of SEIB or CA models is that they cannot be analyzed and that, therefore, models based on mathematically well-known differential equations are preferable to individual-based simulations. However, contrary to the stereotype of mechanistically rich SEIB models being too cumbersome to be validated effectively, these models can be tested in the same way as simpler models. The reason that they can be tested is that they produce a rich array of predictions. We have already mentioned the pattern-oriented modeling approach of Grimm et al. (1996), which uses statistical techniques to compare spatial patterns produced by models with available field data to help calibrate parameters of individual organisms. Here we discuss some of the formal ways in which such model validation can be done, with special reference to the population model of the snail kite (Mooij et al. 2002).

It is important for models to make predictions that are as independent as possible of the data on which the model is based. The snail kite model (Mooij et al. 2002) simulates a population of individual birds, each with its own sex, age and life stage, location on particular wetland, and breeding status. The model is based on observed life cycles, the habitat needs of individuals, and typical breeding phenologies, as well as on the temporally changing conditions of the set of wetland sites that the snail kites utilize. The hypothesis being tested is that the birds are highly mobile and that in the event of drought conditions they have a certain likelihood of reaching another suitable habitat site (probably the closest one available). If no suitable breeding sites exist, they have a probability of reaching a "peripheral" site, on which they can survive but not breed. The model produces a large number of predictions. The most important of these are (1) the distribution of the birds over life stages, (2) the distribution of the birds over the wetlands, (3) the distribution of nests over the wetlands, (4) the ratio between breeding and nonbreeding birds, and (5) the distribution over the nest-stages of breeding birds (these are not all independent). These are all quantitative predictions for which estimates of means and standard deviations are obtainable.

Each of these predictions can be tested against corresponding observed values using a goodness-of-fit approach (Sokal and Rohlf 1981, Chapter 17). Consider the distributions of nests across the wetlands in a given time period, such as a month. The model predictions can be normalized by the total number of nests in the entire modeled region to give the expected fraction of nests in each wetland: $p'_1, p'_2, p'_3, \ldots$, computed by averaging over a set of replicate model simulations. Note that these model values constitute the null model to be falsified in the test. The observed nest frequencies, similarly normalized by the total number of nests observed, are denoted by $p_1, p_2, p_3, \ldots$. The question is whether the deviation of the observed values differs sufficiently from the expected values that the null hypothesis is rejected. This test can be performed by construction of a G statistic, which is constructed as

$$G = 2 \sum np_i \ln(p_i / p'_i), \tag{5.2}$$

where n is the total number of observations, and i is summed over the total number of wetlands, N. The distribution of G can be approximated by the χ^2-distribution with $N - 1$ degrees of freedom. This testing procedure can be carried out for each of the chosen distributions. Failure of any would constitute a rejection of the model or identify a need for recalibration using some of the independent data.

The snail kite model is used not only for testing hypotheses but also for making projections used in management decisions. These models cannot make actual predictions, because they use detailed rainfall data that cannot be predicted in advance. However, based on good up-to-date meteorological information, a small set of likely hydrologic scenarios can be forecast. These scenarios can be used in the SEIB models, and these models can project which water-management strategies will have the highest overall probability of success.

> Point 7. Mechanistically rich models are capable of producing a rich output of predictions that can be compared with monitoring data. It is this richness of predictions, especially the capability of making unexpected predictions, that is the key to testing hypotheses.

> Point 8. Mechanistically rich models operate at the scales at which field investigations take place. Because the unit of observation and the unit of the model are the same, the model output can be analyzed in exactly the same way and with the same tools as the field data.

> Point 9. Statistical methodologies exist that can readily be applied to the output of SEIB models. Because of increased computer speed

in the last decade, intensive replication needed for such testing is not a significant impediment to the thorough analysis of models.

Conclusions

The modeling approach we have described grows directly out of the objective of providing quantitative forms for conceptual models of specific systems, and using these models to test hypotheses. These models tend to be mechanistically rich and often (although not exclusively) tactical in scope, addressing relatively specific hypotheses for specific systems. We appreciate that other approaches to modeling, including some that deliberately focus on few mechanisms and on analytic mathematical tools, are also useful. Minimal models (e.g., Scheffer 1998, Scheffer et al. 2001), for example, have been used to great advantage to elucidate such phenomena as catastrophic shifts in ecosystems in a very general way. We believe that such models must be used with caution, however, because in any given situation many other mechanisms, which may be ignored in a minimal model, may be at work. Another highly important simple modeling approach is that of Markov matrix population models (e.g., Caswell 2000). However, we must remark that to use such models to predict how environmental change will affect a population, one must know the relationships between the survivals and fecundities and environmental conditions. Much of the fundamental, interesting, and difficult biology regarding population behavior resides in determining these relationships, which requires understanding and quantifying the many mechanisms that relate environmental conditions to these survivals and fecundities. We have presented examples above showing the complexity of the relationships between life-cycle parameters and environmental conditions and how simulation approaches, such as SEIB models, may be useful in determining these relationships.

To conclude, in line with our examples of "theory–conceptual model– quantitative model–testing," we regard the first step in theory development to be in recognizing or hypothesizing a set of mechanisms to be operating in a system. Model choice should be focused on providing the best representation for the conceptual model, whether this means differential equations, Markov matrices, SEIB models, complex ecosystem simulations, or some combination of these. Fortunately, more general modeling platforms that allow various alternative model formulations are starting to become available (e.g., Mooij and DeAngelis 1999). In the future, we hope that understanding, not a misplaced emphasis on simplicity, guides modeling in ecology.

Acknowledgments. We acknowledge the many helpful comments of Thorsten Wiegand and Charles Canham. This work was supported in a significant degree by the Department of the Interior's Critical Ecosystem Studies Initiative, a special funding initiative for Everglades restoration administered by the National Park Service, and in part by the U.S. Geological Survey's Florida Caribbean Science Center. This is publication 3052 of the NIOO-KNAW Netherlands Institute of Ecology, Centre for Limnology.

References

Beissinger, S.R. 1995. Modeling extinction in periodic environments: Everglades water levels and snail kite population viability. *Ecological Applications* 5: 618–631.

Beissinger, S.R., and M.I. Westphal. 1998. On the use of demographic models of population viability in endangered species management. *Journal of Wildlife Management* 62: 821–841.

Botkin, D.B., J.F. Janak, and J.R. Wallis. 1972. Some ecological consequences of a computer model of forest growth. *Journal of Ecology* 60: 849–873.

Carpenter, S.R., and J.F. Kitchell. 1992. Trophic cascades and biomanipulation —interface of research and management. *Limnology and Oceanography* 37: 208–213.

Casti, J.L. 1997. *Would-Be Worlds.* New York: John Wiley and Sons.

Caswell, H. 2000. *Matrix Population Models: Construction, Analysis, and Interpretation.* 2nd ed. Sunderland: Sinauer Associates.

Connell, J.H. 1961. The influence of interspecific competition and other factors on the distribution of the barnacle *Chthamalus stellatus. Ecology* 42: 710–723.

Coughenour, M.B. 1993. The SAVANNA landscape model: Documentation and user's guide. Fort Collins: Natural Resource Ecology Laboratory of Colorado State University.

DeAngelis, D.L., B.J. Shuter, M.S. Ridgeway, and M. Scheffer. 1993. Modeling growth and survival in an age fish cohort. *Transactions of the American Fisheries Society* 122: 927–941.

Doak, D.F., and L.S. Mills. 1994. A useful role for theory in conservation. *Ecology* 75: 615–626.

Dreitz, V.J. 2000. The influence of environmental variation on the snail kite in Florida. Ph.D. dissertation, University of Miami.

Fleming, D.M., W.F. Wolff, and D.L. DeAngelis. 1994. The importance of landscape heterogeneity to a colonial wading bird in the Florida Everglades. *Environmental Management* 18: 743–757.

Fu, S., M.L. Cabrera, D.C. Coleman, K.W. Kisselle, C.J. Garrett, P.F. Hendrix, and D.A. Crossley Jr. 2000. Soil carbon dynamics of conventional tillage and no-till agroecosystems at Georgia Piedmont—HSB-C models. *Ecological Modeling* 131: 229–248.

Grimm, V., K. Frank, F. Jeltsch, R. Brandl, J. Uchmanski, and C. Wissel. 1996. Pattern-oriented modeling in population ecology. *Science of the Total Environment* 183: 151–166.

Jeltsch, F., S.J. Milton, W.R.J. Dean, and N. van Rooyen. 1998. Tree spacing and coexistence in semiarid savannas. *Journal of Ecology* 84: 583–595.

Jeltsch, F., S.J. Milton, W.R.J. Dean, N. van Rooyen, and K.A. Maloney. 1996. Modeling the impact of small-scale heterogeneities on tree-grass coexistence in semiarid savanna. *Journal of Ecology* 86: 780–794.

Levin, S.A., B. Grenfell, A. Hastings, and A.S. Perelson. 1997. Mathematical and computational challenges in population biology and ecosystems science. *Science* 275: 334–343.

Levin, S.A., and S. Pacala. 1997. Theories of simplification and scaling of spatially distributed processes. Pp. 271–295 in D. Tilman and P. Kareiva, editors. *Spatial Ecology: The Role of Space in Population Dynamics and Interspecific Interactions.* Princeton: Princeton University Press.

McDermott, D., and K.A. Rose. 2000. An individual-based model of lake fish communities: Application to piscivore stocking in Lake Mendota. *Ecological Modeling* 125: 67–102.

Montoya, R.A., A.L. Lawrence, W.E. Grant, and M. Velasco. 2000. Simulation of phosphorus dynamics in an intensive shrimp culture system: Effects of feed formulation and feeding strategies. *Ecological Modeling* 129: 131–142.

Mooij, W.M., R.E. Bennetts, W.M. Kitchens, and D.L. DeAngelis. 2002. Exploring the effects of drought extent and interval on the Florida snail kite: Interplay between spatial and temporal scales. *Ecological Modeling* 149: 25–39.

Mooij, W.M., and M. Boersma. 1996. An object-oriented simulation framework for individual-based simulations (OSIRIS): *Daphnia* populations as an example. *Ecological Modeling* 93: 139–153.

Mooij, W.M., and D.L. DeAngelis. 1999. Individual-based modeling as an integrative approach in theoretical and applied population dynamics and food web studies. Pp. 551–575 in H. Olff, V.K. Brown, and R.H. Drent, editors. *Herbivores: Between Plants and Predators.* 38th Symposium of the British Ecological Society. Oxford: Blackwell Science.

Nott, M.P. 1998. Effects of abiotic factors on population dynamics of the Cape Sable seaside sparrow and continental patterns of herpetological species richness: An appropriately scaled landscape approach. Ph.D. dissertation, University of Tennessee, Knoxville.

Nott, M.P., O.L. Bass, D.M. Fleming, S.E. Killeffer, N. Fraley, L. Manne, J.L. Curnutt, T.M. Brooks, R.D. Powell, and S.L. Pimm. 1998. Water levels, rapid vegetational changes, and the endangered Cape Sable seaside sparrow. *Animal Conservation* 1: 23–32.

Paine, R.T. 1966. Food web complexity and species diversity. *The American Naturalist* 100: 65–75.

Parton, W.J., D.S. Schimel, C.V. Cole, and D.S. Ojima. 1987. Analysis of factors controlling soil organic matter levels in Great Plains grasslands. *Soil Science Society of America Journal* 51: 1173–1179.

Pastor, J., and W.M. Post. 1986. Influence of climate, soil moisture, and succession on forest carbon and nitrogen cycles. *Biogeochemistry* 2: 3–27.

Possingham, H.P., and J. Roughgarden. 1990. Spatial population dynamics of a marine organism with complex life cycle. *Ecology* 71: 973–985.

Reuter, H., and B. Breckling. 1999. Emerging properties on the individual level: Modeling the reproduction phase of the European robin *Erithacus rubecula*. *Ecological Modeling* 121: 199–219.

Riley, M.J., and H.G. Stefan. 1988. MINLAKE: A dynamic lake water quality simulation model. *Ecological Modeling* 43: 155–182.

Robles, C., and R. Descharnais. 2002. History and current development of a paradigm of predation in rocky intertidal communities *Ecology* 83: 1521–1536.

Roughgarden, J. 1997. Production functions from ecological populations: A survey with emphasis on spatially implicit models. Pp. 296–317 in D. Tilman and P. Kareiva, editors. *Spatial Ecology: The Role of Space in Population Dynamics and Interspecific Interactions*. Princeton: Princeton University Press.

Scheffer, M. 1998. *Community Dynamics of Shallow Lakes*. London: Chapman and Hall.

Scheffer, M., J.M. Baveco, D.L. DeAngelis, E.H.R. Lammens, and B. Shuter. 1995. Stunted growth and stepwise die-off in animal cohorts. *The American Naturalist* 145: 376–388.

Scheffer, M., S. Carpenter, J.A. Foley, C. Folke, and B. Walker. 2001. Catastrophic shifts in ecosystems. *Nature* 413: 591–596.

Schmitz, O. 2000. Combining field experiments and individual-based modeling to identify relevant organizational scale in a field system. *Oikos* 89: 471–484.

Slatkin, M., and D.J. Anderson. 1984. A model of competition for space. *Ecology* 65: 1840–1845.

Sokal, R.R., and F.J. Rohlf. 1981. *Biometry* 2nd ed. San Francisco: W. H. Freeman.

Turner, M.G., Y. Wu, W.H. Romme, and L.L. Wallace. 1993. A landscape simulation model of winter foraging by large ungulates. *Ecological Modeling* 69: 163–184.

Wiegand, T., J. Naves, T. Stephan, A. Fernandez. 1998. Assessing the risk of extinction for the brown bear (*Ursus arctos*) in the Cordillera Cantabrica, Spain. *Ecological Monographs* 68: 539–571.

Wolff, W. 1994. An individual-oriented model of a wading bird nesting colony. *Ecological Modeling* 72: 75–114.

Yodzis, P.P. 1988. The indeterminacy of ecological interactions as perceived through perturbation experiments. *Ecology* 69: 508–515.

6

Modeling for Synthesis and Integration:
Forests, People, and Riparian Coarse Woody Debris

Monica G. Turner

Summary

Many of the most challenging questions in ecosystem science today
are posed at intersections between disciplines, scales, and paradigms. Addressing such questions requires synthesis and integration,
to which quantitative models can make important contributions.
"Synthesis" is derived from the Greek *syntithenai*, meaning "to put
together"; integration is the process through which synthesis is
achieved. Synthetic or integrative quantitative models are developed
to enhance understanding of the system, which can be fostered by
formulating appropriate questions. Four ways in which quantitative
models contribute to integration and synthesis are: (1) facilitating
creative and critical thinking about relationships in a system, (2)
comparing alternative ways of conceptualizing a system, (3) identifying principles or relationships that are general across systems, and
(4) initiating an iterative process in which models and empirical observation constantly inform one another by developing models at the
beginning of a research project. A new simple model integrating
land use, natural disturbance, and the dynamics of riparian coarse
woody debris (CWD) in northern Wisconsin (USA) was developed
as an example. The model simulated relationships among three state
variables: the density of live trees in the riparian forest, the density
of standing snags, and the density of CWD. Four scenarios were
simulated: (1) nominal, (2) single timber harvest, (3) large infrequent wind disturbance, and (4) lakeshore development. Results
suggested that riparian CWD could be enhanced and maintained by
large infrequent disturbances. Shoreline development always reduced riparian CWD, but the relationship was nonlinear. Shoreline
development also had a more persistent negative impact on riparian
CWD than did a single large clearcut. The model illustrated the approaches to question formulation; the combination of data and per-

spectives usually considered separately; and the use of synthetic modeling at the beginning of a research project. Synthesis and integration are critical to scientific progress and to finding solutions to many environmental problems facing society. Ecologists should continue to combine quantitative models with observations and experiments in their search for general understanding of ecological systems.

Introduction

Many of the most challenging questions in ecosystem science today are posed at intersections between disciplines, scales, and paradigms. For example, pressing questions regarding global change require integration of natural and anthropogenic processes over a wide array of spatial and temporal scales (Vitousek et al. 1997, National Research Council 2000). Understanding landscape dynamics requires synthesis of knowledge about the environmental template, natural disturbance regimes, successional change, and past, present, and future patterns of human land use (Turner et al. 2001). Improving our understanding of the complex relationships between the land and water, particularly as it influences eutrophication of aquatic ecosystems, is an important goal of both basic and applied ecological research (Carpenter et al. 1995, 1998; Naiman et al. 1998; Naiman and Turner 2000). Synthesis and integration are required to address these and many other fundamental challenges in ecology, and both are also highly prized in science (Pickett et al. 1994).

Simulation models can make important contributions to synthesis and integration. Many scientists have considered the role of models in science in general and ecology in particular (e.g., Holling 1978, Kitching 1983, Swartzman and Kaluzny 1987, Starfield and Bleloch 1986, Starfield et al. 1994, Grant et al. 1997, Ford 1999). Methods of problem solving, including models, were categorized by Holling (1978) relative to level of understanding and data availability (Figure 6.1). Mathematical models can often be developed and solved analytically if the structure and general dynamics of a system are well understood and there are good data on the important processes occurring within the system. Statistical analyses can be used to search for patterns that will lead to hypotheses about the nature of the underlying processes when good data are available but understanding is lacking. Systems analysis and simulation can be used to investigate hypotheses about how a system works if there are relatively few data but at least some understanding of system structure and dynamics (Figure 6.1). Models were also identified, along with long-term observation, ecosystem experiments, and comparative studies, as one of four key approaches to learning about ecosystems (Carpenter 1998).

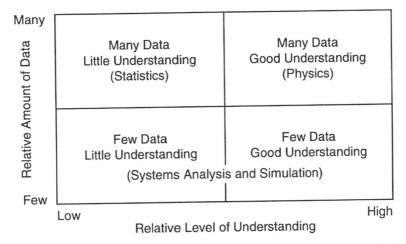

Figure 6.1. Comparison of methods of problem solving in terms of the relative level of understanding and the relative amount of data available about the system. (Modified after Holling 1978; Starfield and Bleloch 1986; and Grant et al. 1997.)

In this chapter, I will emphasize the use of simulation models in synthesis and integration while recognizing the importance of other approaches and models to problem solving. I will begin by defining synthesis and integration, then discuss ways in which questions can be formulated such that the likelihood of achieving synthesis is enhanced. I will next consider four ways in which quantitative models can contribute to synthesis and integration in ecology, using a variety of examples from the literature. Finally, to illustrate these points, I will present a new model developed to explore the dynamics of riparian coarse woody debris for lakes of northern Wisconsin, USA, in response to alternative land-use and disturbance scenarios.

Synthesis, Integration, and Question Formulation

What are synthesis and integration, and how do they differ from other goals of ecological models? The terms are defined in *Webster's New Collegiate Dictionary* (1981) as follows:

> Synthesis: [from Greek *syntithenai*, to put together] 1: a: the composition or combination of parts or elements so as to form a whole, b: the combining of often diverse conceptions into a coherent whole. 2: a: deductive reasoning, b: the dialectic combination of thesis and antithesis into a higher stage of truth.

Deductive reasoning: the deriving of a conclusion by reasoning, specifically, inference in which the conclusion follows necessarily from the premises.

Dialectic combination: any systematic reasoning, exposition, or argument that juxtaposes opposed or contradictory ideas and usually seeks to resolve their conflict; an intellectual exchange of ideas.

Integration: the act or process of integrating, which means 1: forming or blending into a whole. 2: a: uniting with something else, b: incorporating into a larger unit.

These definitions are consistent with those of systems ecology, systems modeling, and simulation modeling (von Bertalanffy 1968, 1969; Watt 1968; Van Dyne 1969; Patten 1971; Ford 1999). In ecology, integration entails the combination of existing data, perspectives, approaches, models, or theories that are apparently disparate (Pickett et al. 1994) and is the process through which synthesis is achieved. Integration may be additive or extractive (Pickett et al. 1994), but it requires the combination of two or more different areas of under-standing to produce new understanding. Integration can occur at any scale or breadth of scope, but many novel attempts in ecology have involved larger scales or levels (e.g., landscape ecology). Integration will be limited if theory in one or more areas is not well developed or if no common currency can be found between the two areas (Pickett et al. 1994).

How can models be developed such that the likelihood of achieving integra-tion and synthesis are enhanced? All models must be question-driven, and mod-els developed for synthesis and integration are no exception. One cannot just "do synthesis" without a clearly stated purpose. Therefore, question identi-fication is a key step in the modeling process—perhaps *the* key step. Model objectives provide the framework for model development, the standard for model evaluation, and the context within which simulation results must be in-terpreted. Thus, stating questions clearly is arguably the most crucial step in the entire modeling process (Grant et al. 1997).

I suggest three general ways to formulate questions to enhance opportunities for integration:

1. Questions posed at the intersection of two or more phenomena require integration of different lines of inquiry. For example, how does x influence, constrain or alter y, where x and y are different ecological phenomena? What are the feedbacks from y to x?

2. Questions exploring a wide range of conditions often require integration and recognize multiple outcomes. For example, what combinations of conditions might produce particular outcomes (i.e., states, flows, or dynamics)? Can multiple pathways produce the same result? Are there thresholds that produce qualitatively different outcomes? A state space can often be used to depict alternative ways

of obtaining particular outcomes, as done by Turner et al. (1993) for the effects of disturbances of different size and frequency for landscape dynamics.

3. Questions about the relative importance of different factors require synthesis of existing knowledge. For example, what is the relative importance of two or more potentially influential drivers on the outcome of a system? Under what conditions are some processes more important than others? Such formulations allow a model to be exercised in an experimental mode (e.g., factorial simulation experiment) and the output analyzed similarly to experimental data. The relative importance of fire size, fire spatial pattern, and winter severity for wintering elk and bison in Yellowstone National Park were evaluated by using an integrative simulation model (Turner et al. 1994). Results from such models can also identify important avenues for empirical study.

Question identification and the subsequent development of the conceptual or qualitative model are perhaps the most intellectually challenging phases of modeling. This phase includes identification of system boundaries, defining model components, identifying the relationships among them, and describing expected patterns of behavior. These are key phases of model development in which opportunities for synthesis and integration can be created.

Quantitative Models in Synthesis and Integration

Simulation models have been widely used in many areas of ecosystem ecology for several decades. In this section, I will highlight four ways in which quantitative models have contributed to synthesis and integration in ecology. Examples are presented to illustrate these uses; these examples are illustrative rather than exhaustive.

Creative and Critical Thinking

Creative and critical thinking about the relationships among parts of a system that might otherwise be considered separately can be represented in a model so that the logical outcome of the representation can be explored. Models of the causes and consequences of land-use change provide excellent examples of the synthesis of subject areas typically considered separately. Understanding land-use changes and their ecological implications presents a fundamental challenge to ecologists, but it requires integrating knowledge from both the social and natural sciences; humans respond to cues from the physical environment and from sociocultural contexts (Riebsame et al. 1994). In addition, forecasting the

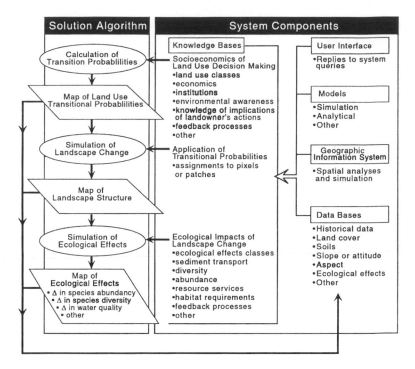

Figure 6.2. Conceptual model of land-use change and ecological responses in temperate forested landscapes of the United States (modified from Lee et al. 1992; reproduced from Turner et al. 2001.)

ecological consequences of land-use change demands integration among traditionally distinct subdisciplines in the natural sciences.

A conceptual model (Figure 6.2) developed by an interdisciplinary group of scientists laid the framework for addressing the causes and consequences of land-use change in temperate forested landscapes in the United States (Lee et al. 1992). This model set the stage for a group of studies in the southern Appalachian Mountains (USA) that quantified land-cover changes (e.g., Turner et al. 1996), related land-cover changes to other ecological response variables (e.g., Harding et al. 1998; Pearson et al. 1998; Wear et al. 1998; Pearson et al. 1999), and explored simulation methods for projecting future landscape patterns (e.g., Wear and Flamm 1993; Flamm and Turner 1994; Wear et al. 1996; Wear and Bolstad 1998). Results demonstrated strong interactions among environmental gradients and social, economic, and aesthetic variables in explaining and prediction land-use patterns. In addition, the likelihood for development to concentrate in riparian areas was suggested (Wear and Bolstad 1998), with important implications for water flow and quality (Wear et al. 1998).

A common theme underlying many studies of land–water interactions is the degree to which upland land uses and their spatial arrangement influence water quality. A simple model of phosphorus (P) transformation and transport for the Lake Mendota watershed, Wisconsin (USA) has provided useful insights into these relationships (Soranno et al. 1996). The watershed of Lake Mendota is dominated by agricultural and urban land uses, and the lake itself has a long history of limnological study (Brock 1985, Kitchell 1992). Soranno et al. (1996) developed a Geographic Information Systems (GIS)-based model of phosphorus loading in which phosphorus-export coefficients varied among land uses. The model was used to compare phosphorus loadings in Lake Mendota under current patterns of land use, presettlement land use, and projected future land use in which the urban area increased nearly twofold. Because rainfall events drive runoff, simulations were conducted for both high- and low-precipitation years. Results demonstrated that most of the watershed did not contribute phosphorus loading to the lake; most P came from a relatively small proportion of the watershed, ranging from 17% of the watershed contributing during low-precipitation years to 50% during high-precipitation years. A six-fold increase in phosphorus loading was estimated to have occurred since settlement.

Land-use models that truly integrate social, economic, and ecological considerations are in their infancy, and there is no consensus on how to best approach this task (Dale et al. 2000). Recent models of land-use change (e.g., Lee et al. 1992; Dale et al. 1993, 1994; Riebsame et al. 1994; Turner et al. 1996; Wear et al. 1996, Wear and Bolstad 1998) often seem simplistic, but they are at the forefront of integration between disciplines that are usually treated separately. Understanding how many diverse factors interact to determine land-use patterns and how ecosystems respond over a range of temporal scales remains a key challenge facing the scientific community and for which models must play an important role.

Alternative Ways of Conceptualizing a System

Seemingly contradictory alternatives, or the implications of alternative ways of conceptualizing a system, can be studied using quantitative models. The same questions or phenomena are often addressed very differently by different modelers. Which models will be most consistent or perform well under what conditions? How much information must be represented in each formulation? Can information be integrated across scales?

The Vegetation/Ecosystem Modeling and Analysis Project (VEMAP) offers an example of the comparative use by many investigators of different models that were developed to address the same general phenomena. In one set of planned comparisons, the responses of net primary productivity (NPP) to doubled CO_2 concentrations (from 335 to 710 ppm) were projected using three biogeochemistry models (Pan et al. 1998). Results found projected increases in NPP of 11% using BIOME-BGC (Running and Coughlan 1988, Running et al. 1989), 5% using CENTURY (Parton 1988), and 8% using the Terrestrial Ecosystem Model (TEM; Raich et al. 1991; Melillo et al. 1993). The models found a negative relationship between precipitation and NPP in all three models, but the relationship between temperature and NPP differed among the models. Each model contained slightly different mechanisms, reflecting, in part, conceptual uncertainty about what controls NPP. Thus, the model comparison project required a highly coordinated, integrated research effort, and results were instrumental in identifying key areas of scientific uncertainty.

Identifying General Principles or Relationships

Models can be used to identify principles or relationships that are general by applying the same model (or suite of models) to different ecological systems. Because of the time required to develop and test complex ecological models, other researchers often apply existing models to new study areas. Models that have been widely used include the JABOWA and FORET models (Botkin et al. 1972; Shugart and West 1977, 1980), CENTURY (Parton et al. 1988), BGC (Running and Coughlan 1988; Running et al. 1989), and many others. Widespread applications of the same model offer opportunities for synthesis of patterns or processes in many different ecosystems or geographic locations within the same ecosystem type. In particular, situations in which the model fails may be most instructive in identifying the limits of applicability of the processes or boundary conditions represented in the model.

One example of widespread application of an ecosystem model is that of CENTURY (Parton et al. 1988), which has been used in many ecosystem types and many locations worldwide. Along a boreal forest transect study in Canada, predictions from CENTURY were tested against field data to explore the sensitivity of carbon dynamics to climate change (Peng et al. 1998; Peng and Apps 1998). The study found that the effects of climate change and enhanced CO_2 concentrations were not simply additive and that they varied along the gradient. Climate change alone would increase total carbon in vegetation but decrease total soil carbon. Increased CO_2 concentrations under current climate would increase total carbon in both vegetation and soil. When both effects were combined, NPP and decomposition both increased, resulting in a decline in soil carbon. Application and testing of the model in varying locations produced qualitatively different insights that could not be obtained by exercising the model at only one location.

Given the widespread use of several influential ecological models, there is a paucity of attempts to synthesize the general knowledge that has been obtained from these many applications. As for many areas in ecology (Baskin 1997), there are numerous opportunities for synthesis and integration of existing modeling results.

Developing Models Early in Research

Models can be used effectively as an exploratory tool at the beginning of a research project, even when data are scarce, to draw inferences about the logical outcomes that follow from the premises. Modeling can then be used to guide empirical study, beginning an ongoing interplay between the model and the data. Developing the conceptual model and exercising it provide a foundation upon which knowledge of the system can build. In this way, the model is not considered to represent final understanding but rather serves as a heuristic device designed to be modified iteratively as research continues. The model presented next illustrates this use of models.

Case Study: Modeling the Dynamics of Riparian Coarse Woody Debris

Coarse woody debris (CWD) is a critical linkage between terrestrial and aquatic ecosystems (Harmon et al. 1986; Murphy and Koski 1989; Gregory et al. 1991; Naiman et al. 2000). Riparian CWD is produced when trees die and fall into a lake, and it provides critical habitat for fishes and other aquatic organisms. In particular, recruitment dynamics of dominant fishes may depend on the physical heterogeneity provided by CWD, which in turn affects entire food webs (Carpenter and Kitchell 1993). The amount of CWD in lakes is determined by the balance between inputs (which depend upon the presence of riparian forest and its composition and structure), decomposition, and removal processes. Tree death may be gradual, as individuals age and senesce, or punctuated by events such as fires, blowdowns, or other natural disturbances that affect large areas and produce pulsed inputs of CWD. The natural dynamics of CWD accumulation are slow, with inputs and decomposition operating over time scales of decades to centuries (Stearns 1951; Hodkinson 1975; Harmon et al. 1986).

Development pressure has increased housing density in many areas that offer appealing natural amenities, such as mountains, coastal areas, and lakes. Land-use development in Vilas County, northern Wisconsin (USA) is one example. Vilas County contains over 1,300 lakes ranging in size from 0.1 to greater than 1,500 ha and covering 16% of the county's surface area. Housing density in Vilas County has increased rapidly since the 1960s, and over half of the new homes were built on the lakeshore (Schnaiberg et al. in press). Furthermore, if the current building rate persists, all undeveloped lakes not in public ownership could be developed within the next twenty years (Wisconsin Department of Natural Resources 1996).

Lakeshore development is negatively related to riparian CWD at the whole-lake scale (Christensen et al. 1997). Humans affect riparian CWD by direct removal to maintain beach areas and boating access. In addition, studies of aesthetic preferences in riverine landscapes have shown preferences for wooded stream channels that do not contain in-channel debris (e.g., Gregory and Davis 1993). Humans also alter the riparian vegetation, for example, by thinning the forest or removing snags, thereby modifying inputs to CWD (e.g., Harmon et al. 1986; Maser and Sedell 1994). The net effect of lakeshore development is the uncoupling of the natural relationship between riparian vegetation and aquatic CWD (Christensen et al. 1997).

The forests surrounding the lakes of Vilas County have changed significantly since European settlement. Much of the northern forest was a hardwood-conifer mix of Eastern hemlock (*Tsuga canadensis*), birch (*Betula* spp.), and maple (*Acer* spp.), with pines (*Pinus* spp.) occurring on the sandy glacial-outwash plains (Curtis 1959). The presettlement disturbance regime was characterized by small gap disturbances along with infrequent, large disturbance events (Canham and Loucks 1984, Frelich and Lorimer 1991). Timbering began in the mid-nineteenth century and increased dramatically during the 1880s with construction of railroads. Most of the forest had been cut by 1900, and timber production declined rapidly due to resource exhaustion (Bawden 1997). Paleoecological studies suggest that this forest harvesting altered the physical structure of the lakes and disrupted the primary producer communities for more than 100 years (Scully et al. 2000). Tourism developed alongside reforestation, beginning with wealthy urban vacationers in the 1920s and evolving to include large numbers of recreational homes and retirees (Gough 1997; Voss and Fuguitt 1979). A long-term and retrospective study of the forest community indicates substantial shifts in composition of riparian forest in Vilas County (Stearns and Likens 2002).

Relatively little is known about the long-term dynamics of production and accumulation of riparian CWD in lakes. A recent modeling study suggested that natural catastrophic disturbances (severe, large-scale events that result in replacement of the riparian forest) could significantly bolster riparian CWD recruitment in streams (Bragg 1997). Compared to undisturbed old-growth forest, large natural disturbances increased the temporal variability and net delivery of CWD, whereas clearcutting reduced both delivery and net amount for many years (Bragg 1997). Natural disturbances in north temperate forests (e.g., large blowdowns in northern Wisconsin occurred in July 1977 and in the Boundary Waters Canoe Area in July 1999) produce large quantities of CWD that may persist in lakes for many centuries.

Question Identification

Predicting the dynamics of CWD requires integrating knowledge of the riparian forest, the natural disturbance regime, human settlement patterns and land-use practices, and the fate of downed wood. A new collaborative, inter-disciplinary study is addressing these issues for lakes in Vilas County using comparative studies, whole-lake experiments, and simulation modeling (see http://biocomplexity.limnology.wisc.edu). To explore the logical outcomes of hypothesized relationships among key components of the riparian system, I developed a prototype of a simulation model that will address the following questions. These questions also illustrate the approaches to question formulation described earlier in this chapter.

1. Is accumulation of riparian CWD a function of small, continuous inputs (e.g., small frequent disturbances), or is it dominated by occasional large pulses (large infrequent disturbances)? I hypothesize that under presettlement natural disturbance regimes, CWD will increase because rates of inputs (both pulsed and gradual) will exceed rates of decomposition and loss to depth by physical transport.

2. What is the effect of timber harvesting and shoreline development on the long-term rhythm of forest development and CWD? I hypothesize that a single timber harvest, such as the turn-of-the-century clearcut, will produce a period of several decades during which the source of CWD will be absent. CWD will decline but then gradually recover. However, with lakeshore development, CWD removal by humans will deplete the long-term resource of CWD while simultaneously reducing the source habitat. Christensen et al. (1997) estimated that it would take approximately 200 years to replace the deficit in CWD density in densely settled lakes. Therefore, I hypothesize that shoreline development will produce long-term declines in CWD abundance that are proportional to the amount of lakeshore developed. Alternatively, particular spatial locations such as sheltered coves may be the primary sources of CWD for the lake as a whole. Loss of riparian forest cover in these critical locations may result in dramatic losses of CWD for the lake, even if other areas of riparian forest remain intact.

3. Under what conditions will CWD fall below critical densities for fish populations? The abundance of CWD required to sustain fish populations over the long term is not known. Detecting such a threshold is complicated because there may be considerable time lags before the effect on fish populations are detected. Depending upon interactions between the slow and fast variables controlling the system, CWD may be pushed below the critical depensation level for fishes. There is evidence from the physical and ecological sciences for nonlinearities in pattern and process related to the thresholds in the area occupied by an entity of interest, such as a land cover type (Stauffer 1985; Stauffer and Aharony 1992; Gardner et al. 1987; Turner et al. 1989; With and King 1997). If gradual changes in riparian land use cause a threshold to be passed either in proportion of lakeshore that is forested or by removal of critical source areas, the fisheries may collapse. Such responses may show strong time lags related to tree life spans and the persistence time of CWD in the lake.

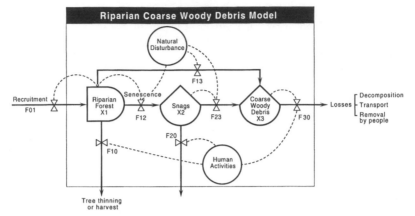

Figure 6.3. Conceptual diagram of the model developed to simulate the dynamics of riparian coarse woody debris (CWD). Flows from compartment i to compartment j are labeled Fij.

Model Structure

The model was purposefully designed to be a simple representation of forest dynamics and the production and accumulation of riparian CWD through time as influenced by natural disturbances and by changes in land use and land cover. Major state variables, flows and controls represented in the model are depicted in the model diagram (Figure 6.3). Spatially, the model considers only the riparian zone of a single lake, defined as the area extending 10 m upland of the lake perimeter; variables and parameters are normalized for 1 km of lake perimeter or 1 ha of riparian shoreline (e.g., 1000 m × 10 m). The model formulation does not vary with lake size. Temporally, the model employs a discrete time step of 1 yr and is run for 200 yr. Initial conditions for the model assume a well-developed forest (not early succession) with live-tree density at steady state.

Three state variables are represented: $X1$, the density of live trees in the riparian forest; $X2$, the density of standing snags in the riparian zone; and $X3$, the density of CWD in the riparian zone. Flows between compartments link these three variables (Figure 6.3). Riparian forest development (F01) is represented deterministically, and these trees may have several fates. They may die and become snags (F12), be blown down, thus bypassing the snag phase and becoming CWD directly (F13), or be removed by humans (F10). Similarly, snags may also be removed by humans (F20) or be converted to CWD when they fall to the ground (F23). Thus, CWD has inputs from the snags and live trees (F23 and F13, respectively). Losses of CWD (F30) occur through decomposition, physical transport to deeper locations in the lake, and direct removal by people.

Changes in the state variables are represented by three equations:

$$dX1 / dt = F01 - F12 - F10 - F13 \qquad (6.1)$$
$$dX2 / dt = F12 - F23 - F20 \qquad (6.2)$$
$$dX3 / dt = F13 - F23 - F30a - F30b - F30c \qquad (6.3)$$

Each flow in the model is represented as follows, with parameter values and sources as shown in Table 6.1.

F01 = RECRUIT * X1, unless time since disturbance or clearcut is less than STAND_RECOV, in which case F01 = 0. Thus, there are no successional processes represented in this initial model. Rather, recruitment of trees into the forest is assumed to occur at a steady rate as a function of the current density of trees.

F12 = X1 * SENES, where tree death is also assumed to be a steady rate that is balancing tree recruitment. However, if a large, infrequent disturbance is simulated, then SENES increases during that event such that 25% of the remaining trees are killed but not toppled and become snags.

F10 = X1 * THINNING * P_DEVEL, representing the thinning of the forest by humans in the presence of shoreline development. A constant rate of thinning is applied until a threshold stand density is reached, after which no more thinning occurs. Thinning is applied only to the proportion of shoreline developed. However, if a clearcut event is simulated, then there is a pulsed event in which a large proportion of the standing forest is removed around the entire lake.

F13 = X1 * TREEFALL. Under nominal conditions, this flow is zero. However, when a large, infrequent disturbance is simulated, there is a pulsed flow of live trees directly to the CWD component, X3.

F23 = X2 * SNAGFALL. Under nominal conditions, snags fall to the ground and become CWD at a steady, relatively low rate. However, if a large infrequent disturbance occurs, snagfall increases during the event.

Table 6.1. Variables and parameters used in the initial model of riparian coarse woody debris dynamics.

Variable	Description	Units	Initial Condition	Source
State Variables				
X1	Density of live trees > 5cm dbh	number/ha	Random value ranging from 1000–2000 ha	Christensen et al. 1996; range of densities for undisturbed lakes.

Table 6.1. *Continued*

Variable	Description	Units	Initial Condition	Source
X2	Density of standing snags	number/ha	100/ha	Christensen et al. 1996, for undisturbed lakes.
X3	Coarse wood debris > 5cm dbh	number/km of shoreline (~ha)	500/ha	Christensen et al. 1996, for undisturbed lakes.

Rates of Natural Processes

Variable	Description	Units	Initial Condition	Source
RECRUIT	Annual recruitment rate of stems > 5cm dbh under steady-state	yr^{-1}	0.005	Set to balance senescence (Frelich and Lorimer 1991, Runkle 1982, 1985); assumed 200-year return interval for gap formation.
SENES	Tree senescence rate	yr^{-1}	0.005 as nominal rate; set to 0.25 with a LID	Set to balance recruitment (Frelich and Lorimer 1991; Runkle 1982; 1985).
SNAGFALL	Rate of snags falling over and becoming CWD	yr^{-1}	0.04 as nominal rate, but set to 0.90 if a large, infrequent disturbance (LID) occurs	Estimated; Christensen et al. (1996) report that ~2.52 pieces of CWD were added per year to the undisturbed lakes; if a major windstorm occurs, this assumes that 90% of the standing snags will fall over.

Table 6.1. *Continued*

Variable	Description	Units	Initial Condition	Source
TREEFALL	Rate at which live trees fall to the ground and become CWD, bypassing the snag stage	yr⁻¹	0.0 as nominal rate, assuming trees die, remain snags for some time, then fall over; 0.75 when a LID occurs	Estimated based on wind-storm events elsewhere (e.g., Turner et al. 1997).
STAND_RECOV	Time for stand to re-cover to its predisturbance density of trees > 5cm dbh	yr	60	Estimated.
DECOMP	Rate at which CWD decomposes	yr⁻¹	0.01 yr⁻¹ as nominal rate, reduced to 50% of this rate following a LID	Estimated; decomposition is very slow, but rates are not known; assumes 100 years.
TRANS-PORT	Rate at which CWD is transported to depth by wind, currents, etc.	yr⁻¹	0.01 yr⁻¹	Estimated; some logs are moved around and transported out of the littoral zone.
Rates of Human Processes				
CWD_REMOV	Rate at which humans remove riparian CWD	yr⁻¹	0.00 yr⁻¹ with no development, 0.10 yr⁻¹ for developed lakeshore	Assumed that humans remove 10% CWD per year from the riparian when development occurs; thus, all CWD is removed within 10 years.

Table 6.1. *Continued*

Variable	Description	Units	Initial Condition	Source
THINNING	Rate at which live trees are thinned from the surrounding forest, occurring only when land-use development has occurred	yr^{-1}	0.0 if there is no development; 0.01 yr^{-1} for areas that are developed if tree density is $\geq$ 800/ha	Christensen et al. (1996) report tree densities of at least 800/ha within the forested areas even of lakes that are developed.
SNAG_ REMOV	Rate at which human cut and remove standing snags	yr^{-1}	0.00 yr^{-1} with no development, 0/01 for developed lakeshore	This assumed that people would remove 10% of the standing dead trees annually, assuming they don't want them to fall on their structures.
P_DEVEL	Proportion of the lake that is developed	Unit less	Varies between 0–1.0	Set by the user for simulation.

F20 = X2 * SNAG_REMOV * P_DEVEL. If shoreline is developed, then this flow represents the removal of snags by people only in the proportion of shoreline that is developed. People remove snags that may endanger their buildings or property. However, if a clearcut is simulated, then F20 = X2 during that event, which assumes that all snags are removed during the clearcut.

Losses of CWD are represented by three flows:

F30a = X3 * DECOMP
F30b = X3 * TRANSPORT
F30c = X3 * CWD_REMOV * P_DEVEL

Decomposition and transport are modeled as constant proportions of the CWD present during a given time period. However, if a large infrequent disturbance occurs, then DECOMP is reduced by 50% because much of the wood will be elevated and not in contact with the soil, thereby slowing decomposition. Direct losses of CWD due to human actions are simulated when some proportion of the shoreline is developed. A constant rate is assumed but applied only to the proportion of shoreline developed.

Scenarios

The model was used to simulate the abundance and temporal dynamics of riparian CWD under four scenarios: (1) nominal, assuming small, frequent gap disturbances that represent the dominant natural disturbance regime; (2) a single extensive timber harvest as occurred during nineteenth-century logging; (3) a single large, infrequent wind disturbance (e.g., a tornado or the 1999 storm in the Boundary Water Canoe Area, Minnesota); and (4) lakeshore development, with proportions of the lakeshore developed varying from 0.10 to 1.0. Small frequent disturbances were modeled to produce gradual inputs and losses of CWD, and stochastic catastrophic disturbances were modeled (e.g., Canham and Loucks 1984; Frelich and Lorimer 1991; Cardille et al. 2001) to produce large pulses of CWD and reset the riparian forest to a pioneer stage. Human settlement influenced the abundance and structure of the riparian forest by reducing the proportion of the lakeshore in riparian forest, reducing tree density, and eliminating standing snags in the remaining forest.

Sensitivity Analysis

A simple sensitivity analysis was conducted to assess model response to variation in the parameter set. Parameter values (Table 6.1) were estimated using existing literature or by assigning reasonable estimates based on expert opinion; none were derived directly from empirical data obtained specifically for this model. Field studies begun during summer 2001 will be used for these purposes in the future. Sensitivity of model results to parameter values was determined by varying selected individual parameter values while holding all other parameter values constant. The initial density of the forest was fixed at 1,500 stems/ha for all sensitivity analyses.

Results

Nominal scenario. Depending on the initial tree density obtained at random, snags and CWD both increased to varying degrees during the simulation. The model was stable, by design. A steady-state condition was generally obtained before 100 yr (Figure 6.4a), with that time being slightly longer for higher initial values of X1 and shorter for the lower values. Higher tree densities in the forest produced a greater abundance of CWD, but CWD always remained at or above initial conditions.

Single extensive timber harvest. Immediately following the single clearcut simulated in model year 50, the density of live trees and snags declined dramatically (Figure 6.4b). After the recovery period, during which I assumed the trees had regrown and achieved a minimum dbh of 5 cm, the density of snags and CWD both began to increase. The CWD recovered from the clearcut scenario in approximately 90 yr. Standing snags took a bit longer to recover. Interestingly, the steady-state density of CWD achieved after the clearcut was always greater than the initial value of CWD.

Large, infrequent disturbance. The occurrence of a single catastrophic blow-down during model year 50 caused a dramatic decrease in the density of live trees and snags and a very large increase in the abundance of CWD (Figure 6.4c). Levels of CWD remained elevated for a relatively long time. Following a decline of CWD for about 150 yr through decomposition and transport to depth, CWD density reached a new steady state that was maintained at or above initial values. Riparian forest and snag density returned to and maintained levels comparable to their initial conditions.

Lakeshore development. CWD always declined with simulated lakeshore development, but the temporal dynamics and steady-state value of snags and CWD varied with the proportion of shoreline development. Furthermore, the relationship between steady-state CWD abundance and shoreline development was nonlinear. If only 10% of the shoreline was developed, CWD declined but stabilized within about 75 yr (Figure 6.5a). With 20% developed, snags and CWD declined quickly, but remained above zero. With 30% developed, CWD was reduced to very low levels within 75 yr, as were snags. As increasing proportions of the lake were developed, the low (near zero) values of CWD were obtained in shorter and shorter time intervals (Figures 6.5b and 6.6). The model still produced stable outcomes (Figure 6.6a) because live forest tree density was maintained well above zero, providing a constant source for CWD and snags. However, CWD density approached zero asymptotically as the proportion of lakeshore developed neared one, and CWD density was very low when 60% or more of the shoreline was developed.

The threshold densities of CWD required to maintain fish populations are not known. However, I examined the conditions under which CWD densities declined to less than 100/ha as a means to explore the potential for threshold dynamics. If the developed shoreline was less than 30%, CWD did not fall below this threshold. If the developed shoreline was greater than or equal to 30%, then this threshold was exceeded in varying amounts of time, ranging from 47 yr with 30% developed to 11 yr with the entire lakeshore developed. These results suggest the potential for complex responses to occur within the system, driven in part by time lags in cause and effect that may be amplified by fish population dynamics.

Model sensitivity. Among the parameters tested, the model was very sensitive to the rate of CWD removal by humans, suggesting that this rate should be estimated with reasonable confidence in future empirical studies. Model results were not sensitive to variation in the decomposition loss parameter or physical transport loss parameter. For example, varying the rate of CWD removal from 0.02/yr to 0.20/yr produced variation from 50 to 175/ha in the steady-state value of CWD. In contrast, variation in the decomposition rate from 0.01 to 0.12/yr produced variation of only 80–90/ha in steady-state CWD.

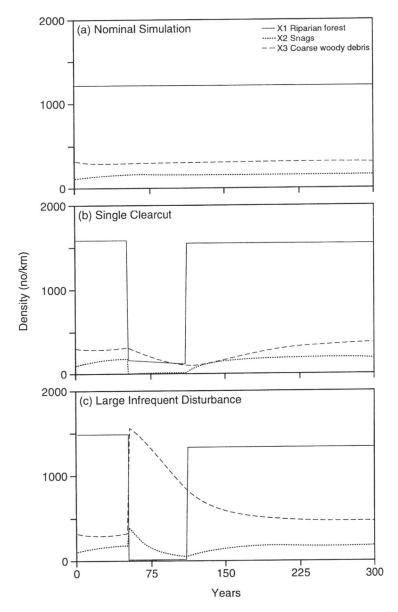

Figure 6.4. Representative simulations from each of three model scenarios: (a) the nominal scenario, which assumes small gap disturbances; (b) the single large clearcut, which occurs during model year 50 and is not repeated; and (c) occurrence of a large, infrequent disturbance (e.g., large blowdown) during model year 50.

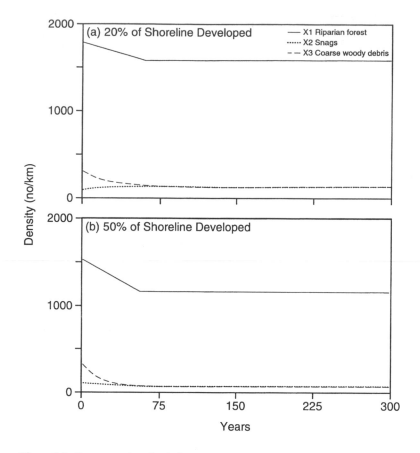

Figure 6.5. Representative simulations of lakeshore development occurring on 20% or 50% of the riparian zone surrounding a lake.

Discussion

The riparian CWD model illustrates several points regarding the use of models for synthesis and integration. First, the questions driving the model were formulated purposefully to require integration of existing data and knowledge. These questions addressed the intersection of different phenomena, allowed exploration of a wide range of conditions that might produce the same or different outcomes, and explored the relative importance of different phenomena for an ecological response. Second, this simple model of riparian CWD dynamics combined data and perspectives that usually are considered separately. These included the riparian forest, disturbance dynamics, human behaviors and

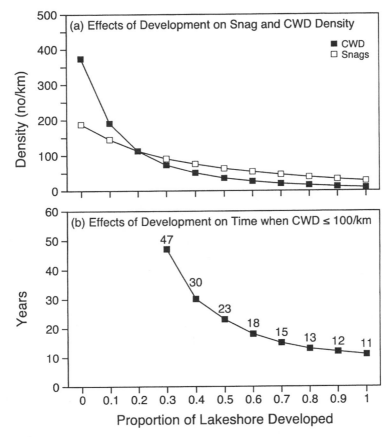

Figure 6.6. (a) Steady-state snag and CWD density for simulated lakeshore development occurring from zero to 100% of a lakeshore. (b) Number of years before simulated CWD abundance drops below 100/km shoreline as a function of the proportion of shoreline developed.

settlement patterns, ecosystem processes (e.g., decomposition), and physical processes (e.g., movement of logs by wind, waves, and ice). Eventually, links to in-lake dynamics and fish population dynamics will also be included. Third, the model was developed at the outset of a large multifaceted research project as comparative and experimental approaches were also beginning. The model predictions can inform the field studies and experiments, and the empirical data will be used to modify the model. Therefore, the model will continue to be integral to the overall research and will facilitate ongoing synthesis of the understanding developed throughout the project.

Clearly, much of the richness of the biological intricacy of the riparian forest ecosystem was excluded from the model. State variables and flows were selected to abstract the essence of hypothesized key interactions, but additional detail or complexity could be added to address other questions or make predictions for particular lakes. For example, successional processes could be modeled, along with variation in forest community composition through time and space. Different stand ages and functional groups of trees (e.g., conifers vs. deciduous trees) will differ in many important characteristics—for example, longevity, susceptibility to disease or windthrow, and decomposition rate—all of which influence CWD recruitment and fate. Other biota, particularly beaver (*Castor canadensis*), can be locally important influences on riparian forest dynamics and CWD, and their effects could be included. The model could also be made spatially explicit, representing the importance of particular landscape positions for CWD, the effects of slope and aspect, or the difference between sheltered and exposed locations. In addition to CWD density, the basal area and geometric complexity of CWD could be simulated. Some parameters, such as physical transport of logs to depth and the rate of CWD removal by humans, will also be influenced by lake type (e.g., seepage or drainage lake) and lake size. Finally, interactions among whole disturbance regimes could be explored rather than simple comparisons among single disturbance events.

Simple as it is, this model still produced interesting dynamics and hypotheses that can be tested empirically. The model suggests that riparian CWD can be maintained by large infrequent disturbances because the pulsed input is substantial and will persist for many decades, even centuries, in the absence of physical removal by humans. This result is consistent with that reported by Bragg (1997), in which natural catastrophic disturbance inflated long-term recruitment of CWD above the level expected in an old-growth forest. The model also suggests nonlinear relationships between lakeshore development and the abundance of riparian CWD, with the potential for thresholds to produce surprises for scientists, managers, and anglers. In addition, despite its acute effects, clearcutting once had a less persistent damping effect on riparian CWD than does moderate to extensive lakeshore development.

Modeling for Synthesis and Integration: Some Take-Home Points

Quantitative models are one of several different approaches to achieving ecological understanding (Carpenter 1998). They can be particularly effective in facilitating the combination of data, perspectives, approaches, or theories that are apparently disparate—that is, integration—to produce new understanding. That models must be question driven, even when synthesis and integration are the goals, cannot be overstated. It is not possible to synthesize everything for all questions, over all scales. Therefore, as with all types of modeling, question identification and selection is key.

Ideally, modeling for synthesis and integration is best begun early and continued throughout a research project. It should be an iterative process in which the models and empirical observation constantly inform one another. Using models as tools for synthesis and integration does not imply that the modeling gets done at the end after the data are all collected and the research project is charged with "synthesizing what has been learned." Modeling should not be an afterthought, an add-on at the end.

Modeling is always a process of abstracting key information and relationships from a system. Synthetic or integrative models require that the *important* processes and details be represented, not that *every* detail or process known about the system must be included. Synthetic models may still be highly abstracted representations of the system. Simple models can, of course, always be expanded as needed based on new questions, but synthetic models need not be encyclopedic.

In conclusion, synthesis and integration are critical to scientific progress and to finding solutions to many environmental problems facing society. Modeling has long been an integral part of ecosystem studies, and ecosystem scientists must strive to strengthen that connection and to exploit the insights that can be derived from modeling (Carpenter 1998; Lauenroth et al. 1998). Ecologists should continue to combine quantitative models with observations and experiments in their search for general understanding of ecological systems.

Acknowledgments. I thank Steve Carpenter, Tim Kratz, and Dan Tinker for helpful discussions of the ideas developed in this chapter and my whole lab group for providing valuable comments as the model was being developed. Funding for this study was provided by the National Science Foundation through the Biocomplexity (Grant No. DEB-0083545) and Long-term Ecological Research Programs (North Temperate Lakes, Grant No. DEB-9632853).

References

Baskin, Y. 1997. Center seeks synthesis to make ecology more useful. *Science* 275: 310–311.

Bawden, T. 1997. The Northwoods: Back to nature? Pp. 450–469 in R.C. Ostergren and T.R. Vale, editors. *Wisconsin Land and Life.* Madison: University of Wisconsin Press.

Botkin, D.B., J.F. Janak, and J.R. Wallis. 1972. Some ecological consequences of a computer model of forest growth. *Journal of Ecology* 60: 849–872.

Bragg, D.C. 1997. Simulating catastrophic disturbance effects on coarse woody debris production and delivery. Pages 148–156 in R. Teck, M. Moeur, and J. Adams, compilers. *Proceedings: Forest Vegetation Simulator Conference.* General Technical Report INT-GTR-373, Ogden, UT: The USDA Intermountain Research Station.

Brock, T.D. 1985. *A Eutrophic Lake: Lake Mendota, Wisconsin.* New York: Springer-Verlag.

Canham, C.D., and O.L. Loucks. 1984. Catastrophic windthrow in the presettlement forests of Wisconsin. *Ecology* 65: 803–809.

Cardille, J.A., S.J. Ventura, and M.G. Turner. 2001. Environmental and social factors influencing wildfires in the upper Midwest, USA. *Ecological Applications* 11: 111–127.

Carpenter, S.R. 1998. The need for large-scale experiments to assess and predict the response of ecosystems to perturbation. Pp. 287–312 in M.L. Pace and P.M. Groffman, editors. *Successes, Limitations, and Frontiers in Ecosystem Science.* New York: Springer-Verlag.

Carpenter, S.R., N.F. Caraco, D.L. Correll, R.W. Howarth, A.N. Sharpley, and V.H. Smith. 1998. Nonpoint pollution of surface waters with phosphorus and nitrogen. *Ecological Applications* 8: 559–568.

Carpenter, S., T. Frost, L. Persson, M. Power, and D. Soto. 1995. Freshwater ecosystems: Linkages of complexity and processes. Chapter 12 in H.A. Mooney, editor. *Functional Roles of Biodiversity: A Global Perspective.* New York: John Wiley and Sons.

Carpenter, S.R., and J.F. Kitchell. 1993. *The Trophic Cascade in Lakes.* Cambridge: Cambridge University Press.

Christensen, D.L., B.R. Herwig, D.E. Schindler, and S.R. Carpenter. 1997. Impacts of lakeshore residential development on coarse woody debris in north temperate lakes. *Ecological Applications* 6: 1143–1149.

Curtis, J.T. 1959. *The Vegetation of Wisconsin.* Madison: University of Wisconsin Press.

Dale, V.H., S. Brown, R. Haeuber, N.T. Hobbs, N. Huntly, R.J. Naiman, W.E. Riebsame, M.G. Turner, and T. Valone. 2000. Ecological principles and guidelines for managing the use of land. *Ecological Applications* 10: 639–670.

Dale, V.H., R.V. O'Neill, F. Southworth, and M. Pedlowski. 1994. Modeling effects of land management in the Brazilian Amazonian settlement of Rondonia. *Conservation Biology* 8: 196–206.

Dale, V.H., F. Southworth, R.V. O'Neill, A. Rosen, and R. Frohn. 1993. Simulating spatial patterns of land-use change in Rondonia, Brazil. *Lectures on Mathematics in the Life Sciences* 23: 29–55.

Flamm, R.O., and M.G. Turner. 1994. Alternative model formulations of a stochastic model of landscape change. *Landscape Ecology* 9: 37–46.

Ford, A. 1999. *Modeling the Environment.* Washington, DC: Island Press.

Frelich, L.E., and C.G. Lorimer. 1991. Natural disturbance regimes in hemlock-hardwood forests of the upper Great Lakes region. *Ecological Monographs* 61: 145–164.

Gardner, R.H., B.T. Milne, M.G. Turner, and R.V. O'Neill. 1987. Neutral models for the analysis of broad-scale landscape pattern. *Landscape Ecology* 1: 5–18.

Gough, R. 1997. *Farming the Cutover: A Social History of Northern Wisconsin, 1900–1940.* Lawrence: University Press of Kansas.

Grant, W.E., E.K. Pedersen, and S.L. Marin. 1997. *Ecology and Natural Resource Management: Systems Analysis and Simulation.* New York: John Wiley and Sons.

Gregory, K.J., and R.J. Davis. 1993. The perception of riverscape aesthetics: an example from two Hampshire Rivers. *Journal of Environmental Management* 39: 171–185.

Gregory, S.V., F.J. Swanson, W.A. McKee, and K.W. Cummins. 1991. An ecosystem perspective of riparian zones. *BioScience* 41: 540–551.

Harding, J.S., E.F. Benfield, P.V. Bolstad, G.S. Helfman, and E.B.D. Jones III. 1998. Stream biodiversity: The ghost of land use past. *Proceedings of the National Academy of Sciences* 95: 14843–14847.

Harmon, M.E., J.F. Franklin, F.J. Swanson, P. Sollins, S.V. Gregory, J.D. Lattin, N.H. Anderson, S.P. Cline, N.G. Aumen, J.R. Sedell, G.W. Lienkaemper, K. Cromack, Jr., and K.W. Cummins. 1986. Ecology of coarse woody debris in temperate ecosystems. *Advances in Ecological Research* 15: 133–302.

Hodkinson, I.D. 1975. Dry weight loss and chemical changes in vascular plant litter of terrestrial origin, occurring in a beaver pond ecosystem. *Journal of Ecology* 63: 131–142.

Holling, C.S. 1978. *Adaptive Environmental Assessment and Management.* New York: John Wiley and Sons.

Kitchell, J.F., editor. 1992. *Food Web Management: A Case Study of Lake Mendota, Wisconsin.* New York: Springer-Verlag.

Kitching, R.L. 1983. *Systems Ecology.* Queensland: Queensland Press, St. Lucia.

Lauenroth, W.K., C.D. Canham, A.P. Kinzig, K.A. Poiani, W.M. Kemp, and S.W. Running. 1998. Simulation modeling in ecosystem science. Pp. 404–415 in M.L. Pace and P.M. Groffman, editors. *Successes, Limitations, and Frontiers in Ecosystem Science.* New York: Springer-Verlag.

Lee, R.G., R.O. Flamm, M.G. Turner, C. Bledsoe, P. Chandler, C. DeFerrari, R. Gottfried, R.J. Naiman, N. Schumaker, and D. Wear. 1992. Integrating sustainable development and environmental vitality. Pp. 499–521 in R.J. Naiman, editor. *New Perspectives in Watershed Management.* New York: Springer-Verlag.

Maser, C., and J.R. Sedell. 1994. *From the Forest to the Sea: The Ecology of Wood in Streams, Rivers, Estuaries, and Oceans.* Delray Beach, FL: St. Lucie Press.

Melillo, J.M., A.D. McGuire, D.W. Kicklighter, B. Moore III, C.J. Vörösmarty, and A.L. Schloss. 1993. Global climate change and terrestrial net primary production. *Nature* 363: 234–240.

Murphy, M.L., and K.V. Koski. 1989. Input and depletion of woody debris in Alaska streams and implications for streamside management. *North American Journal of Fisheries Management* 9: 427–436.

Naiman, R.J., R.E. Bilby, and P.A. Bisson. 2000. Riparian ecology and management in the Pacific coastal forest. *BioScience* 50: 996–1011.

Naiman, R.J., J.J. Magnuson, and P.L. Firth. 1998. Integrating cultural, economic and environmental requirements for fresh water. *Ecological Applications* 8: 569–570.

Naiman, R.J., and M.G. Turner. 2000. A future perspective on North America's freshwater ecosystems. *Ecological Applications* 10: 958–970.

National Research Council. 2000. *Global Change Ecosystems Research.* Washington, DC: National Academy Press.

Pan, Y.D., J.M. Melillo, A.D. McGuire, D.W. Kicklighter, L.F. Pitelka, K. Hibbard, L.L. Pierce, S.W. Running, D.S. Ojima, W.J. Parton, and D.S. Schimel. 1998. Modeled responses of terrestrial ecosystems to elevated atmospheric CO_2: A comparison of simulations by the biogeochemistry models of the Vegetation/Ecosystem Modeling and Analysis Project (VEMAP). *Oecologia* 114: 389–404.

Parton, W.J., J.W.B. Stewart, and C.V. Cole. 1988. Dynamics of C, N, P and S in grassland soils: A model. *Biogeochemistry* 5: 109–131.

Patten, B.C. 1971. A primer for ecological modeling and simulation with analog and digital computers. Pp. 3–121 in B. C. Patten, editor. *Systems Analysis and Simulation in Ecology.* Vol. 1. New York: Academic Press.

Pearson, S.M., A.B. Smith, and M.G. Turner. 1998. Forest fragmentation, land use, and cove-forest herbs in the French Broad River Basin. *Castanea* 63: 382–395.

Pearson, S.M., M.G. Turner, and J.B. Drake. 1999. Landscape change and habitat availability in the Southern Appalachian Highlands and the Olympic Peninsula. *Ecological Applications* 9: 1288–1304.

Peng, C.H., and M.J. Apps. 1998. Simulating carbon dynamics along the Boreal Forest Transect Case Study (BFTCS) in central Canada. 2. Sensitivity to climate change. *Global Biogeochemical Cycles* 12: 393–402.

Peng, C.H., M.J. Apps, D.T. Price, I.A. Nalder and D.H. Halliwell. 1998. Simulating carbon dynamics along the Boreal Forest Transect Case Study (BFTCS) in central Canada. 1. Model testing. *Global Biogeochemical Cycles* 12: 381–392.

Pickett, S.T.A., J. Kolasa, and C.G. Jones. 1994. *Ecological Understanding: The Nature of Theory and the Theory of Nature.* New York: Academic Press.

Raich, J.W., E.B. Rastetter, J.M. Melillo, D.W. Kicklighter, P.A. Steudler, B.J. Peterson, A.L. Grace, B. Moore III, and C.J. Vörösmarty. 1991. Potential net primary productivity in South America: Application of a global model. *Ecological Applications* 4: 399–429.

Riebsame, W.E., W.B. Meyer, and B.L. Turner II. 1994. Modeling land use and cover as part of global environmental change. *Climatic Change* 28: 45–64.

Runkle, J.R. 1982. Patterns of disturbance in some old-growth mesic forests of eastern North America. *Ecology* 63: 1533–1546.

———. 1985. Disturbance regimes in temperate forests. Pp. 17–34 in S.T.A. Pickett and P.S. White, editors. *The Ecology of Natural Disturbance and Patch Dynamics.* New York: Academic Press.

Running, S.W., and J.C. Coughlan. 1988. A general model of forest ecosystem processes for regional applications. I. Hydrologic balance, canopy gas exchange and primary production processes. *Ecological Modeling* 42: 125–154.

Running, S.W., R.R. Nemani, D.L. Peterson, L.E. Band, D.F. Potts, L.L. Pierce, and M.A. Spanner. 1989. Mapping regional forest evapotranspiration and photosynthesis by coupling satellite data with ecosystem simulation. *Ecology* 70: 1090–1101.

Schnaiberg, J., J. Riera, M.G. Turner, and P.R. Voss. In press. Explaining human settlement patterns in a recreational lake district: Vilas County, Wisconsin, USA. *Environmental Management.*

Scully, N.M., P.R. Leavitt, and S.R. Carpenter. 2000. Century-long effects of forest harvest on the physical structure and autotrophic community of a small temperate lake. *Canadian Journal of Fisheries and Aquatic Science* 57(2): 50–59.

Shugart, H.H., and D.C. West. 1977. Development of an Appalachian deciduous forest succession model and its application to assessment of the impact of the chestnut blight. *Journal of Environmental Management* 5: 161–179.

———. 1980. Forest succession models. *BioScience* 30: 308–313.

Soranno, P.A., S.L. Hubler, S.R. Carpenter, and R.C. Lathrop. 1996. Phosphorus loads to surface waters: A simple model to account for spatial pattern of land use. *Ecological Applications* 6: 865–878.

Starfield, A.M., and A.L. Bleloch. 1986. *Building Models for Conservation and Management.* New York: Macmillan.

Starfield, A.M., K.A. Smith, and A.L. Bleloch. 1994. *How to Model It.* Edina, MN: Burgess International Group.

Stauffer, D. 1985. *Introduction to Percolation Theory.* London: Taylor and Francis.

Stauffer, D., and A. Aharony. 1992. *Introduction to Percolation Theory.* 2nd ed. London: Taylor and Francis.

Stearns, F. 1951. The composition of the sugar maple–hemlock–yellow birch association in northern Wisconsin. *Ecology* 32: 245–265.

Stearns, F.W., and G.E. Likens. 2002. One hundred years of recovery of a pine forest in northern Wisconsin. *American Midland Naturalist* 148: 2–19.

Swartzman, G.L., and S.P. Kaluzny. 1987. *Ecological Simulation Primer.* New York: Macmillan.

Turner, M.G., V.H. Dale, and E.E. Everham III. 1997. Fires, hurricanes, and volcanoes: Comparing large-scale disturbances. *BioScience* 47: 758–768.

Turner, M.G., R.H. Gardner, V.H. Dale, and R.V. O'Neill. 1989. Predicting the spread of disturbance across heterogeneous landscapes. *Oikos* 55: 121–129.

Turner, M.G., R.H. Gardner, and R.V. O'Neill. 2001. *Landscape Ecology in Theory and Practice.* New York: Springer-Verlag.

Turner, M.G., W.H. Romme, R.H. Gardner, R.V. O'Neill, and T.K. Kratz. 1993. A revised concept of landscape equilibrium: Disturbance and stability on scaled landscapes. *Landscape Ecology* 8: 213–227.

Turner, M.G., D.N. Wear, and R.O. Flamm. 1996. Land ownership and land-cover change in the Southern Appalachian Highlands and the Olympic Peninsula. *Ecological Applications* 6: 1150-1172.

Turner, M.G., Y. Wu, W.H. Romme, L.L. Wallace, and A. Brenkert. 1994. Simulating winter interactions between ungulates, vegetation, and fire in northern Yellowstone Park. *Ecological Applications* 4: 472–496.

Van Dyne, G.M. 1969. *The Ecosystem Concept in Natural Resource Management.* New York: Academic Press.

Vitousek, P.M., J.D. Aber, R.W. Howarth, G.E. Likens, P.A. Matson, D.W. Schindler, W.H. Schlesinger, and D.G. Tilman. 1997. Human alteration of the global nitrogen cycles: Sources and consequences. *Ecological Applications* 7: 737–750.

von Bertalanffy, L. 1968. *General System Theory: Foundations, Development and Applications.* New York: George Braziller.

―――. 1969. Change or law. Pp. 56–84 in A. Koestler and J.R. Smythies, editors. *Beyond Reductionism: The Alpbach Symposium.* New York: George Braziller.

Voss, P.R., and G.V. Fuguitt. 1979. *Turnaround Migration in the Upper Great Lakes Region.* Final Report to the Upper Great Lakes Regional Commission, Washington, D.C. 70–12, Applied Population Lab, Madison.

Watt, K.E.F. 1968. *Ecology and Resource Management.* New York: McGraw-Hill.

Wear, D.N., and P. Bolstad. 1998. Land-use changes in southern Appalachian landscapes: Spatial analysis and forecast evaluation. *Ecosystems* 1: 575–594.

Wear, D.N., and R.O. Flamm. 1993. Public and private disturbance regimes in the southern Appalachians. *Natural Resource Modeling* 7: 379–397.

Wear, D.N., M.G. Turner, and R.O. Flamm. 1996. Ecosystem management with multiple owners: Landscape dynamics in a southern Appalachian watershed. *Ecological Applications* 6: 1173–1188.

Wear, D.N., M.G. Turner, and R.J. Naiman. 1998. Land cover along an urban-rural gradient: Implications for water quality. *Ecological Applications* 8: 619–630.

Webster's New Collegiate Dictionary. 1981. Springfield, MA: G. and C. Merriam.

Wisconsin Department of Natural Resources. 1996. *Northern Wisconsin's Lakes and Shorelands: A Report Examining a Resource under Pressure.* 18pp.

With, K.A., and A.W. King. 1997. The use and misuse of neutral landscape models in ecology. *Oikos* 79: 219–229.

7

The Role of Models in Prediction for Decision

Roger A. Pielke Jr.

Summary

The processes of science and decision making share an important characteristic: success in each depends upon researchers or decision makers having some ability to anticipate the consequences of their actions. The predictive capacity of science holds great appeal for decision makers who are grappling with complex and controversial environmental issues by promising to enhance their ability to determine a need for and outcomes of alternative decisions. As a result, the very process of science can be portrayed as a positive step toward solving a policy problem. The convergence—and perhaps confusion—of prediction in science and prediction for policy presents a suite of hidden dangers for the conduct of science and the challenge of effective decision making. This chapter, organized as a set of inter-related analytical vignettes, seeks to expose some of these hidden dangers and to recommend strategies to overcome them in the process of environmental decision making. In particular, this chapter will try to distill some of the lessons gleaned from research on modeling, prediction, and decision making in the earth and atmospheric sciences for quantitative modeling of ecosystems. One clear implication is that conventional approaches to modeling and prediction cannot simultaneously meet the needs of both science and decision making. For ecosystem science, there fortunately exists a body of experience in understanding, using, and producing predictions across the sciences on which to develop new understandings of the relationship of science and decision making to the potential benefit of both research and policy.

Introduction: Prediction in Science and Prediction for Decision

The processes of science and decision making share an important characteristic: success in each depends upon researchers or decision makers having some ability to anticipate the consequences of their actions. On the one hand, "[being] predictive of unknown facts is essential to the process of empirical testing of hypotheses, the most distinctive feature of the scientific enterprise" (Ayala 1996). Of course, in science the "unknown facts" in question could lie in the past or the future. "Decision-making," on the other hand, "is forward looking, formulating alternative courses of action extending into the future, and selecting among alternatives by expectations of how things will turn out" (Lasswell and Kaplan 1950).

The predictive capacity of science holds great appeal for decision makers who are grappling with complex and controversial environmental issues because it promises to enhance their ability to determine a need for and outcomes of alternative decisions. As a result, the very process of science can be portrayed as a positive step toward solving a policy problem. The appeal of this "two birds with one stone" line of reasoning is obvious for decision makers who would place the onus of responsibility for problem solving onto the shoulders of scientists. But this reasoning is also seductive for scientists who might wish to better justify public investments in research, as well as for a public that has come to expect solutions because of such investments (Sarewitz and Pielke 1999).

The convergence—and perhaps confusion—of prediction in science and prediction for policy presents a suite of hidden dangers for the conduct of science and the challenge of effective decision making. This paper, organized as a set of interrelated analytical vignettes, seeks to expose some of these hidden dangers and to recommend strategies to overcome them in the process of environmental decision making. In particular, this paper seeks to distill some of the lessons gleaned from research on modeling, prediction, and decision making in the earth and atmospheric sciences for quantitative modeling of ecosystems, the focus of Cary Conference IX. The background materials for the conference noted that "recent years have seen dramatic advancements in the computational power and mathematical tools available to modelers. Methodological advances in areas ranging from remote sensing to molecular techniques have significantly improved our ability to parameterize and validate models at a wide range of spatial scales. The body of traditional, mechanistic, empirical research is also growing phenomenally. Ecosystem science is ripe for major gains in the synthetic and predictive power of its models, and that this comes at a time of growing need by society for quantitative models that can inform debate about critical environmental issues."

This background indicates that the community of ecosystem scientists is following other fields—particularly the atmospheric, oceanic, and earth sciences—down a path of using integrative environmental modeling to advance science *and* to generate predictive knowledge putatively to inform decision making (Clark et al. 2001). This paper distills some of the most important lessons from the other fields that have journeyed down this perilous path, focusing on the use of models to produce predictions for decision.

Modeling for What?

Bankes (1993) defines two types of quantitative models, consolidative and exploratory, that are differentiated by their uses (cf. Morrison and Morgan 1999). A consolidative model seeks to include all relevant facts into a single package and use the resulting system as a surrogate for the actual system. The canonical example is that of the controlled laboratory experiment. Other examples include weather forecasting and engineering design models. Such models are particularly relevant to decision making because the system being modeled can be treated as being closed. Oreskes et al. (1994) define a closed system as one "in which all the components of the system are established independently and are known to be correct" (Oreskes et al. 1994).[1] The creation of such a model generally follows two phases: first, model construction and evaluation; and second, operational usage of a final product. Such models can be used to investigate diagnostics (i.e., What happened?), process (Why did it happen?), or prediction (What will happen?).

An exploratory model—or what Bankes (1993) calls a "prosthesis for the intellect"—is one in which all components of the system being modeled are not established independently or are not known to be correct. In such a case, the model allows for experiments with the model to investigate the consequences for modeled outcomes of various assumptions, hypotheses, and uncertainties associated with the creation of and inputs to the model. These experiments can contribute to at least three important functions (Bankes 1993). First, they can shed light on the existence of unexpected properties associated with the interaction of basic assumptions and processes (e.g., complexity or surprises). Second, in cases where explanatory knowledge is lacking, exploratory models can facilitate hypothesis generation to stimulate further investigation. Third, the model can be used to identify limiting, worst-case, or special scenarios under various assumptions and uncertainty associated with the model experiment. Such experiments can be motivated by observational data (e.g., econometric and hydrologic models), by scientific hypotheses (e.g., general circulation models of climate), or by a desire to understand the properties of the model or class of models independent of real-world data or hypotheses (e.g., Lovelock's Daisyworld).

Both consolidative and exploratory models have important roles to play in science and decision settings (Bankes 1993). However, the distinction between consolidative and exploratory modeling is fundamental but rarely made in practice or in interpretation of research results. Often, the distinction is implicitly (or explicitly) blurred to "kill two birds with one stone" in modeling and predicting for science and policy (Sarewitz and Pielke 1999).[2] Consider, for example, the goal of the U.S. Global Change Research Program, from 1989: "To gain an adequate predictive understanding of the interactive physical, geological, chemical, biological and social processes that regulate the total Earth System and, hence establish the scientific basis for national and international policy formulation and decisions" (CES 1989, 9).[3]

And following from this conflation, most presentations by scientists and the media of the results of national and international climate assessments have sought to imbue the imprimatur of consolidative knowledge upon what are inherently exploratory exercises.[4] Those who conflate the science and policy roles of prediction and modeling trade short-term political or public gain with a substantial risk of a more lasting loss of legitimacy and political effectiveness (Sarewitz et al. 2000).

Thus, one of the most important lessons to be learned from the experiences of other scientific endeavors in which modeling has a potential role to play in research and decision is that one must be clear about the purposes for which the modeling is to be used and carefully examine any assumption that presumes isomorphism between the needs of science and the needs of decision making.

Importance of Uncertainty

Uncertainty, in the view of economist John Maynard Keynes, is the condition of all human life (Skidelsky 2000). Uncertainty means that more than one outcome is consistent with our expectations (Pielke 2001). Expectations are a result of judgment, are sometimes based on technical mistakes and interpretive errors, and are shaped by values and interests. As such, uncertainty is not some feature of the natural world waiting to be revealed but is instead a fundamental characteristic of how human perceptions and understandings shape expectations. Because uncertainty is a characteristic of every important decision, it is no surprise that society looks to science and technology to help clarify our expectations in ways that lead to desired outcomes.

Because decision making is forward-looking, decision makers have traditionally supported research to quantify and even reduce uncertainties about the future. In many cases, particularly those associated with closed systems—or systems that can be treated as closed—understanding uncertainty is a straightforward technical exercise: probabilities in a card game are the canonical example. Two real-world examples include error analysis in engineering and manufacturing, and the actuarial science that underlies many forms of insurance. However, in many other circumstances—particularly those associated with human action—systems are intrinsically open and cannot be treated as closed, meaning that understanding uncertainty is considerably more challenging. In recent decades, many scientists have taken on the challenge of understanding such open systems (e.g., global climate, genetic engineering, etc.), and in the process of securing considerable public resources to pursue this challenge, scientists often explicitly promise to "understand and reduce uncertainties" as input to important societal decisions.

Conventional wisdom holds that uncertainty is best understood or reduced by advancing knowledge, an apparent restatement of the traditional definition of uncertainty as "incomplete knowledge" (Cullen and Small 2000). But in reality, advances in knowledge can add significant uncertainty. For example, in 1990 the Intergovernmental Panel on Climate Change (IPCC) projected a 1.5° to 4.5° C mean global temperature change for 2100 (IPCC 1990). In 2001, after tens of billions of dollars of investment in global-change research, the IPCC projected a 1.4° to 5.8° C temperature change for 2100 (IPCC 2001). Even as the IPCC has become more certain that temperature will increase, the uncertainty associated with its projections has also increased. Why? Researchers have concluded that there are many more scenarios of possible population and energy use than originally assumed and have learned that the global ocean-atmosphere-biosphere system is much more complex than was once thought (IPCC 2001). Ignorance is bliss because it is accompanied by a lack of uncertainty.

The promise of prediction is that the range of possible futures might be narrowed in order to support (and indeed to some degree determine) decision making. By way of contrast, in his Foundation series, science fiction writer Isaac Asimov introduced the notion of "psychohistory." Asimov's psychohistorians had the ability to predict the future with certainty based on complex mathematical models. We know that Asimov's characters lie squarely in the realm of science fiction—there can be no psychohistory such as this. The future, to some degree, will always be clouded. But, experience shows that this cloudiness is variable; we *can* predict some events with skill, and the promise of prediction can be realized. Understanding, using, and producing predictions depends upon understanding their uncertainty. What is it that leads to the uncertainty of earth and environmental predictions? What are the prospects of knowing the uncertainty of specific predictions?

A simple example might prove useful. Consider the poker game known as five-card draw. In a standard fifty-two-card deck, there are 2,598,960 possible five-card poker hands (Scarne 1986). Let us assume that in your hand you hold a pair. What are the chances that by exchanging the other three cards you will draw a third card to match the pair? In this instance you can know with great precision that in 71.428 . . . % of such situations you will fail to improve your hand. Thus, when you exchange three cards you are "uncertain" about the outcome that will result, but you can quantify that uncertainty with great certainty.

This sort of uncertainty is that associated with random processes, that is, one in which each element of a set (in this case a deck of cards) has an equal chance of occurring. Because we know the composition of the deck and the set of possible events (i.e., the relative value of dealt hands), it is possible to calculate precisely the uncertainty associated with future events. Scientists call this *aleatory* uncertainty, and it is studied using mathematical statistics (Hoffman and

Hammonds 1994; Stewart 2000). Such uncertainty, by definition, cannot be reduced. One can never divine what the next card will be, although one can precisely calculate what one's chances are of receiving a particular card. Similarly, in predictions associated with the earth and environmental sciences there is also irreducible uncertainty associated with the nature of random processes.

Let us take the poker example a step further. Assume that you find yourself playing cards with a less-than-honest dealer. This dealer is adding and removing cards from the deck so that the deck no longer has the standard fifty-two cards. The process is no longer *stationary*—it is changing over time. If you were to know the cards added and removed, that is, to have the ability to quantify the changing composition of the deck, to quantify uncertainty, you would simply need to recalculate the probabilities based on the new deck of cards. However, if you were unaware that the deck was changing in its composition, then you could easily miscalculate the uncertainty associated with your options. Similarly, if you were aware that the deck was changing but were not privy to the exact changes, you would be unable to calculate precisely the uncertainty (but would know that the assumption of a standard fifty-two-card deck could be wrong). This sort of uncertainty is called *epistemic* uncertainty and is associated with incomplete knowledge of a phenomenon—and incomplete knowledge of the limits of one's knowledge (Hoffman and Hammonds 1994; Stewart 2000).

Unlike aleatory uncertainty, epistemic uncertainty can be reduced in some cases through obtaining improved knowledge. In the case of the changing deck of cards, uncertainty could be reduced using several methods. For instance, one could carefully observe the outcomes of a large number of hands and record the actual frequencies with which particular hands occur. For instance, if four aces were added to the deck, one would expect to be able to observe the results in the form of more hands with ace combinations. Of course, the more subtle the change, the more difficult it is to detect.[5] The more one understands about the card replacement process, the better understanding one can have about the associated uncertainties. Unless one could discover the pattern underlying the change process (in effect "close" the system, Oreskes et al. 1994), then such theories would be subject to continuous revision as experience unfolds.

But even though epistemic uncertainty can in principle be reduced, if one is dealing with open systems (as is almost always the case for environmental predictions), the level of uncertainty itself can never be known with absolute certainty. Seismologists assigned a probability of 90% to their 1988 prediction of the Parkfield earthquake, but the earthquake never occurred (Savage 1991, Nigg 2000). Were the scientists simply confounded by the unlikely but statistically explicable one-out-of-ten chance of no earthquake? Or, was it because their probability calculation was simply wrong—that is, because the uncertainty associated with the prediction was in fact huge? Similarly, regardless of the sophistication of global climate models, many types of unpredictable events (volcanic eruptions that cool the atmosphere; new energy technologies that reduce carbon emissions) can render today's climate predictions invalid and associated uncertainties meaningless (see, e.g., Keepin 1986).

A central theme that emerges from experience is that important decisions are often clouded by inherent uncertainty, and in many instances, efforts to reduce uncertainty can paradoxically have the opposite effect (Pielke 2001).[6] Frequently, research results in discovery that the vast complexities associated with phenomena that evolve slowly over long periods—like those associated with the integrated earth system—were in fact previously underestimated, thereby having the effect of expanding uncertainties (Sarewitz et al. 2000). In a decision setting, this can have the perverse effect of increasing political controversy rather than reducing it, leading to calls for even more research to reduce uncertainties, while the problem goes unaddressed. No case illustrates this better than global climate change (Sarewitz and Pielke 2000).

One of the most critical issues in using models to develop information for decision making is to understand uncertainty—its sources, and its potential reducibility, as Weber (1999, 43, emphasis) observes:

> If uncertainty is measurable and controllable, then forecasting and information management systems serve a high value in reducing uncertainty and in producing a stable environment for organizations. If uncertainty is not measurable and controllable, then forecasting and predictions have limited value and need to be understood in such context. In short, how we view and understand uncertainty will determine how we make decisions.

Thus, a lesson for any effort that seeks to model open systems to inform decision making, particularly through prediction, is that it is imperative to understand uncertainty, including its sources, potential irreducibility, and relevant experience, in the context of the decision making process. In some cases, such an effort may very well lead to the conclusion that decision making should turn to alternatives to prediction (e.g., examples would include robust strategies insensitive to uncertainties such as trial and error; see e.g., Brunner 2000; Herrick and Sarewitz 2000).

Communicating Uncertainty

Experience shows that neither the scientific community nor decision makers have a good record at understanding uncertainty associated with predictions (Sarewitz et al. 2000). Such understanding is necessary because "the decision making process is best served when uncertainty is communicated as precisely as possible, but no more precisely than warranted" (Budescu and Wallsten 1987, 78). But even in cases where uncertainty is well-understood, as is typically the case in weather forecasting, scientists face challenges in communicating the entirety of their knowledge of uncertainty to decision makers. Often, experts place blame for this lack of communication on the perceived lack of

public ability to understand probabilistic information. The resulting policy prescription is for increased public education to increase scientific literacy (e.g.,
Augustine 1998; Rand 1998; Gibbs and Fox 1999). While improved scientific
literacy has value, it is not the solution to improving communication of information about uncertainty.

Consider the following analogy. You wish to teach a friend how to play the
game of tennis. You carefully and accurately describe the rules of tennis to your
friend, but you speak in Latin to your English-speaking friend. When you get
onto the court, your friend fails to observe the rules that you so carefully described. Following the game, it would surely be inappropriate to criticize your
friend as incapable of understanding tennis and futile to recommend additional
tennis instruction in Latin. But, this is exactly the sort of dynamic observed in
studies of public understanding of scientific uncertainties. For example, Murphy et al. (1980) document that when weather forecasters call for, say, a 70%
chance of rain, decision makers understood the probabilistic element of the
forecast but did not know whether rain has a 70% chance for each point in the
forecast area, or that 70% of the area would receive rain with a 100% probability, and so on.[7] Do you know?

The importance of understanding and communicating uncertainties associated with a prediction product was aptly illustrated in the case of the 1997
flooding of the Red River of the North (Pielke 1999). In February 1997, forecasters predicted that the river would see flooding greater than at any time in
modern history. At Grand Forks, North Dakota, forecasters expected the spring
flood to exceed the 1979 flood crest of 48.8 feet sometime in April. Forecasters
issued a prediction that the flood would crest at 49 feet, hoping to convey the
message that the flood would be the worst ever experienced. But the message
sent by the forecasters was not the message received by decision makers in the
community (cf. Fischoff 1994).

Decision makers in the community misinterpreted both the event being forecast and the uncertainty associated with the forecast. First, the prediction of 49
feet, rather than conveying concern to the public, instead resulted in *reduced*
concern. Locals interpreted the forecast in the context of the previous record
1979 flood, which caused damage, but was not catastrophic. With the 1997
crest expected only a few inches higher than the record set in 1979, many expressed relief rather than concern, that is, "We survived that one OK, how much
worse can a few inches be?" Second, decision makers did not understand the
uncertainty associated with the forecast. Flood forecasts are extremely uncertain, especially forecasts of record floods for which there is no experience.
Forecasters issued a quantitative forecast with a simple qualitative warning
about uncertainty. Hence, many decision makers interpreted the forecast uncertainty in their own terms: Some viewed the forecast as a ceiling, that is, "the
flood will not exceed 49 feet." Others viewed the forecast as uncertain and associated various uncertainties with the forecasts, ranging from 1 to 6 feet. The
historical record showed that flood crest forecasts were, on average, off by
about 10% of the forecast.

On April 22, 1997, at Grand Forks the Red River crested at 54 feet, inundating the communities of Grand Forks, North Dakota, and East Grand Forks, Minnesota, and causing up to $2 billion in damage (current dollars). In the aftermath of the flood, local, state, and national officials pointed to inaccurate flood forecasts as a cause of the disaster. With hindsight, a more reasoned assessment indicates that by any objective measure, the accuracy of the forecasts was not out of line with historical performance. Instead, decision makers *and* scientists failed to understand the meaning of the prediction both in terms of what was being forecast and the uncertainty associated with it.

A significant literature exists on communication of uncertain information, some based on experience in the sciences (e.g., Dow and Cutter 1998; Baker 2000; Glantz 2001) and much more (it seems) from the disciplines of communication, psychology, and sociology (Wallsten et al. 1986; Konold 1989; Hoffrage et al. 2000). The implications of this literature range from the straightforward: "statistics expressed as natural frequencies improve the statistical thinking of experts and nonexperts alike" (Hoffrage et al. 2000) to the more challenging: "probability expressions are interpreted differently by speakers and listeners" (Fillenbaum et al. 1991). However, it is clear that the substantial research on communication of uncertainty has not been well integrated with the research in the earth and environmental sciences that seeks to understand and describe uncertainties relevant to decision making. Understanding and communicating uncertainty by scientists and decision makers alike go hand-in-hand: both are necessary—but not sufficient—for information to contribute systematically to improved decisions.

Understanding Predictability

Consider again the poker example. With perfect knowledge of a card substitution process engineered by a less-than-honest dealer, one would be able to quantify completely and accurately the associated uncertainties in future hands. However, this situation is quite different from most cases that we find in the real world of modeling and prediction in the environmental sciences. In the real world, systems are open and there are fundamental limits to predictability. And, perhaps surprisingly, many scientific efforts to divine the future proceed without an adequate understanding of the limits to predictability. In addition to the aleatory and epistemic uncertainties discussed above, there are a number of other reasons for limits to predictability. Among these are sensitivity to initial conditions, complexity, and human agency.

First, predictability is limited because knowledge of the future depends upon knowing the present, which can never be completely or accurately characterized. For example, weather forecasts depend upon knowing the present state of the atmosphere and then projecting forward its future behavior, based on physical relationships represented in computer models. A result of the dependence on

these "initial conditions" is that small changes in the initial conditions can subsequently lead to large differences in outcomes. Knowledge of initial conditions
is obtained with instruments. In weather prediction, these can include balloons,
radar, satellites, and other instruments that are subject to measurement errors.
But even without such measurement errors, the simple act of rounding off a
decimal can lead to vastly different outcomes. Popularized as the "butterfly
effect," this is a fundamental characteristic of a chaotic system with limited
predictability (Gleick 1987). Scientists have established 10–14 days as the limit
of predictability for weather forecasts. In many other contexts the same limits
hold but are not as well understood. Meteorologists run models repeatedly with
small variations in input data (and sometimes in the model itself) to begin to
understand the sensitivities of model output to initial conditions (e.g., Krishnamurti et al. 1999).

A second factor is that in the environmental sciences, phenomena of interest
to policy makers are often incredibly complex and can result from intercon-
nected human and earth processes. Consider nuclear waste disposal (Metlay,
2000). Predicting the performance of a waste facility 10,000 years into the future depends upon knowing, among a multitude of other potentially relevant
factors, what sorts of precipitation might be expected at the site. Precipitation is
a function of global climate patterns. In addition, global climate patterns might
be sensitive to human processes such as energy and land use. Energy and land
use are functions of politics, policy, social changes, and so on. What at first
seems a narrow scientific question rapidly spirals into great complexity. One
characterization of the concept holds that "a complex system is one whose evolution is very sensitive to initial conditions or to small perturbations, one in
which the number of independent interacting components is large, or one in
which there are multiple pathways by which the system can evolve" (Whitesides and Ismagilov 1999). Scientists are just beginning to understand the implications of complexity for prediction (see Waldrop 1992).

A third factor is the role of human agency. In situations where human decisions are critical factors in the evolution of the future being predicted (that is to
say, almost every issue of environmental policy), the aggregate record of prediction is poor. Ascher (1981) argues that "unless forecasters are completely
ignorant of the performance record, or are attracted solely by the promotional
advantages of the scientific aura of modeling, they can only be attracted to
benefits not yet realized." The poor performance of predictions of societal outcomes is consistent across diverse areas that include energy demand (Keepin
1986), energy supplies (Gautier 2000), population (Cohen 1996), elections
(Mnookin 2001), corporate financial performance (Dreman and Berry 1994),
macro-economics (CBO 1999), and medicine (Fox et al. 1999). To the extent
that modeled outcomes depend upon some degree of accuracy in predicting
factors such as these, predictability clearly will be limited.

And yet, effective decision making cannot occur without some way to anticipate the consequences of alternative courses of action. There is a considerable range of cases in which prediction and modeling contributed to effective

decisions. The practice of insurance and engineering would not be possible without predictive ability. And more relevant to present purposes, the apparently successful response to stratospheric ozone depletion would not have been possible without predictive and diagnostic modeling (Pielke and Betsill 1997). But understanding (and indeed creating) those situations where prediction and modeling serve effective decision making is not straightforward, if simply because questions about the roles of models and prediction in decision making are rarely asked, much less answered.

What Is a "Good" Model?

This section focuses on models as a means to produce predictions for decision making, as well as a social and scientific mechanism that fosters integration of knowledge. The "goodness" of predictions produced from models can be understood from two distinct perspectives: product and process.

Prediction as Product

The first and most common perspective is to view models simply as generators of an information product. Often, when a model is applied to decision problems, it is used to produce a prediction, that is, a "set of probabilities associated with a set of future events" (Fischoff 1994). To understand a prediction, one must understand the specific definition of the predicted event (or events), as well as the expected likelihood of the event's (or events') occurrence. From this perspective, the goal of modeling is simply to develop "good" predictions (Pielke et al. 1999). Three important considerations in the production of "good" predictions are accuracy, sophistication, and experience.

Accuracy. Accuracy is important because "on balance, accurate forecasts are more likely than inaccurate forecasts to improve the rationality of decision making" (Ascher 1979, 6). With a few exceptions, once a forecast is produced and used in decision making, few researchers or decision makers ever look back to assess its skill (Sarewitz et al. 2000). Measuring the skill of a prediction is not as straightforward as it might seem. Consider the case of early tornado forecasts. In the 1880s, a weather forecaster began issuing daily tornado forecasts in which he would predict for the day "tornado" or "no tornado." After a period of issuing forecasts, the forecaster found his forecasts to be 96.6% correct. But others who looked at the forecaster's performance discovered that simply issuing a standing forecast of "no tornadoes" would result in an accuracy rate of 98.2%! This finding suggested that, in spite of the high degree of correct forecasts, the forecaster was providing predictions with little skill—defined as the improvement of a forecast over some naïve standard—and which could result in costs rather than benefits. Simply comparing a prediction with actual events does not provide enough information with which to evaluate its performance. A more sophisticated approach is needed. Thus, predictions should be evaluated in terms of their "skill," defined as the improvement provided by the prediction over a naïve forecast, such as that that would be used in the absence of the prediction.[8]

Sophistication. Decision makers sometimes are led to believe that sophistication of a prediction methodology lends itself to greater predictive skill, that is, given the complexity of the world a complex methodology should perform better. In reality, the situation is not so clear-cut. An evaluation of the performance of complex models has shown that "methodological sophistication contributes very little to the accuracy of forecasts" (Ascher 1981; see also Keepin 1986). A lesson for decision makers is that sophisticated prediction methodologies (or by extension the resources devoted to development of such methodologies) do not necessarily guarantee predictive, much less decision making, successes. Because complex models often require significant resources (computation, humans, etc.), a trade-off invariably results between producing one or a few realizations of a highly complex model and many runs of a simpler, less intensive version of the model. For instance, the U.S. National Assessment of Climate Change used only two scenarios of future climate due to computation limitations (NACC 2000). For many decision makers, having an ability to place modeled output into the context of the entire "model-output space" would have been more useful than the two products that were produced largely without context. This is an example of confusion between consolidative and exploratory modeling.

Experience. In weather forecasts, society has the best understanding of prediction as a product. Consider that in the United States the National Weather Service issues more than 10 million predictions every year to hundreds of millions of users. This provides a considerable basis of experience on which users can learn, through trial and error, to understand the meaning of the prediction products that they receive. Of course, room for confusion exists. People can fail to understand predictions for record events for which there is no experience, as in the Red River case, or even a routine event being forecast (e.g., 70% chance of rain). But experience is essential for effective decision making, and most decision makers have little experience using models or their products. Erev et al. (1993) provide a useful analogy:

> Consider professional golfers who play as if they combine information concerning distance and direction of the target, the weight of the ball, and the speed and direction of the wind. Now assume that we ask them to play in an artificial setting in which all the information they naturally combine in the field is reduced to numbers. It seems safe to say that the numerical representation of the information will not improve the golfer's performance. The more similar are the artificial conditions we create to the conditions with which the

golfers are familiar, the better will be their performance. One can assume that decision making expertise, like golf expertise, is improved by experience, but not always generalized to new conditions.

The importance of experience does not necessarily limit the usefulness of models and their products in decision making, but it does underscore the importance of the decision context as a critical factor in using models (Stewart et al. 1997).

A range of experience illustrates that misunderstandings or misuse of prediction products have presented obstacles to decision makers' efforts to effectively use predictions. Considering the following:

- Global climate change (Rayner 2000). Debate has raged for more than a decade about the policy implications of possible future human-caused changes in climate. This debate has been about "global warming" expressed in terms of a single global average temperature. But global average temperature has no "real-world" meaning, and thus policy advocates have sought to interpret that "event" in different ways, ranging from pending global catastrophe to benign (and perhaps beneficial) change. The issue of uncertainty compounds the issue. As a result, predictive science has been selectively used and misused to justify and advance the existing objectives of participants in the process (Sarewitz and Pielke 2000).

- Asteroid impacts (Chapman 2000). In recent years, scientists have increased their ability to observe asteroids and comets that potentially threaten the earth. In this case, the "event" is clear enough—possible extinction of life on earth if a large asteroid slams into the earth and its prediction seemingly straightforward, uncomplicated by human agency. But scientific overreaction to the discovery of 1997 XF11 and the associated prediction that it could strike the earth on October 26, 2028, illustrate that understandings of uncertainty are critical (Chapman 2000). In this case, hype might have damaged future credibility of scientists who study this threat.

These examples and others illustrate the difficulties associated with understanding prediction as a product (Sarewitz et al. 2000). At the same time, these cases also demonstrate that improving the use of prediction involves more than simply developing "better" predictions, whether more precise, for example, a forecast of a 49.1652 flood crest at East Grand Forks; more accurate, for example, a forecast of a 51 foot crest; or more robust, for example, a probabilistic distribution of various forecast crest levels. While better predictions are in many cases more desirable, better decisions require attention to the broader prediction process. From this standpoint, better predictions may be neither necessary nor sufficient for improved decision making and, hence, desired outcomes. For better decisions, it is necessary to understand prediction as a product in the context of a process.

Prediction as Process

A second perspective is to view modeling as part of a broader prediction process. Included are the participants, perspectives, institutions, values, resources, and other factors that together determine policies for the prediction enterprise, as well as how the prediction enterprise contributes to public demands for action or tools with respect to the issues that they bring to the attention of decision makers. From this perspective, the goal of the prediction enterprise is good decisions. Modeling, due to its (potentially) integrative nature, is an important element of the prediction process.

The successful use of predictions depends more upon a healthy process than just on "good" information (Sarewitz et al. 2000). Weather forecasts have demonstrably shown value not because they are by any means "perfect," but because users of those predictions have successfully incorporated them into their decision routines. The prediction process can be thought of as three parallel subprocesses (Sarewitz et al. 2000):

Research Process   includes the fundamental science, observations, etc. as well as forecasters' judgments and the organizational structure which go into the production of predictions for decision makers.

Communication Process includes both the sending and receiving of information; a classic model of communication is: who, says what, to whom, how, and with what effect.

Choice Process   includes the incorporation of predictive information in decision making. Of course, decisions are typically contingent upon many factors other than predictions.

Often, some persons mistakenly ascribe a linear relation to the processes. Instead, they are better thought of as components of a broader prediction process, with each of the subprocesses taking place in parallel and with significant feedback and interrelations between them.

Peter Drucker has written an eloquent description of the modern organization that applies equally well to the prediction process. "Because the organization is composed of specialists, each with his or her own narrow knowledge area, its mission must be crystal clear . . . otherwise its members become confused. They will follow their specialty rather than applying it to the common task. They will each define 'results' in terms of that specialty, imposing their own values on the organization" (Drucker 1993, 54).

Drucker continues with an apt metaphor. "The prototype of the modern organization is the symphony orchestra. Each of 250 musicians in the orchestra is a specialist, and a high-grade one. Yet, by itself, the tuba doesn't make music; only the orchestra can do that. The orchestra performs only because all 250 musicians have the same score. They all subordinate their specialty to a common task" (Drucker 1993, 55).

In the process of modeling and prediction in support of decision making, success according to the criteria of any subset of the three subprocesses does not necessarily result in benefits to society. Consider the following examples.

- The case of the Red River floods presented earlier illustrates that a technically skillful forecast that is miscommunicated or misused can actually result in costs rather than benefits. The overall prediction process broke down in several places. No one in the research process fully understood the uncertainty associated with the forecast; hence little attention was paid to communication of uncertainty to decision makers. As a result, poor decisions were made and people suffered, probably unnecessarily. Given that Grand Forks will to some degree always depend upon flood predictions, the situation might be improved in the future by including local decision makers in the research process in order to develop more useful products (see Pielke 1999).

- In the case of earthquake prediction, a focus on developing skillful predictions of earthquakes in the Parkfield region of California brought together seismologists with local officials and emergency managers (Nigg 2000). A result was better communication among these groups and overall improved preparation for future earthquakes. In this case, even though the predictions themselves could not be shown to be skillful, the overall process worked because it identified alternatives to prediction that have led to decisions that are expected to reduce the impacts of future earthquakes in this region, such as improving building codes and enforcement, insurance practices, and engineering designs.

- The case of global climate change may be in the early stages of what was documented in the case of earthquakes (Rayner 2000). Policy making focused on prediction has run up against numerous political and technical obstacles. Meanwhile alternatives to prediction have become increasingly visible. The prediction process can be judged successful if the goal of climate policy—to reduce the impacts of future climate changes on environment and society—is addressed, independent of whether century-scale climate forecasts prove to be accurate (Sarewitz and Pielke 2000).

- The case of nuclear waste disposal has also evolved from one in which decision making focused first on developing skillful predictions to one in which decision making focused instead on actions that would be robust under various alternative futures (Metlay 2000). In this case, the policy problem of storing nuclear waste for a very long time (and associated uncertainties) was addressed via decision making by selecting an engineering design that was robust to a very wide range of uncertainties and not by selecting a design based on a specific prediction.

As Robinson (1982, 249) observes, "by basing present decisions on the apparent uncovering of future events, an appearance of inevitability is created that de-emphasizes the importance of present choice and further lessens the probability of developing creative policy in response to present problems . . . [predictions] do not reveal the future but justify the subsequent creation of that future." The lesson for decision makers is that one is in most cases more likely to reduce uncertainties about the future through decision making rather than through prediction.

The criteria for evaluating the "goodness" of a model are thus directly related to the purposes for which a model is to be used. A consolidative model will most likely be evaluated based on the accuracy of its output, whereas an exploratory model could easily succeed even if its results are highly inaccurate (Bankes 1993). Similarly, a model designed to advance understanding should be evaluated by a different set of criteria than a model designed to provide reliable products useful in decision making. For society to realize the benefits of the resources invested in the science and technology of prediction, the entire process must function in a healthy manner, just like the sections of Drucker's orchestra must perform together to make music. Each subprocess of the broader prediction process must be considered in the context of the other subprocesses; they cannot be considered in isolation.

Conclusion: For Better Decisions, Question Predictions

The analytical vignettes presented in this paper begin to highlight some of the shared characteristics of healthy decision processes for the use of model products, particularly predictions. One characteristic is the critical importance of decision makers who have experience with the phenomena being predicted, as well as experience with the predictions themselves. The less frequent, less observable, less spatially discrete, more gradual, more distantly future, and more severe a predicted phenomenon, the more difficult it is to accumulate direct experience. Where direct societal experience is sparse or lacking, other sources of societal understanding must be developed or the prediction process will not function as effectively. Science alone and prediction in particular do not create this understanding.

More broadly, what is necessary above all is an institutional structure that brings together throughout the entire prediction process scientists with those who solicit and use predictions, so that each knows the needs and capabilities of the others. It is crucial that this process be open, participatory, and conducive to mutual respect. Efforts to shield expert research and decision making from public scrutiny and accountability invariably backfire, fueling distrust and counterproductive decisions.

While efforts to predict natural phenomena have become an important aspect of the earth and environmental sciences, the value of such efforts, as judged especially by their capacity to improve decision making and achieve policy goals, has been questioned by a number of constructive critics. The relationship between prediction and policy making is not straightforward for many reasons, among them:

- Accurate prediction of phenomena may not be necessary to respond effectively to political or socioeconomic problems created by the phenomena (for example, see Landsea et al. 1999).

- Phenomena or processes of direct concern to policy makers may not be easily predictable. Likewise, predictive research may reflect discipline-specific scientific perspectives that do not provide "answers" to policy problems, which are complex mixtures of facts and values and which are perceived differently by different policy makers (for example, see Herrick and Jamieson 1995).

- Necessary political action may be deferred in anticipation of predictive information that is not forthcoming in a time frame compatible with such action. Similarly, policy action may be delayed when scientific uncertainties associated with predictions become politically charged (in the issue of global climate change, for example; see Rayner and Malone 1998).

- Predictive information also may be subject to manipulation and misuse, either because the limitations and uncertainties associated with predictive models are not readily apparent, or because the models are applied in a climate of political controversy and high economic stakes.

- Emphasis on predictive products moves both financial and intellectual resources away from other types of research that might better help to guide decision making (for example, incremental or adaptive approaches to environmental management that require monitoring and assessment instead of prediction; see Lee 1993).

These considerations suggest that the usefulness of scientific prediction for policy making and the resolution of societal problems depends on relationships among several variables, such as the timescales under consideration, the scientific complexity of the phenomena being predicted, the political and economic context of the problem, and the availability of alternative scientific and political approaches to the problem.

In light of the likelihood of complex interplay among these variables, decision makers and scientists would benefit from criteria that would allow them to better judge the potential value of scientific prediction and predictive modeling for different types of political and social problems related to earth processes and the environment. Pielke et al. (1999) provide the following six guidelines for the effective use of prediction in decision making:

- Predictions must be generated primarily with the needs of the user in mind. And that user could be another scientist. For stakeholders to participate usefully in this process, they must work closely and persistently with the producers of predictions to communicate their needs and problems.

- Uncertainties must be clearly understood and articulated by scientists, so that users have a chance to understand their implications. Failure to understand uncertainties has contributed to poor decisions that then undermined relations among scientists and decision makers. But merely understanding the uncertainties does not mean that the predictions will be useful. If policy makers truly understood the uncertainties associated with predictions of, for example, global climate change, they might decide that strategies for action should not depend on predictions (Rayner and Malone 1998).

- Experience is a critically important factor in how decision makers understand and use predictions.

- Although experience is important and cannot be replaced, the prediction process can be facilitated in other ways, for example, by fully considering alternative approaches to prediction, such as robust policies insensitive to uncertainties. Indeed, alternatives to prediction must be evaluated as a part of the prediction process.

- To ensure an open prediction process, stakeholders must question predictions. For this questioning to be effective, predictions should be as transparent as possible to the user. In particular, assumptions, model limitations, and weaknesses in input data should be forthrightly discussed. Even so, lack of experience means that many types of predictions will never be well understood by decision makers.

- Last, predictions themselves are events that cause impacts on society. The prediction process must include mechanisms for the various stakeholders to fully consider and plan what to do after a prediction is made.

When the prediction process is fostered by effective, participatory institutions, and when a healthy decision environment emerges from these institutions, the products of predictive science may become even less important. Earthquake prediction was once a policy priority; now it is considered technically infeasible, at least in the near future. But in California the close, institutionalized communication among scientists, engineers, state and local officials, and the private sector has led to considerable advances in earthquake preparedness and a much-decreased dependence on prediction. On the other hand, in the absence of an integrated and open decision environment, the scientific merit of predictions can be rendered politically irrelevant, as has been seen with nuclear waste disposal and acid rain. In short, if no adequate decision environment exists for dealing with an event or situation, a scientifically successful prediction may be no more useful than an unsuccessful one.

These recommendations fly in the face of much current practice where, typically, policy makers recognize a problem and decide to provide resources to science to reduce uncertainty or produce predictions. Scientists then go away and do research to predict natural behavior associated with the problem, and predictions are finally delivered to decision makers with the expectation that they will be both useful and well used. This sequence, which isolates prediction research but makes policy dependent on it, rarely functions well in practice.

Yet, once we have recognized the existence of a prediction enterprise, it becomes clear that prediction is more than a product of science. Rather, it is a complex process, one that includes all the interactions and feedbacks among participants, perspectives, institutions, values, interests, resources, decisions, and other factors that constitute the prediction enterprise. From this perspective, the goal of the prediction enterprise is good decisions, as evaluated by criteria of public benefit. The value of predictions for environmental decision making therefore emerges from the complex dynamics of the prediction process, and not simply from the technical efforts that generate the prediction product (which are themselves an integral part of the prediction process). All the same, it is the expectation of a useable prediction product that justifies the existence of the prediction enterprise. This expectation turns out to be extremely difficult to fulfill.

This chapter has presented only a few of the many considerations that must be understood if scientific modeling and prediction are indeed to fulfill public expectations of the contributions of science in addressing environmental policy problems. There is considerable need for debate and discussion, supported by rigorous knowledge, on the proper role of modeling and prediction in decision making, rather than a simple assumption of what that role should be. However, one clear implication of the considerations presented in this paper is that the belief that modeling and prediction can simultaneously meet the needs of both science and decision is untenable as currently practiced. For ecosystem science, there exists a body of experience in understanding, using, and producing predictions across the sciences on which to build, to the potential benefit of both research and policy.

Acknowledgments. Several sections of this chapter are closely related to two earlier publications: Sarewitz et al. (2000) and Pielke et al. (1999), both of which resulted from the project "Use and Misuse of Predictions in the Earth Sciences," sponsored by the National Science Foundation. I would like to acknowledge gratefully the significant contributions of my collaborators on the prediction project—Dan Sarewitz and Rad Byerly—to the mix of ideas and expressions from those collaborative publications that have found their way into this chapter. This chapter also has benefited a great deal from the comments and suggestions of Rich Conant, Bobbie Klein, Naomi Oreskes, and an anonymous reviewer. An earlier version was drafted in preparation for Cary Conference IX, "Understanding Ecosystems: The Role of Quantitative Models in Observations, Synthesis, and Prediction." Participation in an NSF-sponsored workshop on ecological forecasting (Clark et al. 2001) provided additional motivation for introducing some of the lessons from our prediction project to the ecological community. Of course, all responsibility for errors of fact or interpretation lies solely with the author.

Notes

1. Oreskes (pers. comm.) argues that no model of a natural system can ever be fully specified, and therefore cannot in principle meet the definition of a consolidative model.

2. A consolidative model can be properly used in exploratory fashion, but the real threat to decision making occurs when the opposite occurs. For an extended treatment of such models see Bankes (1993).

3. On the USGCRP, see Pielke 2000a and 2000b.

4. On the interpretation of climate model results, see Trenberth 1997, Edwards 1999, IPCC 2001. On the media's presentation of climate research results, see Henderson-Sellers 1998. On their role in decision, see Sarewitz and Pielke 2000 and Sluijs et al. 1998.

5. In a very similar fashion, some studies of global climate change use such a method to assess whether the storms, temperature, precipitation, etc., of one period differ significantly from that of another period (e.g., Wunsch 1999).

6. A related consideration is that attempts to eliminate uncertainty by changing thresholds for decision (e.g., changing the wind-speed criteria for hurricane evacuation) invariably result in trade-offs between false alarms and misses (i.e., type I and type II errors), with associated societal costs. See Stewart 2000.

7. There is a considerable literature on the use of weather forecasts that supports this line of argument. See in particular the work of Murphy (e.g., Murphy et al. 1980) and Baker (e.g., Baker 2000).

8. The term "skill" is jargon; however the notion of evaluating predictions against a naïve baseline is fundamental to the evaluation of weather forecasts and financial forecasts (such as mutual fund performance). For forecasts that are probabilistic, rather than categorical, the evaluation of skill can be somewhat more complicated, but adheres to the same principles. See Murphy 1997 for a technical discussion of the many dimensions of predictive skill. There are other dimensions of predictive "goodness" that are central to evaluation of its role in decision making—including comprehensibility, persuasiveness, usefulness, authoritativeness, provocative ness, importance, value, etc., for discussion, see Ascher 1979, Armstrong 1999 and Sarewitz et al. 2000.

References

Armstrong, J.S. 1999. *Principles of Forecasting: A Handbook for Researchers and Practitioners.* Netherlands: Kluwer Academic.

Ascher, W. 1979. *Forecasting: An Appraisal for Policymakers.* Baltimore: Johns Hopkins University Press.

———. 1981. The forecasting potential of complex models. *Policy Sciences* 13: 247–267.

Augustine, N. 1998. What we don't know does hurt us. How scientific illiteracy hobbles society. *Science* 279: 1640–1641.

Ayala, F. 1996. The candle and the darkness. *Science* 273: 442.

Baker, E. J. 2000. Hurricane evacuation in the United States. Pp. 306–319 in R. Pielke Jr. and R. Pielke Sr., editors. *Storms.* London: Routledge Press.

Bankes, S. 1993. Exploratory modeling for policy analysis. *Operations Research* 41: 435–449.

Brunner, R. 2000. Alternative to prediction. Chapter 14 in D. Sarewitz, R.A. Pielke Jr., and R. Byerly, editors. *Prediction: Science, Decision Making, and the Future of Nature.* Washington, DC: Island Press.

Budescu, D.V., and T.S. Wallsten. 1987. Subjective estimation of precise and vague uncertainties. Pp. 63–81 in G. Wright and P. Ayton, editors. *Judgmental Forecasting.* Chinchester: Wiley.

CBO (Congressional Budget Office). 1999. Evaluating CBO's record of economic forecasts. ftp://ftp.cbo.gov/14xx/doc1486/fceval99.pdf.

CES (Committee on Earth Sciences). 1989. *Our Changing Planet: A U.S. Strategy for Global Change Research.* A report to accompany the U.S. President's fiscal year 1990 budget. Washington, DC: Office of Science and Technology Policy.

Chapman, C. 2000. The asteroid/comet impact hazard: *Homo sapiens* as dinosaur? Chapter 6 in D. Sarewitz, R.A. Pielke Jr., and R. Byerly, editors. *Prediction: Science, Decision Making, and the Future of Nature.* Washington, DC: Island Press.

Clark, J.S., S.R. Carpenter, M. Barber, S. Collins, A. Dobson, J.A. Foley, D.M. Lodge, M. Pascual, R. Pielke Jr., W. Pizer, C. Pringle, W.V. Reid, K.A. Rose, O. Sala, W.H. Schlesinger, D.H. Wall, and D. Wear. 2001. Ecological forecasts: An emerging imperative. *Science* 293: 657–660.

Cohen, J. 1996. *How Many People Can the Earth Support?* New York: W.W. Norton.

Cullen. A., and M. Small. 2000. Uncertain risk: The role and limits of quantitative assessment. Manuscript.

Dow, K. and S.L. Cutter. 1998. Crying wolf: Repeat responses to hurricane evacuation orders. *Coastal Management* 26: 237–252.

Dreman, D.N., and M.A. Berry. 1994. Analyst forecasting errors and their implications for security analysts. *Financial Analysts Journal.* May/June, 30–41.

Drucker, P. 1993. *Post-capitalist Society.* New York: Harper Collins.

Edwards, P. 1999. Global climate science, uncertainty, and politics: Data-laden models, model-filtered data. *Science as Culture* 8: 437–472.

Erev, I., G. Bornstein, and T.S. Wallsten. 1993. The negative effect of probability assessments on decision quality. *Organizational Behavior and Human Decision Processes* 55: 78–94.

Fillenbaum, S., T.S. Wallsten, B.L. Cohen, and J.A. Cox. 1991. Some effects of vocabulary and communication task on the understanding and use of vague probability expressions. *American Journal of Psychology* 104: 35–60.

Fischoff, B. 1994. What forecasts (seem to) mean. *International Journal of Forecasting* 10: 387–403.

Fox, E., K. Landrum-McNiff, Z. Zhong, N.V. Dawson, A.W. Wu, and J. Lynn. 1999. Evaluation of prognostic criteria for determining hospice eligibility in patients with advanced lung, heart, or liver disease. *Journal of the American Medical Association* 282: 1638–1645.

Gautier, D. 2000. Oil and gas resource appraisal: Diminishing reserves, increasing supplies. Chapter 11 in D. Sarwitz, R.A. Pielke Jr., and R. Byerly, editors. *Prediction: Science, Decision Making, and the Future of Nature.* Washington, DC: Island Press.

Gibbs, W.W. and D. Fox. 1999. The false crisis of science education. *Scientific American.* 1 October, 86–93.

Glantz, M.H. 2001. *Currents of Change: El Niño and La Niña Impacts on Climate and Society.* 2nd ed. Cambridge: Cambridge University Press.

Gleick, J. 1987. *Chaos: Making a New Science.* New York: Penguin Books.

Henderson-Sellers, A. 1998. Climate whispers: Media communication about climate change. *Climatic Change* 4: 421–456.

Herrick, C., and D. Jamieson. 1995. The social construction of acid rain: Some implications for science/policy assessment. *Global Environmental Change* 5: 105–112.

Herrick, C., and D. Sarewitz. 2000. Ex post evaluation: A more effective role for scientific assessments in environmental policy. *Science, Technology, and Human Values* 25: 309–331.

Hoffman, F.O., and J.S. Hammonds. 1994. Propagation of uncertainty in risk assessments: The need to distinguish between uncertainty due to lack of knowledge and uncertainty due to variability. *Risk Analysis* 14: 707–712.

Hoffrage, U., S. Lindsey, R. Hertwig, G. Gigerenzer. 2000. Communicating statistical information. *Science* 290: 2261–2262.

IPCC (Intergovernmental Panel on Climate Change). 1990. *Climate Change: The IPCC Scientific Assessment.* Cambridge: Cambridge University Press.

———. 2001. *Climate Change 2001: The Scientific Basis.* Cambridge: Cambridge University Press.

Keepin, B. 1986. Review of global energy and carbon dioxide projections. *Annual Review of Energy* 11: 357–392.

Konold, Clifford. 1989. Informal conceptions of probability. *Cognition and Instruction* 6(1): 59–98.

Krishnamurti, T.N., C.M. Kishtawal, T. LaRow, D. Bachiochi, Z. Zhang, C.E. Williford, S. Gadgil, and S. Surendran. 1999. Improved skills for weather and seasonal climate forecasts from multi-model superensemble. *Science* 285: 1548–1550.

Landsea, C.L., R.A. Pielke Jr., A. Mestas-Nuñez, and J. Knaff, 1999. Atlantic Basin hurricanes: Indices of climate changes. *Climate Change* 42: 9–129.

Lasswell, H.D., and A. Kaplan. 1950. *Power and Society.* New Haven: Yale University Press.

Lee, K. 1993. *Compass and Gyroscope: Integrating Science and Politics for the Environment.* Washington, DC: Island Press.

Metlay, D. 2000. From tin roof to torn wet blanket: Predicting and observing groundwater movement at a proposed nuclear waste site. Chapter 10 in D. Sarewitz, R.A. Pielke Jr., and R. Byerly, editors. *Prediction: Science, Decision Making, and the Future of Nature.* Washington, DC: Island Press.

Mnookin, S. 2001. It happened one night. *Brill's Content* 4(1): 94–96.

Morrison, M. and M.S. Morgan 1999. Models as mediating instruments. Pp. 10–37 in M.S. Morgan and M. Morrison, editors. *Models as Mediators: Perspectives on Natural and Social Science.* Cambridge: Cambridge University Press.

Murphy, A.H. 1997. Forecast verification. Pp. 19–74 in R. Katz and A. Murphy, editors. *Economic Value of Weather and Climate Forecasts*. Cambridge: Cambridge University Press.

Murphy, A.H., S. Lichtenstein, B. Fischoff, and R.L. Winkler. 1980. Misinterpretations of precipitation probability forecasts. *Bulletin of the American Meteorological Society* 61: 695–701.

NACC (National Assessment on Climate Change). 2000. *Climate Change Impacts on the United States, National Assessment Synthesis Team*. U.S. Global Change Research Program. Cambridge: Cambridge University Press.

Nigg, J. 2000. Predicting earthquakes: Science, pseudoscience and public policy paradox. Chapter 7 in D. Sarewitz, R A. Pielke Jr., and R. Byerly, editors. *Prediction: Science, Decision Making, and the Future of Nature*. Washington, DC: Island Press.

Oreskes, N., K. Shrader-Frechette, and K. Belitz. 1994. Verification, validation, and confirmation of numerical models in the earth sciences. *Science* 263: 641–646.

Pielke, Jr. R.A. 1999. Who decides? Forecasts and responsibilities in the 1997 Red River floods. *Applied Behavioral Science Review* 7: 83–101.

———. 2000a. Policy history of the U.S. Global Change Research Program. Part I. Administrative development. *Global Environmental Change* 10: 9–25.

———. 2000b. Policy history of the U.S. Global Change Research Program. Part II. Legislative process. *Global Environmental Change* 10: 133–144.

———. 2001. Room for doubt. *Nature* 410: 151.

Pielke Jr., R.A., and M.M. Betsill. 1997. Policy for science for policy: Ozone depletion and acid rain revisited. *Research Policy* 26: 157–168.

Pielke Jr. R.A., D. Sarewitz, R. Byerly, and D. Jamieson. 1999. Prediction in the earth sciences: Use and misuse in policy making. *EOS: Transactions of the American Geophysical Society* 80: 9.

Rand, H. 1998. Science, non-science and nonsense: Communicating with the lay public. *Vital Speeches of the Day* 64 (9): 282–284.

Rayner, S. 2000. Prediction and other approaches to climate change policy. Chapter 13 in D. Sarewitz, R.A. Pielke Jr., and R. Byerly, editors. *Prediction: Science, Decision Making, and the Future of Nature*. Washington, DC: Island Press.

Rayner, S., and E.L. Malone, editors. 1998. *Human Choice and Climate Change*. Columbus: Battelle Press.

Robinson, J.B. 1982. Backing into the future: on the methodological and institutional biases embedded in energy supply and demand forecasting. *Technological Forecasting and Social Change* 21: 229–240.

Sarewitz, D., and R.A. Pielke Jr. 1999. Prediction in science and policy. *Technology in Society* 21: 121–133.

———. 2000. Breaking the Global Warming Gridlock. *Atlantic Monthly*. 1 July, 55–64.

Sarewitz, D., R.A. Pielke Jr., and R. Byerly. 2000. *Prediction: Science, Decision Making, and the Future of Nature.* Washington, DC: Island Press.

Savage, J.C. 1991. Criticism of some forecasts of the National Earthquake Prediction Evaluation Council. *Bulletin of the Seismological Society of America* 81: 862–881.

Scarne, J. 1986. *Scarne's New Complete Guide to Gambling.* New York: Simon and Schuster.

Sluijs, J.P. van der, J.C.M. van Eijnhoven, B. Wynne, S. Shackley. 1998. Anchoring devices in science for policy: The case of consensus around climate sensitivity. *Social Studies of Science* 28: 291–323.

Skidelsky, R. 2000. Skidelsky on Keynes. *The Economist.* November 25, 83–85.

Stewart, T. 2000. Uncertainty, judgment and error in prediction. Chapter 3 in D. Sarewitz, R.A. Pielke Jr., and R. Byerly, editors. *Prediction: Science, Decision Making, and the Future of Nature.* Washington, DC: Island Press.

Stewart, T.R., P.J. Roebber, and L.F. Bosart. 1997. The importance of the task in analyzing expert judgment. *Organizational Behavior and Human Decision Processes* 69: 205–219.

Trenberth, K. 1997. The use and abuse of climate models. *Nature* 386: 131–133.

Waldrop, M.M. 1992. *Complexity: The Emerging Science at the Edge of Order and Chaos.* New York: Simon and Schuster.

Wallsten, T.S., D.V. Budescu, A. Rapoport, R. Zwick, and B. Forsyth. 1986. Measuring the vague meanings of probability terms. *Journal of Experimental Psychology* 115(4): 348–365.

Weber, J. 1999. A response to public administration's lack of a general theory of uncertainty: A theoretical vision of uncertainty. *Public Administration Review* 23: 18–45.

Whitesides, G.F., and R.F. Ismagilov. 1999. Complexity in chemistry. *Science* 284: 289–292.

Wunsch, C. 1999. The interpretation of short climatic records with comments on the North Atlantic and Southern Oscillations. *Bulletin of the American Meteorological Society* 80: 245–256.

Part II

Evaluating Ecosystem Models

8

Propagation and Analysis of Uncertainty in Ecosystem Models

Lars Håkanson

Summary

Even if measurement error could be completely eliminated, biological systems are inherently variable, and there are fundamental limits on the predictability of the temporal dynamics of many critical forcing functions. How do we incorporate the important sources of uncertainty in our ecosystem models and determine confidence in model output? How does uncertainty constrain our approaches to prediction? This chapter addresses these questions in a discussion of fundamental concepts in predictive ecosystem modeling, including optimal model size, predictive power, uncertainties in empirical y and x variables (y = target variable to be predicted; x = model variable), uncertainties in model structures, sensitivity analyses, uncertainty analyses (using Monte Carlo techniques), and uncertainties related to step-by-step predictions. Examples are given for aquatic ecosystems using both statistical regression models and dynamical mass-balance models based on differential equations. To meet demands in ecosystem management for generality, practical usefulness, and predictive power, models should not only be as small as possible and driven by readily accessible variables but also be well balanced, which means that the predictions of the target variables do not depend too much on inherent uncertainties associated with just one or a few critical model variables. Such poorly balanced models have generic problems, which means that they will have good predictive power only in a restricted domain.

Background and Aim

The aim of this chapter is to outline the important factors and principles regulating the predictive power of ecosystem models. The first part discusses a key question in predictive ecosystem modeling: What is the maximum predictive success for a given target variable y? It is evident that many factors are involved, including sampling effort, measurement precision, and model structure. The r^2 value will be used here as a standard criterion of predictive success since this is a widely used concept in ecosystem modeling (r^2 = coefficient of determination; r = correlation coefficient). If an ecosystem model is validated, that is, tested against an independent set of data, then the achieved r^2 value will not just depend on the uncertainty in the empirical y value (the uncertainty in the y direction), but also on the structuring of the model, that is, which processes and model variables are accounted for and the empirical uncertainty of the model variables (the uncertainty in the x direction). This chapter will discuss three r^2 values: (1) the theoretically highest reference r^2 (r_r^2), (2) the empirically based highest r^2 (r_e^2), and (3) the highest achieved r^2 when modeled y values are compared to empirical y values (r^2).

"Predictive power" should not be merely two empty words. A scientific definition of the concept is necessary so the meaning is clear. The aim of this chapter is also to discuss a definition of predictive power and the rationale for that definition and to illustrate how the concept can be used for both empirical and dynamical models using radiocesium in lakes as an example.

Optimization will also be addressed; that is, the balance between increasing generality and predictive accuracy as dynamic and empirical/statistical models account for more processes and factors, and the increase in cumulative uncertainty associated with this growth. The apparent predictive power may increase with the number of x variables accounted for in predictive models. However, every x variable and rate in a predictive model has an uncertainty due to the fact that there are always problems associated with sampling, transport, storage, analyses, etc. Uncertainties in x variables may be added or multiplied in the model predictions. The optimal size of predictive models is therefore generally achieved using a (surprisingly) small number of x variables.

I will also focus on systematic predictions where a desired y variable is predicted in several steps in such a way that the model variables used to predict y are themselves predicted variables. The central question is: How is error propagation manifested for step-by-step predictions for regression models based on data from real ecosystems? The last part of the chapter illustrates the practical applicability of sensitivity and uncertainty analyses in contexts of ecosystem modeling, not only in the traditional manner (as a way to critically test models), but also as a tool to structure models.

There are benefits and drawbacks with all types of models. The main disadvantages with empirical models are that they generally only apply under restricted conditions, as set by the range of the x parameters, and that they may give poor insight into the causal mechanisms (see Håkanson 1991). The main disadvantages with dynamic mass-balance models (see Vemuri 1978; Straskraba and Gnauck 1985; Jørgensen and Johnsen 1989) are that they may be difficult and very expensive to calibrate and validate and that they tend to be large (the "elephantiasis problem"), which entails both practical and economical problems, as well as an accumulated uncertainty in the prediction.

The Highest Possible r^2 for Ecosystem Models

Problem Identification

Figure 8.1 illustrates some fundamental concepts related to the question of "the highest possible r^2." Figure 8.1A gives empirical data for the target variable y on both axes, that is, two series of parallel measured data, Emp_1 vs. Emp_2. The scale of interest in this context is the ecosystem scale, which means that the data emanate from several sampling occasions from an entire lake for defined time intervals. There are uncertainties for all such mean values. One standard way of quantifying such uncertainties is by means of the coefficient of variation, CV (CV = SD/MV; SD = standard deviation, MV = mean value). It is evident that the CV value of a given target variable, for example, the chlorophyll a concentration in contexts of eutrophication or the Hg content in fish muscle in contexts of toxic contamination, is *"not"* a constant but a variable. The CV value for within ecosystem variability is always related to very complex climatological, biological, chemical, and physical conditions, which means that simplifications are often requested. "Everything should be as simple as possible, but not simpler," according to Albert Einstein. It has been shown that it is often possible to define a characteristic CV value for a given variable, like 0.35 for total P and 0.25 for Hg concentration in fish and chlorophyll (see Håkanson and Peters 1995). Such characteristic CV values are very useful, for example, in uncertainty analysis of models using Monte Carlo simulations.

For simplicity, it will be assumed here that the samples emanate from normal frequency distributions. CV values expressed in this manner are by far the most commonly expressed statistic for this purpose in the ecological literature (see Whicker 1997). When more data become available, this assumption may be challenged, since it is well known that many water chemical variables are not normally but rather *log* normally distributed (see Håkanson and Peters 1995). Alternative methods of expressing the relative variability could be used (see Gilbert 1987), such as the geometric standard deviation, the range, and/or the ratio between maximum and minimum values. However, for the present purpose, this simplification seems justified. It is not likely that other statistical measures of uncertainty will substantially change the general conclusions given later in this chapter. By definition, CV is also largely independent of n, the number of data used to determine MV and SD, if n is large enough ($n > 6$).

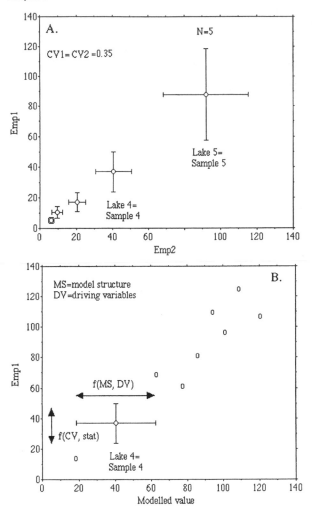

Figure 8.1. Illustration of some fundamental concepts related to the question of "the highest possible r^2" of ecosystem models. A. Empirical data for the target variable y on both axes, that is Emp₁, vs. Emp₂ for five lakes (1 to 5). B. Empirical data on the y axis and modeled values on the x axis. (Figure modified from Håkanson 1999a).

Figure 8.1 illustrates a case when a CV of 0.35 has been used for all empirical data on both axes. The uncertainty associated with the given target variable is illustrated by the uncertainty bands. This uncertainty will evidently influence the result of the regression, including the r^2 value and the confidence intervals.

If the CV for y is large, one cannot expect a model to predict y well. Figure 8.1B illustrates a normal hypothetical model validation when modeled values are put on the x axis. The empirical uncertainty associated with y remains the same on the y axis but the uncertainty in the x direction is related to the uncertainty associated with the model structure and the uncertainty of the model variables (x_i). Generally, one can expect the model uncertainty to be larger, or much larger, than the uncertainty in the y direction. This is also illustrated in Figure 8.1B for one of the data points.

This means that the r^2 value in the regression in Figure 8.1B is likely to be lower than the r^2 value obtained in the Emp$_1$-Emp$_2$ comparison. The crucial question is how high an r^2 value is it possible to obtain? This can only be answered by structuring the model in the best possible manner, by accounting for the most important processes and model variables, and by omitting the relatively unimportant processes and variables that can increase the uncertainty of the model (wider uncertainty bands, CI) more than the predictive success (a larger r^2 value).

Figure 8.1 illustrates a rather simple scenario for empirical/statistical regression models that produce one y value for one ecosystem (such as a lake). For a more general perspective, one must also include dynamic models, which yield time-dependent predictions of y, that is, time series of y, where the data in the time series are not independent of each other (see Figure 8.2). Empirical data of the target y variable and the empirical uncertainties associated with y (the CV value is also in this case set to 0.35) are given on the y axis with the series of model-predicted data (which can be expected to be even more uncertain). Figure 8.2 illustrates a situation where the model predictions fall within the empirical confidence bands for all months except for June, July, and August, that is, a situation where the model structure is systematically inadequate. By changing (excluding or including) equations or model variables, it may be possible to obtain a better fit between empirical data and modeled values for the summer period. The fit can be expressed by the r^2 value and also by the difference between modeled values and empirical data. Such a simple comparison is given in Figure 8.3A. One can note that the r^2 value is very low in this example (0.012) in spite of the fact that the model gives rather accurate predictions. The mean difference (Diff = $(M - E)/E$; M = modeled value, E = empirical data) is just 0.11 or 11%. This illustrates a well-known drawback with the use of r^2; r^2 depends on the amount of data and the range of the data in the regression. This is illustrated in Figure 8.3B. The data from Figure 8.3A is given by the cluster called Lake 3. In contexts of ecosystem modeling, when the ultimate aim is to have a general model covering the entire range of the target variable y, it is evident from Figure 8.3A that the r^2 value is not an adequate statistic for time series of dependent data. In such cases, one can preferably use, for example, the Diff value. The r^2 value could, on the other hand, be preferred for the entire data set. Figure 8.3B illustrates that this hypothetical model yields very good predictions ($r^2 = 0.979$) over the entire range.

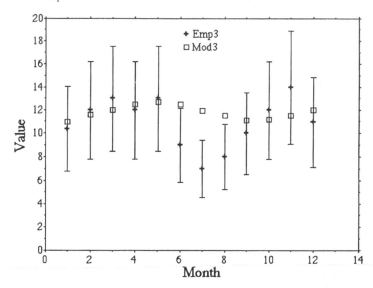

Figure 8.2. A comparison between parameter values for one empirical data series with uncertainty intervals (CV = 0.35) and modeled values in a time series from a dynamic model. (Modified from Håkanson 1999a).

Figure 8.3C highlights an important aspect related to the empirical uncertainties in y and the model structure. A time series from another lake ecosystem (called Lake 6) has been included with the three lakes given in Figure 8.3B. One can note that the model structure of this (hypothetical) dynamic model cannot cope with the conditions prevailing in Lake 6. The r^2 value has dropped to 0.645. A comparison between two data sets for all four lakes for y (Emp$_1$ vs. Emp$_2$) gave an (hypothetical) r^2 value of 0.92 under defined statistical conditions (CV for $y = 0.35$, six samples were taken from each lake, $n = 6$, and four lakes were included in the study, $N = 4$). This example illustrates a case when it would be possible to improve the model significantly, from $r^2 = 0.645$ to an r^2 higher than 0.9, if a better model structuring is used. Since there are many statistical considerations involved with the r^2 value from the Emp$_1$-Emp$_2$ comparison, it would be valuable to have access to a more stable reference r^2 value than r_e^2.

The Sampling Formula and Uncertainties in Empirical Data

If the within ecosystem variability (CV) is large, many samples must be analyzed to obtain a given level of certainty in the mean value. There is a general formula that is derived from the basic definitions of the mean value, the standard deviation, and the Student's t value and that expresses how many samples

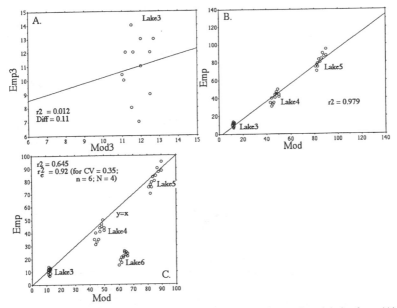

Figure 8.3. A theoretical comparison between empirical data and modeled values. (A) A regression between empirical data (Emp₃ on the y axis) and modeled values (Mod₃ on the x axis). The r^2 value for these 12 data is 0.012 and the Diff value 0.11 (Diff is defined in the text). (B) A comparison between data series for three lakes (Lake 3, 4, and 5). The overall r^2 value is 0.979. (C) The same three data as given in (B) and a new data series from Lake 6. The model gives poor predictions for Lake 6 but good predictions for the other three lakes, so the model structure can be improved. A direct comparison between two empirical data series (Emp₁ vs. Emp₂) gave an r^2 value (r_e^2) of 0.92, so the poor results for Lake 6 are not due to poor empirical data but to a poor model structure. (Modified from Håkanson 1999a).

are required (*n*) in order to establish a mean value with a specified certainty (from Håkanson 1984):

$$n = (1.96 \cdot CV/L)^2 + 1 \qquad (8.1)$$

where L is the level of error accepted in the mean value. For example, $L = 0.1$ implies 10% error, so the measured mean will be expected to lie within 10% of the expected mean with the probability assumed in determining t. Since one often determines the mean value with 95% certainty ($p = 0.05$), the t value is often 1.96 (or about 2).

If the CV is 0.35, about 50 samples are required to establish a lake-specific mean value for the given variable provided that one accepts an error of $L = 10\%$. If one accepts a larger error, for example, $L = 20\%$, fewer samples would be required. Since most variables in most lakes have CV's between 0.1 to 0.5, one can calculate the error in a typical estimate. If $n = 5$ and CV = 0.33 then L is about 33%. Because few studies take more samples, this calculation has profound implications about the quality of our knowledge of ecosystems. It shows that for most lake variables, existing empirical estimates are only rough measures of the lake-typical mean value. This is especially so for total P, and to a lesser extent for Secchi depth, conductivity, and pH.

One reason for these high CV values has to do with the fact that there are large differences in analytical reliability for different variables (see Håkanson and Peters 1995). The average CV related to analytical procedures is only about 0.025 for the determination of Hg in fish. Lake pH can also be determined with great reliability. The average CV value for pH is about 0.02. This error represents the combined effects of errors in all phases of sampling, sample preparation, and analysis. Determinations of conductivity, hardness (CaMg) and the Ca concentration are also generally highly reliable. Color (Col), Fe concentration, total P concentration, and alkalinity have higher methodological CV's (0.15–0.2).

It is important to remember this uncertainty in the empirical data when one derives empirical models or calibrates and validates dynamic models. Since one rarely has very reliable empirical data, one cannot expect to obtain models that explain all variability. The uncertainty in the empirical determinations of lake variables for which relatively few samples have been analyzed may produce marked divergences between modeled and empirical data. In such cases, wide divergences may not necessarily indicate errors and deficiencies in the models but could reflect deficiencies in the empirical base.

In summary, very many factors (from methods of sampling and analysis to chemical and ecological interactions in the lake) influence the empirical estimates of water quality variables (pH, alkalinity, total P, color, etc.) used to characterize entire lakes at the time scale of weeks to years. Since many variables fluctuate greatly, it is often difficult in practice to make reliable, representative, lake-typical empirical estimates. Data from specific sites and sampling occasions may not represent the prevailing, typical conditions in the lake very well at all.

Empirically Based Highest r^2: r_e^2

A first step to determine the highest possible r^2 of a predictive model is, as already mentioned, to compare two empirical samples. The parameters in these two samples should be as time and area compatible as possible; they should also be sampled, transported, stored and analyzed in the same manner. To illustrate the basic approach used to determine r_e^2, data from 70 Swedish lakes (from Håkanson et al. 1990) will be used for the target operational variable for Hg research, the Hg concentration in fish for human consumption (here, values in mgHg/kg muscle of 1-kg pike, abbreviated as Hgpi). At least 4 fish are in-

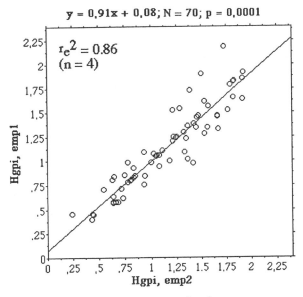

$$y = 0.91x + 0.08; N = 70; p = 0.0001$$

$$r_e^2 = 0.86$$
$$(n = 4)$$

Figure 8.4. Determination of "empirically highest r^2" (r_e^2) for the Hg concentration in 1-kg pike (Hgpi) from a regression between two parallel empirical samples (4 fish in each sample, i.e., 8 fish from 70 lakes). (Figure modified from Håkanson 1999a.)

cluded in each sample, that is, at least 8 fish per lake. The highest empirical r_e^2 is 0.86 (see Figure 8.4).

The following sequence of values for r_e^2 (from Håkanson and Peters 1995) have been determined from two sets of mean values (Emp$_1$ and Emp$_2$) each representing 6 samples from different months in 1986 from 25 Swedish lakes.

Variable	temp	Total P	Secchi	Fe	Ca and pH	Alk, color and CaMg
r_e^2	0.76	0.85	0.90	0.95	0.96	0.99

There are generally high r_e^2 values between the two series. Temperature gave, as expected, the lowest value (0.76), and color one of the highest (0.99). These data also indicate something about the temporal variability of these variables: that it is important to establish representative values for larger areas and longer periods of time (like annual or monthly mean values). The more changeable a variable is, the more difficult it will be to establish representative empirical data of the given variable.

The important point here is that one should never hope to explain all of the variation in any ecological variable. It is interesting to note that total P, which is a fundamental state variable in practically all lake contexts, is one of the most variable measures tested ($r_e^2 = 0.85$).

Theoretically Highest Reference r^2: r_r^2

From a statistical point of view, an equation has been derived which gives the theoretically highest r^2 value as a function of (1) the number of samples (n_i) for each y_i value in the regression, (2) the number of data points in the regression (N), (3) the standard deviations related to all individual data points, (4) the standard deviation of all points in the regression and (5) the range of the y variable (Håkanson 1999a). The r_r^2 value is defined as

$$r_r^2 = 1 - 0.66 \cdot CV^2 \tag{8.2}$$

where CV is the characteristic within-lake variability for the given y variable. The equation is valid for actual (nontransformed) y values. One cannot expect to obtain r^2-values higher than the r_r^2 values, so these values may be used as reference values for the following models. It can be noted that, e.g., r_r^2 for the Secchi depth is 0.98.

Defining Predictive Power

Figure 8.5A illustrates two curves, one based on empirical data, the other on modeled values. One can see that there is almost perfect agreement between the two curves. One way to quantify the fit between empirical and modeled y is to do a regression. The r^2 value, the intercept, and the slope of the regression line will reveal the fit. The r^2 value and the slope should be as close to 1 as possible (Figure 8.5B), and the intercept should ideally cross the origin. Would this model work also for other systems? If the answer is yes, then we may have a very useful predictive model. However, one can safely assume that the r^2 value and the slope will not be equally high in all cases. There will be situations when the model will yield a poor prediction, a low r^2, and a slope much lower or higher than 1. Such a spread of values indicates the uncertainty of the model in predictions. For each validation, one can determine the r^2 value and the slope between empirical and modeled y. One can also determine the coefficient of variation for r^2 (CV_r^2). If the model generally has a high predictive power, then the CV_r^2 should be small. From these arguments on r^2, slope and CV_r^2, Håkanson (1997) gave a general definition of predictive power (PP) as:

If $\alpha = 1$, then PP $= R^2/((1.1 - \alpha) \cdot CV_r^2)$
If $\alpha > 1$, then PP $= R^2/((1.1 - 1/\alpha) \cdot CV_r^2)$ (8.3)

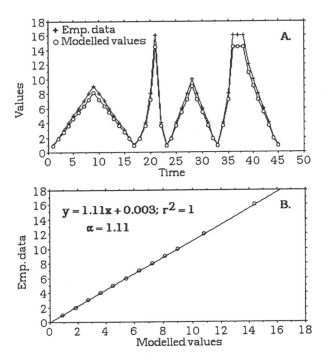

Figure 8.5. (A) Illustration of a very good correspondence between empirical data and modeled values. (B) The same data illustrated by regression analysis. The fit is almost perfect, the r^2 value is 1.00, but the slope is 1.1, which is higher than the ideal 1.00, indicating that the model consistently underestimates the empirical data. (Figure modified from Håkanson 1997.)

where

- R^2 is the *mean r^2* of all model validations: The higher R^2 the higher PP.

- The slope, α, of the regression line may be smaller and larger than 1. If the slope is smaller than 1, one can quantify the influence on PP by the factor $1.1 - \alpha$. Since α may be equal to 1, and since division by zero is not allowed, 1.1 is used instead of just 1. This means that the slope factor is always larger than 0.1. If the slope is larger than 1, one simply sets it to 1/slope. This means that a slope of 0.5 will give the same factor as a slope of 2, namely $1.1 - 0.5 = 0.6$ or $1.1 - 1/2 = 0.6$.

- CV_r^2 is the coefficient of variation for the r^2 values obtained in the empirical tests.

One can safely assume that in practice CV_r^2 will never be zero for ecosystem models; neither are models likely to yield r^2 values of 1.00. Very good models may give r^2 values of about 0.95. CV_r^2 values lower than 0.1 ought to be rare. With this definition, PP will generally be lower than 100. Models yielding PP higher than 10 would be very good since this means that, for example, $R^2 = 0.9$, $CV_r^2 = 0.1$, and $\alpha = 1.1$ Models giving PP lower than 1 would be useless for all practical purposes. Such models have a poor fit (a low r^2 or a slope much diverging from 1 or both) and great uncertainties (a high CV_r^2).

This expression of predictive power (Eq. 8.1) should be regarded like most predictive models for complex ecosystems: A tool that accounts, not for every conceivable situation and factor, but for the most important factors in a simple and useful manner. Here, the fit between modeled values and empirical data is given by the value of r^2 and the slope factor. The fit may, however, be expressed in many alternative ways. Instead of the mean or median r^2, one could use the adjusted r^2. All expressions related to such r^2 values would depend on the number of data pairs (n), the range and the transformation of the x and y variables. Logarithmic x and y variables will give different r^2 values than non-transformed variables. Instead of using this definition of the slope factor, one could use other alternatives and also include the intercept. Methods and guidelines for this have been discussed by Håkanson and Peters (1995).

It should be stressed that the uncertainty (CV_r^2) is determined *independently* of the fit. One should not use expressions related to the confidence interval of the regression line for the uncertainty since such measures are directly related to the r^2 value. In this approach, the model uncertainty may be expressed in two ways: either by Monte Carlo simulations or from repeated validations which enable the determination of CV_r^2 from the obtained r^2 values between modeled values and empirical data.

This definition should not be used in an uncritical manner and PP values determined for different models for different purposes may not be directly comparable. In all modeling situations, it is the responsibility of the modeller to define and explain the presuppositions of the models and their applicability, so that the net result is more clarity and less confusion.

Predictive Power of Empirical Models

Consider the concentration of radiocesium in pike from 14 Swedish lakes in 1988 (two years after the Chernobyl disaster). Many lake variables (including K concentration, pH, total P and color) could influence the bio-uptake of radiocesium and the Cs concentration in pike. The concentration of ^{137}Cs in water in 1987 was the most important x variable (Håkanson 2000). It explains statistically about 78% ($r^2 = 0.778$) of the variability in the y variable (Cspi88, Cs in

Bq/kg ww pike in 1988) among these 14 Swedish lakes. The other three factors were K concentration, the percent of open land in the catchment (OL%), and lake total P. Lake total P is, as already pointed out, a highly variable parameter. Its coefficient of variation (CV) was, on average, 38%. The corresponding CV for K is only about 12%, for Cs in water it was about 26% (Andersson et al. 1991), and for OL% it was much smaller—on the order of 1% (CV = 0.01; Nilsson 1992). With this information, one can use independent Monte Carlo simulations to estimate the uncertainty (CV) in the y variable. PP attains a maximum value for $n = 3$. By accounting for total P in this stepwise regression analysis, one increases r^2, but decreases PP. The reason for this is that one adds an uncertain variable, which contributes more to the model uncertainty (CV) than to the r^2 value. The net result is a model with a lower PP. One should also note that empirical regression analysis automatically yields a slope close to 1. The PP values of these empirical models are very high: PP > 35 for all four models. The maximum PP is not obtained for the largest model. For the other models, the highest PP may very well be obtained for other model sizes. It should be remembered that empirical models are basically meant to be used within given ranges of applicability.

Predictive Power of Dynamic Models

Dynamic models derive from a causal analysis of ecological and biological fluxes. Dynamic models are used mostly to study complex interactions and time-dependent variations within defined ecosystems. If dynamic models are to be used in practice, for example, to quantify fluxes, amounts, and concentrations of energy, carbon, or contaminants in lakes, the parameters that govern the transport among the various compartments have to be known, simulated, or guessed. In dynamic modeling, dimensional analysis (of each parameter) is very important. Dynamic models are often difficult to calibrate and validate, and they tend to grow indefinitely. If dynamic models are not validated, they may yield absolutely worthless predictions. As is the case for any model, the presuppositions ("traffic rules") of the model must always be clearly stated.

In this section, the relationship between predictive power (PP) and model size (n) will be examined for three dynamic models for radiocesium in lakes. The results can be compared to the results from the empirical model for radiocesium.

1. A small model (Håkanson 1991). It has only three compartments: water, prey, and predatory fish; 6 model variables; and 5 lake-specific variables. The total number of driving variables (n) is thus 11. Note that there is no catchment area, no sediments, no food web, and no partition coefficient (Kd) in this model.

2. The VAMP model (see Håkanson et al. 1996). It has 10 compartments, 21 model variables, and 13 lake-specific variables.
The model size is given by $n = 34$.

3. The generic model (see IAEA 2000). It is a traditional model with
9 compartments, 27 model variables, and 9 lake-specific variables, which gives $n = 36$.

These models have been tested using the data for the VAMP lakes (see
IAEA 2000), which vary in size (from 0.042 to 1147 km^2), mean depth (from
1.7 to 89.5 m), precipitation (from 600 to 1840 mm/year), pH (from 5.1 to 8.5),
K concentration (from 0.4 to 40 mg/l), and primary productivity (from 0.8 to
350 g C/m$^2 \cdot$ year).

The results concerning predictive power of the dynamic models are summa-
rised in Table 8.1. One can note that the smallest model yields the highest PP,
the biggest model, the generic model, the lowest PP. In this case, one obtains
the best predictive power with a relatively small model that accounts for the
most important processes, no more, no less. Big models with many uncertain
rates and model variables give lower values of PP. One should also note that
the PP values obtained by these dynamic models for radiocesium are much
lower than those from the empirical models. To conclude: within their range of
applicability, empirical models often provide better predictive power than dynamic models in ecosystem contexts.

Table 8.1. Predictive power for three dynamic models for cesium in lakes.

	Model Variable n_1	Lake-specific Variable n_2	$n = n_1 + n_2$	R^2	CV_r^2	Slope, MV	$1.1-\alpha$	PP
Small model	6	5	11	0.65	0.427	1.12	0.21	7.42
VAMP model	21	13	34	0.65	0.428	0.85	0.25	6.12
Generic model	27	9	36	0.56	0.602	1.15	0.23	3.99

The Optimal-Size Problem

Is it possible to quantitatively assess optimal model size, that is, the balance
between an increasing generality as dynamic and empirical models account for
more processes (more x variables) and an increasing in predictive uncertainty

associated with this growth? Every process, state variable, and compartment added to a dynamic model entails a certain error, since there is always an uncertainty linked to the method of sampling, transport, storage, handling, analysis, and data processing. This problem has been addressed in several contexts (see Peters 1986, 1991). It is evident to many modelers that the risks of predictive failure, as determined from both a decreasing accuracy in the prediction of y and increasing uncertainty limits (confidence or tolerance limits) around the predicted y value, will increase if more and more x parameters (like compartments, boxes, factors, rates, processes, etc.) are accounted for in the model.

It should be stressed that due to the great complexity of ecosystems, ecosystem models constructed for the purpose of describing and understanding interactions, food webs, fluxes of contaminants, etc., must be extensive! However, it is quite another issue with models for specific *predictions* of just one or two target y variables. Even if "everything depends on everything else" in ecosystems, it is clear that all x variables could not have the same predictive weight for one specific y variable. In predictive modeling, there seem to exist two ways of assessing that influence: by empirical/statistical methods (correlations, etc.) or by sensitivity and uncertainty analyses.

The Predictive Accuracy of Models

In the balance between predictive accuracy and accumulated uncertainty, accuracy will be treated first. Figure 8.6 illustrates the relationship between r^2 and model size (number of variables) for

1. A simple *biological* variable: Hgpe, the mean lake Hg content in perch fry in mg/kg wet weight based on data from about 10 fish per lake from 25 lakes (25 = N) (data from Håkanson et al. 1990). Many different parameters from the catchment areas (like percentage of rocks, lakes, and mires), the bathymetric map of the lakes (mean depth, Dm; and dynamic ratio, $\sqrt{area}$ / Dm, and water-chemistry variables (like pH and color) were tested in deriving this model for Hgpe.

2. An *abiotic* (chemical) variable: RHg, the mean reactive Hg from water samples in ng/l based on about 10 samples per lake from 25 lakes (data from Håkanson et al. 1990). The model for RHg was derived in the same manner as that for Hgpe.

3. Mean annual lake pH (pH12; 12 as in 12 months) predicted from *random parameters* (as given by a random data generator); pH12 was determined empirically from monthly samples from 25 lakes.

It is evident that these three selected variables are not typical for all biotic and abiotic lake variables; they have been used here as examples to illustrate some important principles. The following conclusions may be drawn from Figure 8.6.

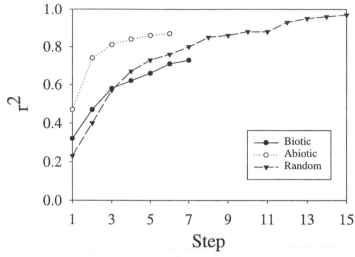

Figure 8.6. The relationship between r^2 and model size for a biotic variable, an abiotic variable, and a random variable. See text for description of the variables.

- These results indicate what ought to be generally true, namely that it should be more difficult to predict biological variables than abiotic variables. As soon as biology becomes a factor, predictions often get difficult because beside the abiotic variables, one must then also expect uncertainties related to consumption, predation, turnover, metabolic characteristics, etc. For the simple biological parameter, the Hg concentration in fish, 7 variables are needed to obtain an r^2 value of 0.73. For the abiotic variable, 6 variables are needed to get an r^2 value of 0.87. It should also be noted that the value of RHg in a given lake depends on many complex biological processes (see Lindqvist et al. 1991; Meili 1991).

- It is possible to obtain very high r^2 values in models based entirely on random parameters.

One can see that the r^2 values increase as the number of x variables accounted for in the models increases and that the curves for the biological variable (Hgpe) and the random parameter prediction of pH12 (for $n = 25$) are very close to one another. But this is totally dependent on the choice of n. The curve for $n = 35$ is significantly different from the curve for $n = 25$. The reason for this is simply that it is much more unlikely to obtain high r^2 values for random x parameters if n is large. To conclude: the predictive power, expressed for

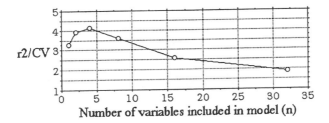

Figure 8.7. Hypothetical calculation of the optimal number of x variables in a predictive model. The r^2 value, a measure of predictive accuracy, should be as high as possible. CV = a measure of the accumulated model uncertainty should be as small as possible; n = the number of parameters or x variables in the model. (Figure modified from Håkanson 1995.)

example by the r^2 {xe "-"}value, generally increases with the number of x variables accounted for in the predictive model, but differently for biological and chemical variables (and for different modeling presuppositions). This has been exemplified here with regression models, but the same principles apply also to dynamical models.

The Accumulated Uncertainty

From any statistical textbook (see Neter et al. 1988, or Helsel and Hirsch 1992), it is easy to demonstrate that there exists a mathematical and exact way of defining the uncertainty of additive models, since the standard deviation, SD, may be written as $SD_y = SD_x \cdot \sqrt{n}$, where SD_y = the standard deviation of the y variable; SD_x = the standard deviation of all the x variables. From this, one can conclude that the uncertainty factor increases significantly as n increases.

The Optimal Size

To determine the optimal size (the number of x variables) of a predictive model, one can combine the predictive accuracy (r^2) and the uncertainty factors (SD or CV) in several ways. From Figure 8.7, one can note that the optimal size for predictive models under the given conditions is generally achieved for surprisingly small n. The reason is that the predictive power (r^2) increases rapidly when n is small and the increasing accumulated error (CV or SD) decreases the ratio for higher n.

These results support Albert Einstein's well-known admonition that "Everything should be as simple as possible, but not simpler!"

Step-by-Step Predictions

In predictive ecosystem modeling and practical environmental management, it is often necessary to make predictions in several steps. For example, key limnological variables, like total P, pH, and color could be predicted from readily available map parameters describing the catchment area (bedrocks, soils, land use, etc.) and the lake morphometry (area, volume, mean depth, etc.). These predicted variables might then be used in other models to predict various measures of the state of the fish, bottom fauna, etc. (Peters 1986). Models based on this process of step-by-step predictions (Håkanson and Peters 1995) are an important and integral part of lake management and research.

The concept of step-by-step prediction should not be confused with stepwise multiple regression analysis (Draper and Smith 1966). There are several problems with step-by-step models, and the aim of this section is to discuss them using data from lakes and regression analyses. The mathematical/analytical aspects of error propagation are treated in basic textbooks (Blom 1984) and are often based on Gauss's approximation equation. Error propagation is also discussed in contexts of, for example, modeling using Geographical Information Systems (GIS; see Goodchild et al. 1993, and Burrough and McDonnell 1998).

To highlight a typical problem in step-by-step predictions, a simple and typical case study will be given. It concerns the task of predicting lake morphometric variables. Morphometric parameters are usually important and useful x variables in lake models—they are predictors, not predicted variables. Nevertheless, there are lakes where no depth measurements have been made, so the mean depth and the volume cannot be known. From simple topographical maps, one can easily determine the lake area the shoreline length, the size of the catchment area and its relief (RDA = $dH/ADA^{0.5}$; RDA = relief of drainage area; dH = the max. height difference in the drainage area; ADA = area of drainage area in m^2; see Håkanson and Peters 1995). But it is not possible to calculate characteristics like lake volume (Vol), maximum depth, and mean depth (Dm = Vol/area), unless one has access to a bathymetric map. This example concerns the prediction of such characteristics.

Table 8.2A illustrates a regression between lake area and volume. The transformations that yield the most normal frequency distributions for this lake type have been used (Håkanson and Peters 1995), namely log(area) and log(1000 · Vol), where area is in km^2 and volume in km^3. Log(1000 · Vol) is used instead of log(Vol) for convenience. The regression line is: log(1000 · Vol) = 0.605 + 1.142 · log(area). The r^2 value for these 95 lakes is 0.85. This means that lakes with large areas can be predicted to have large volumes, which sounds plausible and is consistent with previous observations (Straskraba 1980). Table 8.2A shows that the r^2 value may be further increased by accounting for the relief of the catchment (RDA): the higher the relief, the larger the mean depth and the larger the volume. The r^2 value for the empirical model with these two x variables is 0.87.

Table 8.2. Regressions for lake morphometric data from ninety-five Swedish lakes and their catchments.

Model Parameters	r^2	Model
A. Prediction of lake volume		
Lake area	0.85	$\log(1000 \cdot \text{Vol}) = 0{,}605 + 1{,}142 \cdot \log(\text{area})$
Lake area + catchment relief (RDA)	0.87	$\log(1000 \cdot \text{Vol}) = 0{,}134 + 1{,}224 \cdot \log(\text{area}) + 0{,}332 \cdot \log(\text{RDA})$
B. Prediction of lake maximum depth (Dmax) from empirical data		
Lake volume (Vol)	0.47	$\log(\text{Dmax}) = -0{,}601 + 1{,}499 \cdot (1000 \cdot \text{Vol})^{\wedge}0{,}1$
Lake volume, area	0.87	$\log(\text{Dmax}) = -4{,}202 + 4{,}558 \cdot (1000 \cdot \text{Vol})^{\wedge}0{,}1 - 1{,}008 \cdot \log(\text{area})$

Source: Håkanson 1996b.
Note: Lake volume (Vol) is in km^3; F > 4

If one has access to empirical data on lake volume (as for these 95 lakes), it is possible to develop models predicting the maximum depth (Dmax in m). This is shown by two regression models in Table 8.2B. From lake area and volume, one can predict maximum depth; $r^2 = 0.87$.

The crucial question here is what will happen to the prediction of Dmax if one uses predicted values for lake volume instead of empirical data? This involves a step-by-step prediction: First the lake volume is predicted from the lake area and the catchment relief, then Dmax is predicted from the predicted volume. The r^2 value drops from 0.87 to 0.34. This is a typical feature of step-by-step predictions. There is often a considerable loss in predictive accuracy at each step.

The reason for the drastic decrease in predictive power can be statistically explained by uncertainty analysis (Monte Carlo simulations).

Sensitivity and Uncertainty Analyses

The aim of this section is to give only a few examples of two very useful methods for sensitivity and uncertainty tests. Additional discussion of this topic for aquatic ecosystem models can be found in Håkanson and Peters (1995). All model variables (such as rates and distribution coefficients) and all model uncertainties will not be of equal importance in predicting a target *y* variable. Sensitivity and uncertainty tests are very useful to rank such dependencies. Sensitivity analysis involves the study, by modeling and simulation, of how the change in one rate or variable of a model influences a given prediction while everything else is kept constant. This type of analysis is commonly used in ecosystem modeling (see Hinton 1993; Hamby 1995; IAEA 2000).

In sensitivity analyses, it is common to repeat the calculations for all interesting model variables to try to produce a ranking of the factors influencing the target variable. The basic idea is to identify the most sensitive part of the model, that is, the part that is most decisive for the model prediction. In a simple sensitivity analysis, one would vary all model variables using the same uncertainty factor. However, it is evident that it is *not* realistic to apply the same uncertainty for all model variables. There are major differences among model variables in this respect. Morphometric parameters can often be determined very accurately, while some model variables, like rates and distributions coefficients, cannot be empirically determined at all for real ecosystems but, rather, have to be estimated from laboratory tests or theoretical derivations. This means that the values used for such model variables are often very uncertain. Table 8.3 gives a compilation of typical, characteristic CV values for many types of variables used in aquatic ecosystem models. Note that about 70% of the data in a frequency distribution fall within the values given by MV ± SD, and that 95% of the data are likely to fall within MV ± 2 · SD. From this table, one can note that model variables, such as rates and distribution coefficients generally can be given CV values of 0.5. The highest expected CV values are for certain sedimentological variables. In the following uncertainty tests, the CV values given in Table 8.3 will be used.

Table 8.3. Compilation of characteristic CV values for within-lake variability based on individual samples for different types of lake variables.

	Coefficient of Variation
Catchment variables	
Catchment area (ADA)	0.01
Percent outflow areas (OA)	0.10
Fallout of radiocesium (Cssoil)	0.10
Mean soil type or permeability factor (SP)	0.25
Lake variables	
Lake area (Area)	0.01
Mean depth (Dm)	0.01
Volume (Vol)	0.01
Maximum depth (Dmax)	0.01
Theoretical water retention time (Tw)	0.10

Table 8.3. *Continued*

	Coefficient of Variation
Water chemical variables	
pH	0.05
Conductivity (cond)	0.10
Ca concentration (Ca)	0.12
Hardness (CaMg)	0.12
K concentration (K)	0.20
Color (col)	0.20
Fe concentration (Fe)	0.25
Total P concentration (TP)	0.35
Alkalinity (alk)	0.35
Sedimentological variables	
Percent ET areas (ET)	0.05
Mean water content for E areas	0.30
Mean water content for T areas	0.20
Mean water content for A areas	0.05
Mean bulk density for E areas	0.10
Mean bulk density for T areas	0.10
Mean bulk density for A areas	0.02
Mean organic content for E areas	0.50
Mean organic content for T areas	0.50
Mean organic content for A areas	0.10
Mean TP-concentration for E areas	0.50
Mean TP-concentration for T areas	0.75
Mean TP-concentration for A areas	0.35
Mean metal concentration for E areas	0.50
Mean metal concentration for T areas	0.25
Mean metal concentration for A areas	0.20
Lake management variables	
Secchi depth (Sec)	0.15
Chlorophyll *a* concentration (Chl)	0.25
Hg concentration in fish muscle	0.25
Cs concentration in fish	0.22
Cs concentration in water	0.30
Cs concentration in sediments	0.60
Climatological variables	
Annual runoff rates	0.10
Annual precipitation	0.10
Temperature	0.20

Table 8.3. *Continued*

	Coefficient of Variation
Model variables	
Fall velocities	0.50
Age of A sediments	0.50
Age of ET sediments	0.50
Diffusion rates	0.50
Retention rates	0.50
Bioconcentration factors	0.50
Feed habit coefficients	0.50
Distribution (= partion) coefficients	0.50

Source: Håkanson 1999c.
Note: E = areas of fine sediment erosion; T = areas of fine sediment transport; and A = areas of fine sediment accumulation.

Uncertainty Tests Using Monte Carlo Techniques

Two main approaches to uncertainty analysis exist, analytical methods (Cox and Baybutt 1981; Worley 1987) and statistical methods, like Monte Carlo techniques (Rose et al. 1989). In this section, only Monte Carlo simulations will be discussed.

Uncertainty tests using Monte Carlo techniques may be done in several ways: with uniform CV values, or more realistically, with characteristic CV values (Table 8.3). Monte Carlo simulation is a technique to forecast the entire range of likely observations in a given situation; it can also give confidence limits to describe the likelihood of a given event. Uncertainty analysis (which is a term for this procedure) is the same as conducting sensitivity analysis for all given model variables at the same time. A typical uncertainty analysis is carried out in two steps (Håkanson 1999c). First, all the model variables are included with defined uncertainties, and the resulting uncertainty for the target variable is calculated. Then, the model variables are omitted from the analysis one at a time.

Figure 8.8 shows the results from such an uncertainty analysis for a version of the "classical" ELS-model (effect-load-sensitivity; see Figure 8.9) using data on total phosphorus (TP, µg/l) from Lake S. Bullaren, Sweden. This figure also shows the results for an often-used operational target variable for lake eutrophication studies, maximum algal volume (AV, mm^3/l), as well as results from sensitivity analyses, which can be directly compared to the results for the uncertainty analyses.

One can note that the most crucial uncertainty component for the predicted AV value is the value used for Cin, the tributary concentration of total phosphorus (TP). The values selected for the settling velocity for TP (v), the exponent for the lake water retention rate ($1/Tw^{Expo}$, regulating lake outflow of TP) or the specific runoff rate (SR, regulating lake inflow of TP) influence lake TP

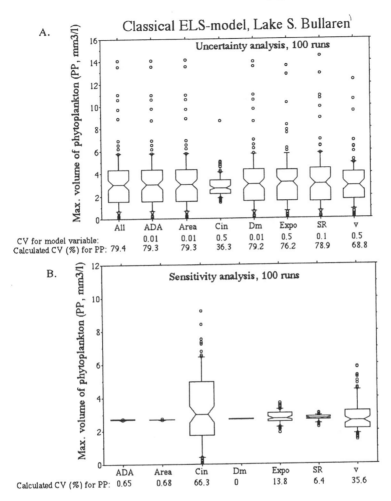

Figure 8.8. (A) Uncertainty analyses using the "classical" ELS-model (effect-load-sensitivity) for algal volume (AV) in Lake S. Bullaren. CV values for the model variables are given as well as calculated CV values for AV. (B) Corresponding results from sensitivity analyses. (Figure modified from Håkanson 1999c.)

concentrations, and hence also the AV predictions, much less. The uncertainties associated with the morphometric data (CV = 0.01) do not affect the predictions in any significant manner. The figure also gives the calculated CVs for the target variable, AV. These CVs can be used to rank the influence the model variables have, under the given conditions, on the predictions of AV.

The "classical" ELS-model

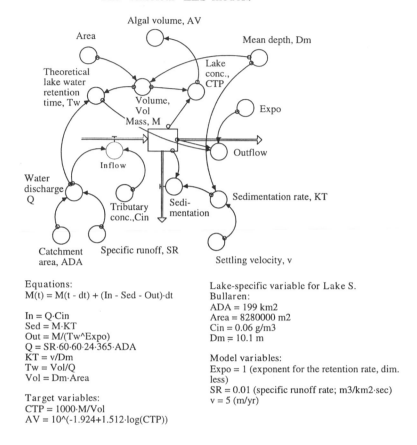

Equations:
$M(t) = M(t - dt) + (In - Sed - Out) \cdot dt$

$In = Q \cdot Cin$
$Sed = M \cdot KT$
$Out = M/(Tw\text{\textasciicircum}Expo)$
$Q = SR \cdot 60 \cdot 60 \cdot 24 \cdot 365 \cdot ADA$
$KT = v/Dm$
$Tw = Vol/Q$
$Vol = Dm \cdot Area$

Target variables:
$CTP = 1000 \cdot M/Vol$
$AV = 10\text{\textasciicircum}(-1.924 + 1.512 \cdot \log(CTP))$

Lake-specific variable for Lake S.
Bullaren:
$ADA = 199$ km2
$Area = 8280000$ m2
$Cin = 0.06$ g/m3
$Dm = 10.1$ m

Model variables:
Expo = 1 (exponent for the retention rate, dim. less)
SR = 0.01 (specific runoff rate; m3/km2·sec)
v = 5 (m/yr)

Figure 8.9. The "classical" ELS-model. Equations, target variables, lake-specific variables, and model variables. (Figure modified from Håkanson 1999c.)

The results given in Figure 8.8 are indicative of a typically poorly balanced model. The calculated uncertainties for these 100 runs for the target variable AV are also given, and together with the box-and-whisker plots, they demonstrate that this model, and all similar traditional lake eutrophication models of the Vollenweider- and OECD-type (Vollenweider 1968; OECD 1982) are poorly balanced and highly dependent on the reliability of the data for Cin. This is nothing new. It is the reason why so much effort has been devoted to developing a good understanding of the processes regulating the tributary flow

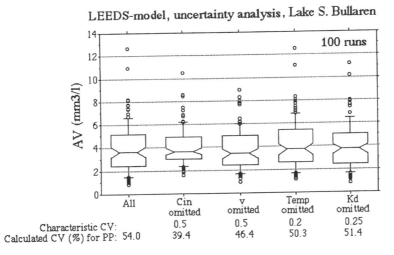

Figure 8.10. Monte Carlo simulations using the LEEDS-model for Lake S. Bullaren. The box-and-whisker plots (giving medians, quartiles, percentiles, and outliers) show the results for the four model variables contributing most to the uncertainty in the target variable, AV. The characteristic CV values are given, as well as calculated CVs for maximum algal volume (AV). (Figure modified from Håkanson 1999c.)

of phosphorus and the efforts to identify sources for TP fluxes from catchments (see Dillon and Rigler 1974; 1975; Nichols and Dillon 1978; Chapra and Reckhow 1979; 1983). A well-balanced model should *not* be too dependent on the uncertainty related to a single variable. Instead, all the box-and-whisker plots should look more alike and the CV values for y should *not* change too much if a model variable is omitted in the uncertainty analysis.

From these uncertainty tests, it is clear that the value for the tributary concentration, Cin, is very important in the "classical" ELS model for lake eutrophication studies. It is often important in other models for lake eutrophication (and in models for other substances than P). This is exemplified again for predictions of algal volume (AV) using the more comprehensive LEEDS model (see Håkanson 1999c for a model description) Figure 8.10 presents box-and-whisker plots for the LEEDS model when uncertainty in all 15 variables is included, and for four cases where uncertainty in a single critical variable is omitted. The main message from Figure 8.10 is that LEEDS is a well-balanced model. Cin is, however, still the most important model variable for the uncertainty in the AV predictions, but CV for AV does not decrease more than from 54.0% to 39.4% when Cin is omitted.

Concluding Remarks

All model approaches, and all approaches to science, have drawbacks and limitations. In dynamic models, one may say that scientific knowledge is used to set the processes, but guesses may be necessary to set the rates. In empirical regression models, science is used to obtain the constants and guesses are made about the processes. Both approaches have their places.

Statements on models (from Håkanson and Peters 1995):

- Models are built and validated with empirical data. Empiricism enters at many steps from start to finish in modeling, but empirical data, and any knowledge based on empirical data, are uncertain. Accumulated uncertainties in the models will cause accumulated uncertainties in model predictions.

- The predictive power of a model is not governed by the strength of the model's strongest part, but by the weakness of its weakest part.

- Big models are simple to build but hard to validate. Small models are hard to build and simple to validate.

- Small size is necessary, but not sufficient, for utility and predictive power; so useful models must be small. Small models should be based on the most fundamental processes, but that is far easier to say than it is to accomplish.

- Scientific knowledge does not lie in the model alone nor in the empirical data alone, but in their overlap as validated, predictive models.

- The key issue is not to verify, but to falsify a model, and thereby to determine its limitations.

- It is important to predict mean values, but it is equally important to predict the confidence interval around the mean.

Acknowledgments. I would like to thank an anonymous reviewer for many constructive remarks and suggestions.

References

Andersson, T., L. Håkanson, H. Kvarnäs, and Å. Nilsson. 1991. *Measures against High Levels of Radiocesium in Lake Fish*. Final report from the Liming-Mercury-Cesium Project. (in Swedish, Åtgärder mot höga halter av radioaktivt cesium i insjöfisk. Slutrapport- för cesiumdelen av projektet Kalkning-kvicksilver-cesium). Statens strålskyddsinst. SSI Report 91–07. Stockholm.

Beck, M.B., and G. Van Straten, editors. 1983. *Uncertainty, System Identification, and the Prediction of Water Quality*. Heidelberg: Springer.

Blom, G. 1984. *Sannolikhetsteori Med Tillämpningar*. Lund: Studentlitteratur.

Burrough, P.A., and R.A. McDonnell. 1998. *Principles of Geographical Information Systems*. Oxford: Oxford University Press.

Chapra, S.C., and K. Reckhow. 1979. Expressing the phosphorus loading concept in probabilistic terms. *Journal of Fisheries Research Board of Canada* 36: 225–229.

———. 1983. Engineering approaches for lake management. Vol. 2. *Mechanistic Modeling*. Woburn: Butterworth.

Cox, D.C., and P. Baybutt. 1981. Methods for uncertainty analysis: A comparative survey. *Risk Analysis* 1: 251–258.

Dillon, P.J., and F.H. Riegler. 1974. A test of a simple nutrient budget model predicting the phosphorus concentration in lake water. *Journal of Fisheries Research Board of Canada* 31: 1771–1778.

———. 1975. A simple method for predicting the capacity of a lake for development   based on lake trophic status. *Journal of Fisheries Research Board of Canada* 32: 1519–1531.

Draper, N.R., and H. Smith. 1966. *Applied Regression Analysis*. New York: Wiley.

Gilbert, R.O. 1987. *Statistical Methods for Environmental Pollution Monitoring*. New York: Van Nostrand Reinold.

Goodchild, M.F., B.O. Parks, and L.T. Steyaert, editors. 1993. *Environmental Modeling with GIS*. New York: Oxford University Press.

Håkanson, L. 1981. *A Manual of Lake Morphometry*. Heidelberg: Springer.

———. 1984. Sediment sampling in different aquatic environments: Statistical aspects. *Water Resources Research* 20: 41–46.

———. 1991. *Ecometric and Dynamic Modeling: Exemplified by Caesium in Lakes after Chernobyl*. Heidelberg: Springer-Verlag.

———. 1995. Optimal size of predictive models. *Ecological Modeling* 78: 195–204.

———. 1997. Modeling of radiocesium in lakes: On predictive power and lessons for the future. Pp. 3–45 in G. Desmet et al., editors. *Freshwater and Estuarine Radioecology*. Amsterdam: Elsevier.

————. 1999a. On the principles and factors determining the predictive success of ecosystem models, with a focus on lake eutrophication models. *Ecological Modelling* 121: 139–160.

————. 1999b. Error propagations in step-by-step predictions: Examples for environmental management using regression models for lake ecosystems. *Environmental Modeling and Software* 14: 49–58.

————. 1999c. *Water Pollution: Methods and Criteria to Rank, Model and Remediate Chemical Threats to Aquatic Ecosystems.* Leiden: Backhuys Publishers.

————. 2000. *Modeling Radiocesium in Lakes and Coastal Areas: New Approaches for Ecosystem Modellers.* A textbook with Internet support. Dordrecht: Kluwer Academic.

Håkanson, L., P. Andersson, T. Andersson, Å. Bengtsson, P. Grahn, J-Å. Johansson, H. Kvarnäs, G. Lindgren, and Å. Nilsson. 1990. *Measures to Reduce Mercury in Lake Fish.* Final report from the Liming-Mercury-Cesium project. Swedish Environmental Protection Agency, PM 3818, Stockholm.

Håkanson, L., J. Brittain, L. Monte, R. Heling, U. Bergström, and V. Suolanen. 1996. Modeling of radiocesium in lakes: The VAMP model. *Journal of Environmental Radioactivity* 33: 255–308.

Håkanson, L., and H. Johansson. 2000. Models to predict the distribution coefficient (dissolved and particulate) for phosphorus in lakes. Manuscript. Institute of Earth Science, Uppsala University.

Håkanson, L., and R.H. Peters. 1995. *Predictive Limnology: Methods for Predictive Modelling.* Amsterdam: SPB Academic.

Hamby, D.M. 1995. A comparison of sensitivity analysis techniques. *Health Physics* 68: 195–204.

Helsel, D.R., and R.M. Hirsch. 1992. *Statistical Methods in Water Resources.* Amsterdam: Elsevier.

Hinton, T.G. 1993. Sensitivity analysis of ecosys-87: An emphasis on the ingestion pathway as a function of radionuclide and type of disposition. *Health Physics* 66: 513–531.

IAEA (International Atomic Energy Agency). 2000. *Modelling of the Transfer of Radiocesium from Deposition to Lake Ecosystems.* Report of the VAMP Aquatic Working Group, Vienna, IAEA-TECDOC-1143.

Jørgensen, S.E., and J. Johnsen. 1989. *Principles of Environmental Science and Technology.* 2d ed. Amsterdam: Elsevier.

Lindqvist, O., K. Johansson, M. Aastrup, A. Andersson, L. Bringmark, G. Hovsenius, L. Håkanson, Å. Iverfeldt, M. Meili, and B. Timm. 1991. Mercury in the Swedish environment. *Water, Air and Soil Pollution* 55: 261.

Meili, M. 1991. Mercury in boreal forest lake ecosystems. *Acta University Upsaliensis 336.* Thesis, Uppsala University, Sweden.

Neter, J., W. Wasserman, and G.A. Whitmore. 1988. *Applied Statistics.* 3d ed. Boston: Allyn and Bacon.

Nicholls, K.H., and P.J. Dillon. 1978. An evaluation of phosphorus-chlorophyll-phytoplankton relationships for lakes. *Int. Revue ges. Hydrobiology* 63: 141–154.

Nilsson, Å. 1992. Statistical modelling of regional variations in lake water chemistry and mercury distribution. Thesis, Umeå University, Sweden.

OECD (Organisation for Economic Co-operation and Development). 1982. *Eutrophication of waters. Monitoring, Assessment, and Control*. Paris: OECD.

Peters, R.H. 1986. The role of prediction in limnology. *Limnology and Oceanography* 31: 1143–1159.

———. 1991. *A Critique for Ecology*. Cambridge: Cambridge University Press.

Rose, K.A., R.I. McLean, and J.K. Summers. 1989. Development and Monte Carlo analysis of an oyster bioaccumulation model applied to biomonitoring. *Ecological Modeling* 45: 111–132.

Straskraba, M. 1980. The effects of physical variables on freshwater production: analyses based on models. Pp. 13–84 in E.D. LeCren and R.H. Lowe-McConnell, editors. *The Functioning of Aquatic Ecosystems*. Cambridge: Cambridge University Press.

Straskraba, M., and A. Gnauck. 1985. *Freshwater ecosystems. Modelling and simulation*. Vol. 8 of Developments in Environmental Modelling. Amsterdam: Elsevier.

Vemuri, V. 1978. *Modeling of Complex Systems*. New York: Academic Press.

Vollenweider, R.A. 1968. *The Scientific Basis of Lake Eutrophication, with Particular Reference to Phosphorus and Nitrogen as Eutrophication Factors*. Technical Report DAS/DSI/68.27. Paris: OECD.

Welch, P.S. 1948. *Limnological Methods*. Toronto: Blakiston.

Whicker, F.W. 1997. Measurement Quantities and Units. Preprint from the international workshop on "Measuring Radionuclides in the Environment: Radiological Quantities and Sampling Designs. Bad Honnef, Germany.

Worley, B.A. 1987. *Deterministic Uncertainty Analysis*. Oak Ridge National Laboratory Report ORNL-6428. Oak Ridge.

9

Bayesian Approaches in Ecological Analysis and Modeling

Kenneth H. Reckhow

Summary

Bayesian analysis provides a normative framework for use of uncertain information in decision making and inference. From a practical perspective, Bayes Theorem has a logical appeal in that it characterizes a process of knowledge updating that is based on pooling precision-weighted information.

For years however, Bayesian inference was largely ignored or even discredited in favor of frequentist inference; among the reasons were computational difficulties and the formal use of subjective probabilities in applications of Bayes Theorem. In recent years, new computational approaches (e.g., Markov chain Monte Carlo) have greatly reduced the first problem, while the general recognition of the role of expert judgment in science has at least lessened resistance with respect to the second problem. Beyond that, Bayesian approaches facilitate certain analyses and interpretations that are often important to scientists.

For example, the growing recognition of the value of combining information or "borrowing strength" in ecological studies, as new information is acquired to augment existing knowledge, is one of several reasons why interest in Bayesian inference continues to increase. Many currently used analytic techniques, such as random coefficients regression, multilevel models, data assimilation, and the Kalman filter are focused on this theme; all of these techniques reflect the basic framework of Bayes Theorem for pooling information.

Most ecologists initially are taught that probabilities represent long-run frequencies: A consequence of this perspective is that probabilities have no meaning in a single unique or nonreplicated analysis. Scientists often ignore this constraint and interpret probabilities to suit the particular analysis. Related confusion sometimes arises in classical hypothesis testing and in the interpretation of p values. Bayesian inference provides appealing options in these situations.

Collectively, these developments and perspectives have resulted in an increase in the application of Bayesian approaches in ecological studies, a number of which are noted here. Specific examples dealing with combining information, hypothesis testing, and Bayesian networks are discussed in more detail. In sum, it seems reasonable to make the judgmental forecast that Bayesian approaches will continue to increase in use in ecology.

Introduction

Might the science of ecology be richer and further advanced if research were conducted within a Bayesian framework? This question may not be as absurd as it first appears. After all, Bayes Theorem can be viewed as a logical way to combine information or pool knowledge, so one might reasonably argue that application of that theorem is the appropriate way to integrate new research findings with existing knowledge. Correspondingly, failure to use the theorem may perhaps mean that new scientific knowledge is combined with existing knowledge in an ad hoc, judgmental way. So, over time, one might reasonably expect that the logical, structured approach of Bayes Theorem would advance scientific knowledge to a greater degree than would informal judgmental approaches.

In consideration of that perspective, the objective of this chapter is to present the case for Bayesian analysis as the basis for scientific inference. The next section begins with a description of Bayes Theorem, supported by a discussion of its interpretation, application, and controversies. Following that, several sections are devoted to specific applications in ecology, with examples illustrating the merits associated with a Bayesian approach. The chapter ends with a concluding observation on the role of judgment in scientific analysis.

Bayesian Inference

Bayes Theorem lies at the heart of Bayesian inference: it is based on the use of probability to express knowledge and the combining of probabilities to characterize the advancement of knowledge. The simple, logical expression of the theorem stipulates that, when combining information, the resultant (or posterior) probability is proportional to the product of the probability reflecting a priori knowledge (the prior probability) and the probability representing newly acquired knowledge (the sample information, or likelihood). Expressed more formally, Bayes Theorem states that the probability for y conditional on experimental outcome x (written $p(y|x)$) is proportional to the probability of y before the experiment (written $p(y)$) times the probabilistic outcome of the experiment (written $p(x|y)$):

$$p(y \mid x) \propto p(x \mid y)p(y) \tag{9.1}$$

To fix ideas, suppose an ecologist is interested in the reduction in chlorophyll a in a lake associated with a 30% reduction in phosphorus concentration. She could use existing data from similar lakes to develop a simple chlorophyll-phosphorus regression model and predict the reduction in chlorophyll for the lake of interest. Alternatively, she could conduct dilution experiments on the lake, collecting new data to estimate the quantity of interest. Adopting a third option, a Bayesian ecologist would use "the best of both worlds" by combining the estimates using Bayes Theorem. In the language of that theorem, the regression model would yield the prior probability, since this estimator exists *prior* to the collection of new data, and the posterior probability would represent the revised estimate based on both prior knowledge and new experimental evidence.

At first glance, it seems hard to argue against this seemingly rational quantitative strategy for updating scientific knowledge. Indeed, one might ask why all ecologists aren't Bayesians? There are a number of reasons. Certainly, the most important is that virtually all ecologists still learn probability and statistics from a classical, or frequentist, perspective, and Bayes Theorem is at best a minor topic within that curriculum. Beyond that, Bayesian inference has been widely regarded as subjective and thus not suitable for objective scientific analysis. The problem with this perspective is that most science is hardly the objective pursuit that many choose to believe.

Consider the judgments we make in a scientific analysis. Implicit (or explicit) in the ecologist's lake study on phosphorus and chlorophyll are judgments about the adequacy of the existing lakes' data, the merits of dilution experiments, and the truth of the model relating phosphorus to chlorophyll. There are no purely scientific, absolutely correct, choices here; these represent "gray areas" about which reasonable scientists would disagree. Yet, these also represent judgments that *must* be made by the ecologist in order to carry out the lake study. Ordinary scientists are not unique in their reliance on judgment, however. Tanur and Press (2001) examine the scientific methods in the work of some of the most distinguished scientists (e.g., Galileo, Newton, Darwin, Einstein) in history, noting the substantial role of subjectivity in their work.

Further, consider how most scientists address the revision of scientific knowledge in light of their own new contributions. In some cases, the scientist simply states the conclusions of his work, not attempting to quantitatively integrate new findings with existing knowledge. When integration is attempted in the concluding section of a research paper, it is typically a descriptive subjective assessment of the implication of the new knowledge. Bayesian inference has the potential to make combining evidence more analytically rigorous. It is ironic that the subjectivity of Bayesian analysis would be its undoing.

Applications in Ecology

Bayesian analysis can be understood and applied in science in several different "ways." In perhaps the most fundamental sense, Bayesian inference provides a probability-based, normative approach for updating scientific knowledge based on new information. Less comprehensive but more common, Bayesian analysis in mathematical models supports parameter estimation and yields predictive distributions for quantities of interest. For inferential statements, Bayesian statistical analysis, unlike classical or frequentist statistical analysis, is compatible with the statements that scientists are inclined to make following research as they directly relate to the quantity/parameter of interest (e.g., a reaction rate) rather than to the value of a test statistic (e.g., a t statistic).

As recently as ten years ago, one could probably list all of the Bayesian applications in ecology on a single page; that is clearly no longer true. Thus, no attempt is made here to identify all ecology papers using Bayes Theorem; instead, a representative set is listed below.

For ecologists, it is likely that the most compelling arguments in favor of conducting science in a Bayesian framework result from performance. Simply put, does the Bayesian analysis contribute to better scientific inference? To help assess this, the next four subsections outline ecological applications of Bayes Theorem. These subsections describe four different types of Bayesian analyses; the discussions are largely self-contained so that the particular approaches and applications can be understood without reading all subsections.

Combining Information: An Empirical Bayes Analysis

In ecological studies, investigators often use existing scientific knowledge to specify hypotheses or models and then collect data at a site of interest to test the hypotheses or fit the models. If collateral data from nearby or similar sites exist, it is common practice to use this information to make a judgmental assessment of the support for and against the model/hypothesis but otherwise not to incorporate these collateral data into the analysis in a formal way.

For example, consider the situation in which a state agency has maintained a statewide surface-water-quality monitoring network, and a local community is interested in using some of these data to assess trends in selected contaminants at sites within its jurisdiction. The common practice is to use the data at each site for a site-specific trend analysis, while using data from other nearby sites only in a comparative analysis or discussion. This approach persists despite the fact that if variability in water quality at a site is high, a long record of single-site observation is required to be confident in a conclusion concerning change over time at that site.

A seemingly natural question of interest might be whether collateral data at nearby sites can contribute to the site-specific analysis other than in a comparative study. The answer often is "yes," as a consequence of exploiting the commonality (or exchangeability) among sites. On the one hand, each field site has unique features associated with forcing functions (e.g., watershed conditions and pollutant inputs) and with response functions (e.g., water depth and hydraulic conditions). However, ecological science includes common principles that should lead us to expect similarity in ecosystem response to stresses, and implied in a discussion of response at other nearby sites is often an expectation that these sites have something in common with the site of interest.

As a result, it should often be possible to improve (i.e., reduce inferential error) the single-site analysis by "borrowing strength" from other similar sites. This may be accomplished using an empirical Bayes approach (Maritz and Lwin 1989) or another similar information pooling method (Draper et al. 1992). In empirical Bayes inference, collateral information (which in the above example is the assessment of trends at the other similar sites) is used to construct a "prior" probability model that characterizes this information. Using Bayes Theorem, the prior probability is then combined with a probability model for the trend at the site of interest. In many instances, combining information using empirical Bayes methods yields smaller interval estimates and thus stronger inferences than would result if this information was ignored.

The strategy of "borrowing strength" from other similar analyses is an attribute shared by several statistical methods. Bayesian inference (Box and Tiao 1973; Berger 1985; Gelman et al. 1995), empirical Bayes methods (Martz and Lwin 1989; Carlin and Louis 2000), and the classical method of random coefficients regression (Swamy 1971; Reckhow 1993) all have this characteristic. Bayesian inference, of course, results from the application of Bayes Theorem, which provides a logical framework for pooling information from more than one source. Empirical Bayes (EB) methods also use Bayes Theorem, but otherwise they are more classical (or frequentist) than Bayesian (see Morris 1983, including comments) in that they involve estimators and consider classical properties. In the typical parametric empirical Bayes problem, we wish to simultaneously estimate parameters $\mu_1, \ldots, \mu_p$ (e.g., p means). The EB prior for this problem is often *exchangeable*; that is, the prior belief for each of the $i = 1, \ldots, p$ parameters to be estimated does not depend on the particular value of i (the prior belief is the same for each parameter). With exchangeability, the prior model is assumed to describe a simple underlying relationship among the μ_j, and Bayes Theorem is used to define the EB estimators for the posterior parameters.

Exchangeability in the empirical Bayes application is a particularly useful concept for simultaneous parameter estimation with a system that has a hierarchical or nested structure (Bryk and Raudenbush 1992). Examples of these systems are plentiful. For instance, cross-sectional lake data may arise from individual lakes (at the lowest level of the hierarchy) that are located within ecoregions (at the next level of the hierarchy). Alternatively, individual stream stations may be nested within a stream segment or nested within a watershed. This nestedness implies a structure for the linkage of separate sites or systems that could be exploited in a hierarchical model.

Empirical Bayes descriptions and applications are less common than are Bayesian analyses in the statistics and ecology literature. While most textbooks on Bayesian inference have sections treating EB problems, they tend not to be emphasized, perhaps because they have frequentist attributes and do not require a "true" prior probability. A few books focus on Empirical Bayes methods, notably Martz and Lwin (1989) and Carlin and Louis (2000); beyond that, the previously cited references on similar frequentist methods (hierarchical models and random coefficients regression) should be useful. Other related methods of interest include the Kalman filter (e.g., Collie and Walters 1991) and data assimilation (e.g., Robinson and Lermusiaux 2000). Ecological applications of interest include Reckhow (1993 and 1996), Solow and Gaines (1995), and Whiting et al. (2000).

Incorporating Expert Judgment: A Bayesian Analysis

The conventional approach for the determination of parameters for process-based ecological simulation models is for the modeler to use his judgment in selecting parameters that are consistent with any available data as well as with tabulations of accepted coefficient values (e.g., Zison et al. 1978). Experienced modelers appear to have greater understanding of the built-in correlations and sensitivities, and thus these scientists are usually able to achieve better fits. Models are ultimately judged adequate based on a visual comparison of predictions and observations; formal mathematical optimization is not usually involved.

Parameter estimation in empirical models has traditionally been undertaken using classical optimization methods such as maximum likelihood or least squares. Judgment is typically involved in the specification of the model but not in the actual estimation algorithm.

The difference in approach between these two categories of models has occurred for a few reasons. In some cases, process modelers have believed that the model equations are theoretically correct, and the model parameters are physically measurable quantities that are simply measured in the field and inserted into the model. In other cases, the process models, with large numbers of parameters, were not identifiable without imposing constraints on most of the parameters (and effectively estimating these parameters using expert judgment and literature tabulations). In contrast, empirical models, with one or only a few parameters, are identifiable from the available data, and the empirical modeler believes that the model is more credible if an optimality criterion is used to estimate the parameters.

In situations where available data and expert judgment permit, it often should be to the advantage of the modeler to use both data and judgment in the estimation of model parameters with Bayes Theorem. For example, Reckhow (1988) demonstrated how expert judgment could improve a model of fish population response to acid deposition in lakes (Reckhow et al. 1987) when the knowledge of an expert is elicited and formally incorporated into the model using Bayes Theorem. This example illustrates a number of key features of expert elicitation, so it is useful here to present it in some detail. In Reckhow's study, an expert (Dr. Joan Baker) in fish response to acidification was interviewed to elicit a prior probability for the model parameters. The model was a logistic regression model with the form

$$p(\text{Presence}) = 1/(1 + e^{-\beta x}) \tag{9.2}$$

where p(Presence) is the probability of species presence, β represents the model parameters, and x represents the predictor variables (pH and calcium).

Since with a statistical model scientific experts are more likely to think in terms of the variables (pH, calcium, and species presence/absence) rather than in terms of the model parameters, a predictive distribution elicitation approach (Winkler 1977; Winkler et al. 1978) was used to determine the prior. For this procedure, the expert (Dr. Baker) was given a set of predictor variables and then asked to give her estimate of the median response. A frequency perspective was thought to facilitate response; thus a typical question was "Given 100 lakes that have supported brook trout populations in the past, and if all 100 lakes have pH = 5.6 and calcium concentration = 130 μeq/L, what number do you now expect continue to support the brook trout population?" This question was repeated 20 times with a variety of pH-calcium pairs to yield 20 predicted responses. Twenty was chosen to provide some redundancy to improve characterization of the prior probability yet not burden the expert with time-consuming questions. The pH-calcium pairs were not randomly selected but rather were chosen to resemble the sample data matrix.

The expected response provided by Dr. Baker does not provide a crucial measure of error. Thus, it was assumed that the errors in the conditional response were approximately normally distributed, and additional questions were posed to Dr. Baker to determine fractiles of the predictive distribution, conditional on pH and calcium. A typical question was "For pH = 5.1 and calcium = 90 μeq/L, you estimated that 55 lakes supported brook trout populations. If the odds are 3:1 that the number of lakes (of 100 total lakes) supporting brook trout is greater than a particular value, what is that value?" This question yields the 25th percentile, and other similar questions provide other percentiles. These fractiles were assessed for six conditional y [p(Presence)] distributions, producing six estimates for standard error that are conditional on an assumed known underlying variance (estimated from the data). A thorough description of this probability elicitation and the complete Baycsian analysis can be found in Reckhow (1988).

Of course, the success of a modeling approach is based on how well objectives are achieved. In this case, the model was to be used for prediction, so two prediction-error criteria were examined, and the Bayes model was compared to a classical maximum likelihood model on both the data set used for fitting and a set-aside data set. Based on the error criteria, the Bayes model was nearly comparable to the maximum likelihood model on the fitting data and was far superior on the set-aside data. It seems possible that the judgmental prior adds a level of robustness to the model, conceivably due to the broad knowledge in the mind of the expert.

Perhaps because of the controversial nature of subjective probability, or the relative inaccessibility of technical guidance, there are relatively few papers in ecology presenting a Bayesian analysis using elicitation of a judgmental prior probability. Of particular note, Anderson (1998) presents a thoughtful discussion of Bayesian analysis in ecology from a cognitive science perspective; she makes a strong case for the frequency approach to elicit a subjective prior probability. Crome et al. (1996) and Wolfson et al. (1996) also discuss the elicitation process and conducted studies involving expert elicitation of the prior probability. Finally, a number of additional references may be useful to guide probability elicitation, including Morgan and Henrion 1990, Cooke 1991, and Meyer and Booker 1991.

Hypothesis Testing

Many studies in ecology, both experimental and observational, are designed to assess what may be referred to as a "treatment effect." The treatment effect can pertain to such things as the influence of various factors on growth rate of an organism, the effect of a pollution control strategy on ambient pollutant concentrations, or the effect of a newly created herbicide on animal life. It is common practice in these situations for the scientist to obtain data on the treatment effect and use hypothesis testing to assess the significance of the effect.

In classical or frequentist statistical analysis, hypothesis testing for a treatment effect is often based on a point null hypothesis (which should actually be used only if considered appropriate; see Berger and Sellke 1987). Typically, the point null hypothesis is that there is no effect; it is often stated in this way as a "straw man" (Wonnacott and Wonnacott 1977) that the scientist expects to reject on the basis of the data evidence. To test the null hypothesis, data are obtained to provide a sample estimate of the effect of interest and then to compute an estimate of the test statistic. Following that, a table for the test statistic is consulted to assess how unusual the observed value of the test statistic is, *given* (assuming) that the null hypothesis is true. If the observed value of the test statistic is unusual, that is, if it essentially incompatible with the null hypothesis, then the null hypothesis is rejected.

In classical statistics, this assessment of the test statistic is based on the sampling distribution for the test statistic. The sampling distribution is a probability density function that is hypothetical is nature. In effect, it is a smoothed histogram for the test statistic plotted for a large number of hypothetical samples with the same sample size. Inference in classical statistics is based on the distribution of estimators and test statistics in many (hypothetical) samples, despite the fact that virtually all statistical investigations involve a single sample. This hypothetical sampling distribution provides a measure of the frequency, or probability, that a particular value, or range of values, for the test statistic will be determined for a set of many samples. In classical statistics, we equate this long-run frequency to the probability for a particular sample before that sample is taken.

There are two problems with this approach that are addressed through the use of Bayesian statistical methods. The first is that the hypothesis test is based on a test statistic that is at best indirectly related to the quantity of interest—the truth (or probability of truth) of the null hypothesis. The p value commonly reported in hypothesis testing is the probability (frequency), given that the null hypothesis is true, of observing values for the test statistic that are as extreme, or more extreme, than the value actually observed; in other words:

$$p(\text{test statistic equals or exceeds } k | H_0 \text{ is true}) \tag{9.3}$$

The scientist, however, is interested in the probability of the correctness of the hypothesis, given that he has observed a particular value for the test statistic; in other words:

$$p(H_0 \text{ is true} | \text{test statistic} = k) \tag{9.4}$$

Classical statistical inference does not provide a direct answer to the scientist's question; Bayesian inference does.

The second problem relates to the issue of "conditioning," which concerns the nature of the sample information in support of the hypothesis. Bayesian hypothesis tests are conditioned only on the sample taken, whereas classical hypothesis tests are conditioned on other hypothetical samples in the sampling distribution (more extreme than that observed) that could have been selected but were not. The Bayesian approach, of course, uses more than the sample, as it also incorporates prior information. However, the prior probability, while judgmental, does relate to the hypothesis of interest, whereas the sampling distribution relates to logically irrelevant, hypothetical samples. Clearly, the Bayesian approach is more focused on the problem of interest to the ecologist.

The statement of the hypothesis, the issue of conditioning, and the selection of the prior probability are discussed in a number of references that address Bayesian hypothesis testing. Hilborn and Mangel (1987) discuss hypothesis testing using Bayes Theorem and present several interesting examples. Berry (1996) provides an excellent introduction: Although not concerned with ecology, his detailed example (focusing on basketball free-throw shooting) is a nice illustration of selection of the prior and Bayesian updating. Of particular note, Edwards et al. (1963) and Berger and Sellke (1987) compare p values with Bayesian posterior probabilities, demonstrating that p values tend to overstate the sample evidence against the null hypothesis. Reckhow (1990) illustrates this tendency in an example concerning acidification of lakes.

Bayesian (Probability) Network Analysis

The analyses described in the previous sections are all relatively simple in scope. That is, they consist of univariate analyses or single equation models. Many interesting ecological problems, however, are multivariate and involve complex relationships among variables. To address these issues, probability networks, or Bayes nets, have become one of the most interesting and potentially useful current research directions for the application of Bayesian methods in ecology.

It is common modeling practice to develop a flow diagram consisting of boxes and arrows to describe relationships among key variables in an aquatic ecosystem and to use this diagram as a device to guide model development and explanation. In ecology, this graphical model is typically used to display the flow of materials or energy in an ecosystem. For a probability network, however, the "flows" indicated by the arrows in a graphical model do not represent material flow or energy flow; rather, they represent conditional dependency. Thus, while the probability modeling approach of interest here typically begins with a graphical model, in this case the presence or absence of an arrow connecting two boxes specifies conditional dependence or independence, respectively. An example is given in Figure 9.1 for eutrophication in the Neuse Estuary in North Carolina (Borsuk 2001).

Probability networks like that in Figure 9.1 can be as simple or as complex as scientific needs, knowledge, and data allow. The relationships may reflect direct causal dependencies based on process understanding or a statistical, aggregate summary of more complex associations. In either case, the relationships are characterized by conditional probability distributions that reflect the aggregate response of each variable to changes in its "up-arrow" predecessor, together with the uncertainty in that response. In that regard, it is important to recognize that the conditional independence characterized by absence of an arrow in a Bayes net graphical model is not the same as complete independence, a feature that is likely to be quite rare in systems of interest. For example, ecological knowledge suggests that many of the variables in Figure 9.1 are likely to be interrelated or interdependent. However, the arrows in that figure indicate that conditional on *sediment oxygen demand* and *duration of stratification*, *oxygen concentration* is independent of all other variables. This means that once *sediment oxygen demand* and *duration of stratification* are known, knowledge of other variables does not change the probability for *oxygen concentration* [*p*(oxygen concentration)].

Conditional probability relationships may be based on either (1) observational or experimental data or (2) expert scientific judgment. Observational data that consist of precise measurements of the variable or relationship of interest are likely to be the most useful, and least controversial, information. Unfortunately, appropriate and sufficient observational data may not always exist. Experimental evidence may fill this gap, but concerns may arise regarding the applicability of this information to the natural, uncontrolled

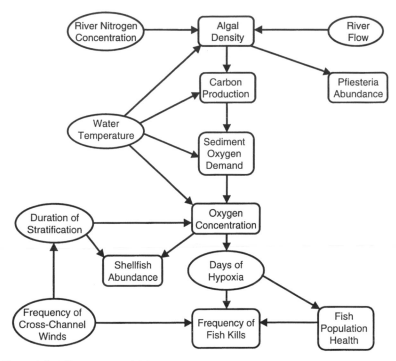

Figure 9.1. A Bayes net model for eutrophication in the Neuse Estuary.

system, and appropriate experimental data may also be limited. As a consequence, the elicited judgment of scientific experts may be required to quantify some of the probabilistic relationships. Of course, the use of subjective judgment is not unusual in water-quality modeling. Even the most process-based computer simulations rely on subjective judgment as the basis for the mathematical formulations and the choice of parameter values. Therefore, the explicit use of scientific judgment in probability networks should be an acceptable practice. In fact, by formalizing the use of judgment through well-established techniques for expert assessment, the probability network method may improve the chances of accurate and honest predictions.

The Bayes net presented in Figure 9.1 as a graphical model has recently been used for development of the total maximum daily load (TMDL) for nitrogen in the Neuse Estuary to address water quality standard violations (NC DENR 2001). Several features of the Bayes net improved the modeling and analysis for the Neuse nitrogen TMDL. To facilitate explanations and to enhance credibility, the underlying structure of the Neuse probability network (reflected in the variables and relationships conveyed in the graphical model) is designed to be consistent with mechanistic understanding. The relationships in this model are probabilistic: they were estimated using a combination of optimization techniques and expert judgment. The predicted responses are also probabilistic, reflecting uncertainty in the model forecast. This is important, as it allows computation of a "margin of safety" which is required by the U.S. Environmental Protection Agency for the assessment of a total maximum daily load. In addition, the Bayes net model can be updated (using Bayes Theorem) as new information is obtained; this capability supports adaptive implementation, which is becoming an important strategy for successful ecological management (Collie and Walters 1991; NRC 2001).

Applications of Bayes nets are becoming more common in the ecological literature. Varis and colleagues (e.g., Varis et al. 1990; Varis 1995) were among the first to apply Bayes nets and the closely related technique of influence diagrams. Reckhow, Borsuk, and Stow (see Reckhow 1999, Borsuk 2001, Borsuk et al. 2001, 2002) present additional details documenting the Neuse Estuary Bayes net models described above. Other interesting recent applications of Bayes nets include Rieman et al. (2001) and Haas (2001). Software, both commercial and free, is available for development and application of Bayes nets; some examples are Genie (www2.sis.pitt.edu/~genie), Netica (www.norsys .com), and Bayesware (www.Bayesware.com).

Other Bayesian Literature of Interest

Recent developments in Bayesian methods are significant; thus, a paper on this topic written five years from now would probably focus on different techniques and have more ecological analyses to report. Of particular significance, the important advancements in computation afforded by Markov chain Monte Carlo (MCMC; see Spiegelhalter et al. 1993; 1995; Congdon 2001; and Meyer and Millar 1999) have greatly expanded the type and complexity of problems amenable to Bayesian analysis. Limited observational data may still prevent Bayesian analysis with large process models, but MCMC methods have removed the computational hurdle. The prospect of the next generation of quantitative ecologists trained in the use of MCMC methods and armed with user-friendly software such as WinBUGS (Spiegelhalter et al. 1995) portends major changes in ecological analysis and modeling.

While MCMC methods represent the most significant recent advancement for Bayesian analysis, there have been other new developments in Bayesian methods. As just one example, nonparametric Bayesian analysis (Lavine and Mockus 1995), which makes use of MCMC methods, allows flexible characterization of a functional relationship, similar to smoothing techniques in exploratory analyses. Qian and Reckhow (1998) and Stow and Qian (1998) present examples of nonparametric Bayesian regression in ecology.

Finally, this is not the first, nor should it be expected to be the last, discussion paper on the merits of Bayesian analysis in science. Of particular note is an excellent set of papers that appeared in *Ecological Applications* (Dixon and Ellison 1996); within this collection, the papers by Ellison (1996) and Dennis (1996) are recommended for additional insights on the strengths and limitations of Bayesian analysis in ecology.

Concluding Thought

As a final point, it must be acknowledged that judgment plays a key role in essentially all areas of statistical and process modeling in ecology. In mechanistic models, judgment is utilized in the specification of equations and in the selection of parameters. In statistical models, judgment is required for the selection of predictor variables and functional relationships. Despite some claims, judgment-free analysis is rare for predictive ecological assessments. Further, if one believes that current scientific knowledge has information to contribute, then the judgmental aspects of Bayesian and empirical Bayes methods are not only desirable but also important for the conduct of good scientific analysis.

Acknowledgments. Mark Borsuk and Craig Stow provided references and useful comments during many fruitful discussions. Mark's work contributed significantly to the section on Bayes nets.

References

Anderson, J.L. 1998. Embracing uncertainty: The interface of Bayesian statistics and cognitive psychology. *Conservation Ecology* 2(1): 2. http://www.consecol.org/vol2/iss1/art2.

Berger, J.O. 1985. *Statistical Decision Theory and Bayesian Analysis.* 2d ed. New York: Springer-Verlag.

Berger, J.O., and Sellke, T. 1987. Testing a point null hypothesis: The irreconcilability of p values and evidence (with discussion). *Journal of the American Statistical Association* 82: 112–122.

Berry, D.A. 1996. *Statistics: A Bayesian Perspective.* Belmont, CA: Duxbury.

Borsuk, M.E. 2001. A Graphical Probability Network Model to Support Water Quality Decision Making for the Neuse River Estuary, North Carolina. Ph.D. dissertation, Duke University.

Borsuk, M.E., D. Higdon, C.A. Stow, and K.H. Reckhow. 2001. A Bayesian hierarchical model to predict benthic oxygen demand from organic matter loading in estuaries and coastal zones. *Ecological Modelling* 143: 165–181.

Borsuk, M.E., C.A. Stow, and K.H. Reckhow. 2002. Predicting the frequency of water quality standard violations: A probabilistic approach for TMDL development. *Environmental Science and Technology* 36(10): 2109–2115.

Box, G.E.P., and G. Tiao. 1973. *Bayesian Inference in Statistical Analysis.* London: Addison-Wesley.

Bryk, A.S., and S.W. Raudenbush. 1992. *Hierarchical Linear Models.* Newbury Park, CA: Sage Publishing.

Carlin, B.P., and T.A. Louis. 2000. *Bayes and Empirical Bayes Methods for Data Analysis.* 2d ed. Boca Raton, FL: Chapman and Hall/CRC.

Collie, J.S., and C.J. Walters. 1991. Adaptive management of spatially replicated groundfish populations. *Canadian Journal of Fisheries And Aquatic Science* 48: 1273–1284.

Congdon, P. 2001. *Bayesian Statistical Modelling.* Wiley Series in Probability and Statistics. Chichester, UK: John Wiley and Sons.

Cooke, R.M. 1991. *Experts in Uncertainty, Opinion and Subjective Probability in Science.* New York: Oxford University Press.

Crome, F.H.J., M.R. Thomas, and L.A. Moore. 1996. A novel Bayesian approach to assessing impacts of rain forest logging. *Ecological Applications* 6: 1104–1123.

Dennis, B. 1996. Discussion: Should ecologists become Bayesians? *Ecological Applications* 6(4): 1095–1103.

Dixon, P., and A.M. Ellison. 1996. Introduction: Ecological applications of Bayesian inference. *Ecological Applications* 6(4): 1034–1035.

Draper, D., D.P. Gaver Jr., P.K. Goel, J.B. Greenhouse, L.V. Hedges, C.N. Morris, J.R. Tucker, and C.M. Waternaux. 1992. *Combining Information: Statistical Issues and Opportunities for Research.* Washington, DC: National Academy Press.

Edwards, W., H. Lindeman, and L.J. Savage. 1963. Bayesian statistical inference for psychological research. *Psychological Review* 70: 193–242.

Ellison, A.M. 1996. An introduction to Bayesian inference for ecological research and environmental decision-making. *Ecological Applications* 6(4): 1036–1046.

Gelman, A., J. Carlin, H. Stern, and D.B. Rubin. 1995. *Bayesian Data Analysis.* London: Chapman and Hall.

Haas, T.C. 2001. A web-based system for public-private sector collaborative ecosystem management. *Stochastic Environmental Research and Risk Assessment* 15: 101–131.

Hilborn, R., and M. Mangel. 1997. *The Ecological Detective.* Princeton, NJ: Princeton University Press.

Lavine, M.L., and A.J. Mockus. 1995. A nonparametric Bayes method for isotonic regression. *Journal of Statistic Planning Inference* 46: 235–248.

Maritz, J.S., and T. Lwin. 1989. *Empirical Bayes Method.* 2d Ed. London: Chapman and Hall.

Meyer, M., and J. Booker. 1991. *Eliciting and Analyzing Expert Judgment: A Practical Guide.* London: Academic Press.

Meyer, R., and R.B. Millar. 1999. BUGS in Bayesian stock assessments. *Canadian Journal of Fisheries and Aquatic Science* 56: 1078–1086.

Morgan, M.G., and M. Henrion. 1990. *Uncertainty.* New York: Cambridge University Press.

Morris, C. 1983. Parametric empirical Bayes inference: Theory and applications. *Journal of the American Statistical Association* 78: 47–65.

NC DENR (North Carolina Department of Environment and Natural Resources). 2001. *Phase II of the Total Maximum Load for Total Nitrogen to the Neuse River Estuary, North Carolina.* Raleigh, NC: NC DENR Division of Water Quality.

NRC (National Research Council). 2001. *Assessing the TMDL Approach to Water Quality Management.* Committee to Assess the Scientific Basis of the Total Maximum Daily Load Approach to Water Pollution Reduction, Water Science and Technology Board. Washington, DC: National Academy of Sciences.

Qian, S.S., and K.H. Reckhow. 1998. Modeling phosphorus trapping in wetlands using nonparametric Bayesian regression. *Water Resources Research* 34: 1745–1754.

Reckhow, K.H. 1988. A comparison of robust Bayes and classical estimators for regional lake models of fish response to acidification. *Water Resources Research* 24: 1061–1068.

———. 1990. Bayesian inference in non-replicated ecological studies. *Ecology* 71: 2053–2059.

———. 1993. A random coefficient model for chlorophyll-nutrient relationships in lakes. *Ecological Modelling* 70: 35–50.

———. 1996. Improved estimation of ecological effects using an empirical Bayes method. *Water Resources Bulletin* 32: 929-935.

———. 1999. Water quality prediction and probability network models. *Canadian Journal of Fisheries and Aquatic Sciences* 56: 1150–1158.

Reckhow, K.H., R.W. Black, T.B. Stockton Jr., J.D. Vogt, and J.G. Wood. 1987. Empirical models of fish response to lake acidification. *Canadian Journal of Fisheries and Aquatic Science* 44: 1432-1442.

Rieman, B.E., J.T. Peterson, J. Clayton, P. Howell, R. Thurow, W. Thompson, and D. Lee. 2001. Evaluation of the potential effects of federal land management alternatives on the trends of salmonids and their habitats in the Interior Columbia River Basin. *Forest Ecology and Management* 5501: 1–20.

Robinson, A.R., and P.F.J. Lermusiaux. 2000. Overview of data assimilation. Harvard University Reports in Physical/Interdisciplinary Ocean Science. Report no. 62. Harvard University. Cambridge, MA.

Solow, A.R., and A.G. Gaines. 1995. An empirical Bayes approach to monitoring water quality. *Environmetrics* 6(1): 1–5.

Spiegelhalter, D.J., A.P. David, S.L. Lauritzen, and R.G. Cowell. 1993. Bayesian analysis in expert systems (with discussion). *Statistical Science* 8: 219–283.

Spiegelhalter, D.J., A. Thomas, N. Best, and W.R. Gilks. 1995. *BUGS: Bayesian Inference Using Gibbs Sampling*. Version 0.50. Technical report. Medical Research Council Biostatistics Unit. Institute of Public Health. Cambridge University.

Stow, C.A., and S.S. Qian. 1998. A size-based probabilistic assessment of PCB exposure from Lake Michigan fish consumption. *Environmental Science and Technology* 32: 2325–2330.

Swamy, P.A.V.B. 1971. *Statistical Inference in a Random Coefficient Regression Model*. Berlin: Springer Verlag.

Tanur, J.M., and S.J. Press. 2001. *The Subjectivity of Scientists and the Bayesian Approach*. New York: Wiley.

Varis, O. 1995. Belief networks for modeling and assessment of environmental change. *Environmetrics* 6: 439-444.

Varis, O., J. Kettunen, and H. Sirvio. 1990. Bayesian influence diagram approach to complex environmental management including observational design. *Computational Statistics and Data Analysis* 9: 77–91.

Whiting, D.G., D. Tolley, and G.W. Fellingham. 2000. An empirical Bayes procedure for adaptive forecasting of shrimp yield. *Aquaculture* 182: 215–228.

Winkler, R.L. 1977. Prior distributions and model-building in regression analysis. Pp. 233–242 in A. Aykac and C. Brumat, editors. *New Developments in the Applications of Bayesian Methods*. North Holland: Amsterdam.

Winkler, R.L., W.S. Smith, and R.B. Kulkarni. 1978. Adaptive forecasting models based on predictive distributions. *Management Science* 24: 977–986.

Wolfson, L.J., J.B. Kadane, and M.J. Small. 1996. Bayesian environmental policy decisions: Two case studies. *Ecological Applications* 6: 1056–1066.

Wonnacott, T.H., and R.J. Wonnacott 1977. *Introductory Statistics*. New York: John Wiley and Sons.

Zison, S.W., W.B. Mills, D. Diemer, and C.W. Chen. 1978. *Rates, constants, and kinetic formulations in surface water modeling*. Report of the U.S. Environmental Protection Agency. EPA-600/3-78-105. Environmental Protection Agency. Athens, GA.

10

Model Validation and Testing: Past Lessons, Present Concerns, Future Prospects

Robert H. Gardner and Dean L. Urban

Summary

There is extensive literature on model validation and testing in eco-system science. In spite of this solid foundation, a survey of the current literature reveals that the practice of model validation and testing has not matured over the past twenty-five years. Few exemplars can be found that rigorously explain the rationale behind model formulation, the methods of parameter estimation, and the procedures for model testing. It is more common to find that the problems and pitfalls of validation and testing are poorly understood, inadequately executed, or entirely ignored. Among the reasons for this are the accelerating pace of technologies for model development has not been matched by the parallel development of methods for model analysis and testing; a trend in increased model complexity often requires the inclusion of critical variables that are unmeasured or unmeasurable (e.g., dispersal in population and metapopulation models); and the short duration of most research projects provides insufficient time and resources for adequate testing of model performance. These challenges are particularly daunting for models that are implemented spatially over large regions (e.g., landscape simulators). These difficulties are further exacerbated by the fact that no single method is suitable for testing all possible models and modeling objectives.

Here we review a heuristic approach to model evaluation and present a general method for constructing informative model–data comparisons. Our premise is that models never work in their entirety (they always simplify reality to some extent), but even inadequate models can be used effectively so long as their behavior is well understood. This approach builds a richer appreciation for *how* a model matches data and the circumstances under which the model performs best. This fuller understanding of model–data comparisons allows client users to choose cases where the model can be used confidently or to take advantage of model bias to obtain conservative results. We believe that a generalized approach to a broad variety of models, including complex ecosystem simulators, must be developed. There are pressing needs for formal research on model construction, evaluation, and testing that should be a high priority for all ecosystems scientists.

Introduction

Even the most casual observations indicate that ecological systems are complex and difficult to describe. Because they have many components whose interactions are uncertain, their behavior may vary widely from time to time and place to place. The development of an abstract representation of an ecosystem (i.e., a model) is probably the best, first step in understanding their dynamics (Caswell 1988). The complexity of these models may range from simple, idealized conceptualizations to detailed, quantitative formalizations. However, no matter what degree of complexity is involved, model usefulness can only be assessed by objective evaluation of design and performance. The relevant criteria for design are usually based on qualitative assessments of the adequacy of model structure (i.e., the equations) as representations of relevant ecological variables and interactions, while criteria for performance usually revolve around the quantitative comparison of predictions against independently gathered information. As simple as this scheme appears, the philosophical and practical issues of verifying design and testing predictions remains largely unresolved in spite of extensive emphasis on these issues since the early days of ecological modeling (Levins 1966; Goodall 1972; Wiegert 1975; Caswell 1976; Overton 1977).

Semantic inconsistencies in the ecological literature have been an important source of confusion regarding the design, testing, and application of ecological models (Rykiel 1996). These inconsistencies may be traced, in part, to the frequent borrowing by ecologists of techniques and technologies from other disciplines. For example, early developments in ecosystem modeling relied heavily of the development of analog computers and mathematical techniques designed by electrical engineers (e.g., Olson 1963; Funderlic and Heath 1971; Child and Shugart 1972). This heritage explains why early diagrams of ecosystem models are similar to the wiring diagrams used to program analog computers—a formalism still familiar to users of Stella® modeling software (HPS 2000). However, the process of borrowing from other fields has often eroded the generality and rigor of the original applications. For instance, the term "sensitivity analysis," originally defined in engineering as the partial differential of model response to changes in model parameters (Tomovic and Karplus 1963), is now generally used as a term describing the analysis of model response to any "realistic" set of parameter perturbations (Rose and Swartzman 1981). Although these two techniques share the same term, the methods of analysis and the conclusions that may be drawn are quite distinct (Gardner et al. 1981; Dale et al. 1988).

Establishing the usefulness of models as a means of improving our understanding or predicting ecosystem dynamics requires the use of objective measures of performance. Discussions of this process, often referred to as "model validation," have been surrounded with an inordinate degree of confusion (for general reviews see Oreskes et al. 1994, Rykiel 1996). Based on etymological arguments, a "valid" model denotes that all model behaviors totally correspond to observed system behavior (e.g., Mankin et al. 1975). Because all models are simplifications of reality and are thus, to some degree, incorrect, model validation in the strictest sense is impossible (Oreskes et al. 1994). Faced with this impossibility, resourceful individuals have simply redefined the term "validation" to mean "the comparing of model output . . . with actual historical data" (Miller et al. 1976; Aber 1997; Wilder 1999). Even looser definitions may be found, including what Rykiel (1996) has termed "face validity"—the increased level of confidence in model performance derived from extensive experience and use. These problems might seem comical were it not for the fact that modelers continue to comply with demands of empiricists to "validate models" by comparison with field data (Rykiel 1996). Sadly, such comparisons often involve no more than simple graphical comparison of predictions with available data. Although such graphs may seem visually informative, such comparisons can be misleading because output that "looks good" may actually fail to represent critical aspects of system dynamics. For example, Urban et al. (2000) illustrated a strikingly good "visual test" of their model against independent data, but a conservative randomization test revealed that model output was often significantly different from the data.

Considerable confusion regarding model testing may also arise from the failure to recognize that methods and conclusions will vary as a function of the degree (or stage) of model development. To illustrate this relationship, Table 10.1 provides a heuristic overview of six stages of model development from model conceptualization (stage 1) to application (stage 6). Model testing occurs throughout this process, but only during the final stages (5 and 6) is it appropriate to statistically compare predictions against measurements. Earlier stages focus instead on measures of model structure (stage 1), parameter sensitivity (stage 2), model adequacy (stage 3), and hypothesis testing (stage 4). Performing tests remains an important challenge, but relevant techniques are available (Table 10.2).

The problem of model evaluation and testing has escalated because of recent advances in the acquisition of geospatial data and the modeling of ecosystem dynamics at landscape-to-global scales. Although the consequences of errors and uncertainties for spatial estimates have been considered (e.g., Henderson-Sellers et al. 1985; Turner et al. 1989; Costanza and Maxwell 1994; Cherrill and McClean 1995; Burrough et al. 1996) and issues of prediction across scales recognized (e.g., Hall et al. 1995; Kemp et al. 2001), methods for model testing

Table 10.1. Key stages in model development, testing and application.

Stage of Development[1]	Tools for Analysis	Information Derived
1. Conceptualization and selection	Mathematical and graphical analysis	Class of dynamics defined
2. Parameter estimation	Calibration[2]	Adequacy of model representation quantified
3. Parameter refinement	Formal sensitivity analysis,[3] uncertainty analysis	Important parameters and processes identified
4. Model evaluation	Preliminary experiments, evaluation of uncertainties	Alternative hypotheses, scenarios may be tested
5. Validation	Statistical comparison of model predictions with independent data	Reliability of model predictions established
6. Application	Simulation of relevant management and/or policy scenarios, synthesis of results	Understanding of system dynamics

[1] These stages of model development are presented for heuristic purposes only. The development of most models is iterative: failures at any stage usually result in model redevelopment.

[2] Calibration refers to a wide range or curve-fitting exercises designed to choose the best model formulations and parameter sets.

[3] Sensitivity analysis is defined in the strict sense as the partial derivative of model response to changes in model parameters (Gardner 1984). Other forms of parameter perturbation experiments may be useful for identifying model dynamics over a broad range of conditions.

which can simultaneously consider errors in both data and models have been primarily based on developments within the remote sensing arena (Campbell 1996). Many issues remain unresolved as, once again, numerical techniques are "borrowed" by ecologists to be applied to these situations. This chapter evaluates this new interface, attempts to define the key issues around the use and application of spatial data for model testing, and presents a general framework for testing and evaluation of spatial models.

Table 10.2. Key tools for model testing during development and application.

Tool for Analysis	References
Loop analysis (model structure) Goodness of fit tests based on regression[1]	Lane 1986; Giavelli et al. 1988 Leggett and Williams 1981; Power 1993; Janssen and Heuberger 1995; Smith and Rose 1995; Haefner 1996
Goodness of fit based on likelihoods, Bayesian methods[2]	Gelman et al. 1995; Hilborn and Mangel 1997; Burnham and Anderson 1998
Sensitivity analysis[3]	Gardner et al. 1981; Caswell and Trevisan 1994; Haefner 1996; Klepper 1997
Error or uncertainty analysis[4]	Hakanson 1996, Monte et al. 1996; Hwang et al. 1998
Spatial error analysis	Costanza and Maxwell 1994; Henebry 1995

[1] These methods are general and can be used to fit equations for components of a model, or to compare output from a model to data for verification (calibration) or validation using independent data.
[2] See also Reckhow (this volume).
[3] Many papers on sensitivity analysis also treat uncertainty or error analysis as well.
[4] Error propagation techniques can also be used to incorporate uncertainty into model forecasts.

Comparing Models against Data

The process of evaluating model performance would seem to be a simple and straightforward problem. Simulation results and empirical measurements are summarized and the degree of agreement statistically assessed; if significant agreement occurs, then model performance is satisfactory, while lack of agreement indicates that the model has failed. However, for any given model–data comparison, ambiguities may exist that make the outcome uncertain and inconclusive. These ambiguities often revolve around three general questions: (1) Do the data for testing provide an unbiased, comprehensive measure of system behavior? (2) Is the nature of the comparison consistent with the objectives for which the model was developed? and (3) Will the method of model–data comparison provide a robust measure of agreement?

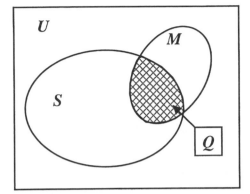

Figure 10.1. Venn diagram of the universe of possible behaviors of an ecosystem (U), the domain of model predictions (M), and the region of observed system behavior (S). The cross-hatched region (Q) represents the overlap between M and S.

Mankin et al. (1975) developed an intuitive graphical scheme that sheds light on many of the issues. They used Venn diagrams (Figure 10.1) to represent the potential interactions among U, the universe of possible outcomes of ecosystem behavior; S, the set of ecological dynamics actually observed; and M, the set of model behaviors that one wishes to evaluate. The region over which M and S intersect defines the domain of Q, the region of agreement between the model and observed system behavior. When the domain of M and S are specified, two important ratios can be estimated: First, the ratio of Q to S that estimates model *adequacy*, the degree to which the model "explains" system behaviors. Adequacy increases as Q increases because a larger proportion of S is reproduced by the model simulations, M. The second quantity is model *reliability*, the ratio of Q to M. Reliability increases as the portion of M outside of S decreases. Therefore, the extent to which Q is "nonempty" establishes the degree to which the model is useful for predicting system behavior. It is interesting to note that this scheme avoids the term "valid" in favor of two more specific quantities, *adequacy* and *reliability*. Under this scheme the process of model-data comparison is an ongoing effort with successive (and continuous) comparisons establishing the domains of M, S, and Q.

The use of Venn diagrams is appealing because it makes it easy to envision the use of models for a variety of purposes (Table 10.1). For example, a model wholly contained within S is completely reliable because it never makes a prediction that is either novel or unprecedented. Such a model is perfectly adequate and reliable—and also perfectly boring. By contrast, excitement is sometimes generated when model performance is less reliable; that is, a model that makes reasonable predictions that fall outside the domain of S. This is a familiar experience in disciplines that have predicted and then discovered new planets or subatomic particles. Indeed, one may argue that science progresses best when "wrong" predictions (i.e., M not within S) are taken seriously, resources are devoted to new data collection efforts, and predictions are verified by extending the domain of S. This excitement, of course, comes at the cost that these predictions might actually be wrong—a risk that is acceptable only for specific, exploratory applications. Applied problems, such as environmental risk assessment or biological conservation, require greater reliability in model predictions.

The construct developed by Mankin et al. (1975) also provides insight into the issues raised above regarding potential ambiguities in model–data comparisons. The first two issues involve the certainty with which the boundaries around S and M are described. When either boundary is uncertain estimates of adequacy (Q/S) or reliability (Q/M) will be imprecise. The clarity of the boundaries of M are determined by the evaluation of model assumptions and by the adequacy of parameter estimation to simulate S. The boundaries of S are, of course, determined by the nature of the data used to describe system dynamics. Therefore, because data are required to define both M and S, these data must satisfy two criteria: (1) The set of behaviors being described empirically must match the objectives and assumptions used to develop the model. For instance, a model of ecological succession might assume constant (or average) environmental conditions. If significant climatic events (e.g., droughts or floods) have occurred that modify patterns of species composition, then the data describing these responses should not be used to test model performance (i.e., define the domain of Q). Of course the problem is that it is very difficult to obtain data that meet all (or even most) of the model assumptions and objectives. Nevertheless, it is contingent on the investigator to recognize that data that fail to satisfy this criterion are inadequate for model testing. (2) The data for testing must also be independent of the data used to parameterize the model. This is a familiar topic that revolves around the problem of a "fitted" model simply regenerating the data from which it was parameterized (Rykiel 1996; Rose and Smith 1998). One must also be certain that the use of novel data does not measure behaviors of S that are outside the domain originally used to develop and calibrate the model. Data-splitting avoids these problems (Rossi et al. 1993; Deutsch and Journel 1998) by providing a consistent set of observations for model parameterization and testing.

The third source of ambiguity revolves around the uncertainties associated with the selection and application of statistical tests designed to measure model performance. Because model development typically involves trade-offs among precision (e.g., a detailed focus on specific ecological processes), complexity (e.g., consideration of multiple processes and state variables), and generality (e.g., extension of results to other conditions or ecosystems), it is important to explore the implications of these trade-offs as they affect model adequacy and reliability. Therefore, it is not surprising that a large number of potential tests have been suggested and applied (Table 10.2). The diversity of tests makes the selection of an appropriate method difficult—and hence conclusions to be drawn from such tests—confusing and uncertain. A rather subtle, but dangerous, problem also exists: close scrutiny from specific model–data comparisons may lose sight of the larger issue of how well a model performs over a broad range of conditions and simulation scenarios. This is, of course, the classical case of losing sight of the forest by concentrating too much on individual trees.

Table 10.3. The confusion matrix for binary event classifications.

		Predicted classes		
		0	1	
Observed classes	0	A	C	$O_{(-)}$
	1	B	D	$O_{(+)}$
		$P_{(-)}$	$P_{(+)}$	N

Source: Adapted from Vayssières et al. (2000).
Note: Class 0 and Class 1 indicate that the event of interest did or did not occur, respectively. The symbols in the table indicate: A, the number of cases predicted as Class 0 that are in Class 1; B, the number of cases predicted as class 0 that are in class 1, etc.; $P_{(-)}$ and $P_{(+)}$ are the number of cases predicted as class 0 and class 1, respectively; $O_{(-)}$ and $O_{(+)}$ are the number of cases observed in class 0 and class 1, respectively; and $N = A + B + C + D$ = the total number of comparisons. Lower case symbols in the text indicate values normalized that are normalized to N, for instance: $a = A/N$, $p_{(-)} = P_{(-)}/N$, etc.

Simple Comparisons

The simplest type of comparison between models and data involves the classification of results into discrete categories. The difficulty with this scheme is that most ecosystem measurements and predictions involve continuous rather than discrete variables (e.g., measures of respiration, photosynthesis, ecosystem productivity, etc.). The transformation from continuous to discrete can be effected via a quantitative criterion used to simultaneously evaluate and place model and data results into two classes: a "negative" state where results fall below the criterion and a "positive" state where results are above the criterion. Such binary classifications are clear and remarkably flexible, allowing them to be broadly applied. A binary classification rule yields four possible outcomes: (1) the response of interest (i.e., values above the criterion) was neither observed nor predicted; (2) observations showed a positive response while the model failed to predict a similar result; (3) model predictions were positive but observations negative; and (4) model predictions and measurements agreed that a positive response occurred. These four situations are illustrated in Table 10.3 along with the marginal totals: the number of observed negative and positive outcomes, $O_{(-)}$ and $O_{(+)}$, respectively, and the number of negative and positive outcomes predicted by the model, $P_{(-)}$ and $P_{(+)}$, respectively. This matrix, often referred to as a contingency table or the "confusion" matrix (Campbell 1996), completely describes a criterion-based model–data comparison.

Table 10.4. A confusion matrix for a hypothetical example where the frequency of positive events observed is small and model fails to predict these events.

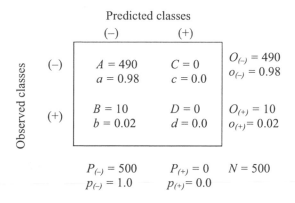

Predicted classes

	(−)	(+)	
(−)	$A = 490$ $a = 0.98$	$C = 0$ $c = 0.0$	$O_{(-)} = 490$ $o_{(-)} = 0.98$
(+)	$B = 10$ $b = 0.02$	$D = 0$ $d = 0.0$	$O_{(+)} = 10$ $o_{(+)} = 0.02$

$P_{(-)} = 500$ $P_{(+)} = 0$ $N = 500$
$p_{(-)} = 1.0$ $p_{(+)} = 0.0$

(Observed classes — left vertical label)

Note: Values in upper case are the observed counts for the hypothetical example presented in the text. The lower case symbols are the relative frequencies (i.e., $a = A/N$, etc.).

Many indices can be developed from the information summarized in the confusion matrix. For instance, the relative number of true positives, $d = D/O_{(+)}$, the relative number of true negatives, $a = A/O_{(-)}$, and the relative error rate, $e = (B + C)/N$, all measure aspects of model performance. However, caution should be used as it is easy to construct a meaningless test from the confusion matrix. To illustrate, imagine that an outbreak of a forest insect has occurred in 2% of 500 measured sites (the criterion for an outbreak is a 25% defoliation of trees in each forest stand). Further imagine that a set of model simulations have been produced to anticipate this situation (Table 10.4). A superficial inspection of Table 10.4 indicates that model results are quite similar to the measured values: The number of "true negatives" equals 490, the exact number of observed sites that were measured in that category (i.e., $O_{(-)} = 490$, Table 10.3); the "true positives" ($D = 0$) are incorrect but the error rate, $(B + C)/N$, is only 0.02, or 2% of the total. One might conclude that model provides a sufficiently accurate overview of what actually occurred. However, further reflection shows a fallacy in this conclusion—surely a model *that predicts nothing at all* cannot be regarded as a satisfactory predictor!

A more robust measure of model performance is provided by the kappa statistic (Lynn et al. 1995; Monserud and Leemans 1992, Pontius 2000). Kappa measures the level of agreement between sets of observations corrected for chance and normalized to the maximum possible (i.e., the perfect predictor). Thus, kappa is equal to $(P_i - P_c)/(1.0 - P_c)$, where P_i is the sum of the proportions correctly predicted and Pc is the sum of proportions expected by chance alone (Pontius 2000). The kappa statistic has the useful property that values near 0 indicate results that may be achieved by chance alone, while those greater than 0 indicate results with significant departures from chance. For the binary confusion matrix in Table 10.4, $P_i = (a + d)$ and $P_c = (O_{(-)} \times P_{(-)} + (O_{(+)} \times P_{(+)})$. Using the calculation for kappa we find that $(0.98 - 0.98) / (1.0 - 0.98) = 0.0$, a result quite clearly not different from random. Although we need not test the significance of this kappa statistic, methods have been developed to do so (Pontius 2000).

One shortcoming of the kappa statistic is that it is based only on the agreement between model and data and, thus, is unable to identify where the comparison may have failed. For instance, a simple rearrangement of the "failures" (i.e., switching the values of b and c in Table 10.3) has no effect on the value of the kappa statistic. This deficiency can be addressed by the estimation of two quantities: *sensitivity* (Se) and *specificity* (Sp). *Sensitivity*, which measures the relative number of true positives predicted, is estimated as $a/(a + c)$; while *specificity*, which measures the relative accuracy of estimating true negatives, is estimated as $d/(b + d)$. The sum of Se and Sp behaves in a manner similar to the kappa statistic: when (Se + Sp) are approximately equal to 1.0 the results are no more predictive than a random model; when (Se + Sp) < 1.0 the results are less predictive than a random model; when (Se + Sp) ≥ 1.0 the results indicate a level of agreement between model and data that is greater than chance (Pearce and Ferrier 2000; Vayssières et al. 2000). For our example (Table 10.4), (Se + Sp) = 1.0, a value that, similar to kappa = 0.0, indicates results not different from chance alone. However, the rearrangements of b and c alters the values of Se and Sp (without changing the sum), providing insight into changes that are not considered by the kappa statistic. The responsiveness of *Se* and *Sp* is logically consistent with concepts illustrated in the Venn diagram (Figure 10.1): b measures values in S that are not in Q while c measures values in M not in Q. However, because Mankin et al. (1975) do not explicitly consider a of the confusion matrix (i.e., events outside the region of both S and M, Table 10.4) sensitivity (Se) can only be approximated by $(U - S - M) / (U - S)$ and specificity (*Sp*) by Q/S.

Interesting possibilities for model testing occur when the criterion for comparison is allowed to vary. This happens, of course, when we are not certain what constitutes a "positive" event. For instance, in the above insect defoliation example, the insect may be ubiquitous and always defoliating trees to some extent. If this is the case, then it is useful to compare model performance when the criterion for defoliation varies. For each value of the criterion a unique confusion matrix can be produced, with Se and Sp estimated for each criterion value used. The multiple outcomes of the family of confusion matrices can then be plotted (Figure 10.2) and optimum model performance determined. The resulting curve produced by this family of decision criteria has been termed a receiver-operator characteristic (ROC) curve in the signal-detection literature (Vayssières et al. 2000; Pontius and Schneider 2001).

Once again numerical methods developed independent of ecosystem studies prove to be useful for ecological applications. The ROC curves have two distinct advantages: (1) ROCs evaluate models against data across a range of criteria; and (2) each comparison uses the same criterion to simultaneously evaluate the model and the data. The key is, of course, the sliding criteria that may seem new and unfamiliar. However, many habitat classification models are based on logistic regressions that perform binary classifications of sites into "habitat" or "nonhabitat." The binary nature of the outcome is deceiving because the logistic regression actually predicts a continuous response of the probability of a site being a given habitat type. By convention, we usually "decide" the classification with an arbitrary criterion that results exceed the probability 0.5 for a positive event to be recorded and the site labeled as suitable habitat. Other classification methods, including two-group discriminant functions, also collapse continuous probabilities into discrete binary outcomes. However, it is often reasonable to use other values for this decision threshold. If we reduce the probability from 0.5, more samples will be classified as habitat and fewer as nonhabitat, resulting in a rearrangement of the map. Reducing the threshold to the extreme value of 0 results in all sites being classified as habitat, as the example in Table 10.4 illustrates. Reciprocally, if we raised the threshold to a much higher probability, fewer actual habitat samples would be classified as such, but also fewer nonhabitat samples would be misclassified as habitat. In the extreme, we could classify everything as nonhabitat and never make a mistake with a nonhabitat sample.

These two extremes represent the endpoints of a ROC curve (Figure 10.2), and also illustrate the trade-offs in calibrating or tuning a classification model to a particular goal. In fact, the ROC curve describes the shape of this tuning curve. Two important points should be kept in mind: (1) There exists a tuning of the model that will maximize its classification accuracy (the outermost point of the convex ROC curve), which might not be located at the default threshold value. (2) Any point along the ROC curve can be a "valid" tuning of the model for specific applications. For example, if we predict potential habitat for an endangered species whose exact requirements are uncertain, then it may be desirable to tune the model to overpredict habitat area and thus provide a map indicating all potential areas that might merit special management consideration. Conversely, if we wish to highlight only the very best habitat for active protection, a more restrictive tuning might be preferred. In either case, the most useful tuning of the model is not the optimal tuning. The key point is that, for each model application, a specific criterion for model–data comparison must be formulated and consistently applied.

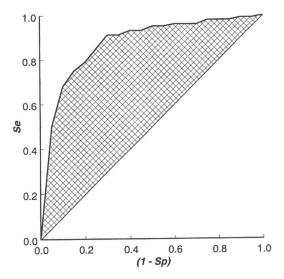

Figure 10.2. An example ROC curve generated by comparing model predictions against system observations. Sensitivity (Se), plotted on the ordinate, is the fraction of true positives while reliability (1–Sp), plotted on the abscissa, is measured as 1.0 – ratio of true negatives. The straight line under the curve is formed by the line Se + Sp = 1.0 and represents the frequencies of positive and negative results generated by a random model. The crosshatched region indicates the degree of departure of the model results from the random model. Models that behave better than random would lie in the upper left quadrant of the figure. A perfect model would have provided only "true" and never "false" results, and thus Se = 1.0 and (1–Sp) = 0.0. (Figure adapted from Vayssières et al. 2000.)

Generalizing the ROC Protocol

The examples we have used to construct the ROC curve (Figure 10.2) involve a single criterion and thus constitute a one-dimensional comparison between model and data. However, many ecological problems exist that require multi-dimensional criteria to evaluate model performance. For instance, variability in space might be described by a variety of attributes (e.g., geology, topographic position, community type) display a variety of states (e.g., different land-cover types, degrees of species persistence, etc.), or involve temporal as well as spatial changes (e.g., rate of invasive spread, time to recovery from disturbance, etc.). A simple, elegant method of comparison for multidimensional contrasts would appear possible using the ROC approach, but as far as we can determine, multidimensional extensions to ROCs have not been attempted. We offer the following suggestions in the hopes of stimulating such developments.

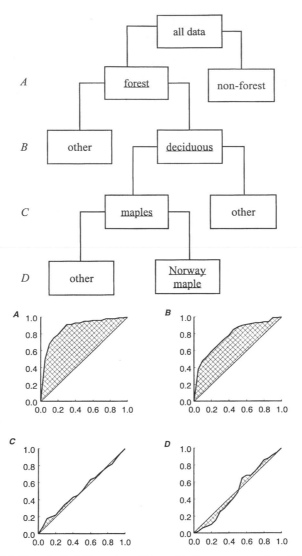

Figure 10.3. Hypothetical illustration of hierarchically arranged series of ROC analyses. The top figure shows the progressive subsetting of data into *A*: forest species; *B*: deciduous species; *C*: all maple species; and *D*: only Norway maple. The ROC curves for each level of discrimination are given in the bottom figure. The area under the curve is greatest for *A* (forests), and remains high for *B* (deciduous).The ROC curves for *C* and *D* indicate that this model is not useful for discriminating maples from other deciduous species or Norway maples from other maple species.

A Family of ROCs. The simplest situation involving multidimensional criteria is where several states, rather than the binary states of the above examples, may be observed and predicted. This occurs, for example, in land-cover maps with multiple cover types defined. Separate ROCs may be developed for each cover type by iterative map reclassifications, with classes being either "the cover type of interest" or "all other cover types." A family of ROC curves will be produced, one for each cover type. The danger of this approach is a lack of independence between tests, with type II errors undefined. Potential areas for research involve the development of methods (e.g., orthogonal comparisons)what would specify and control errors due to multiple comparisons.

Multivariate Metrics. The binary transformation of a continuous variable can be generalized to the multivariate case by collapsing a multivariate index into a binary outcome according to specified threshold. For example, one might compute a multivariate index of species composition, perhaps using ordination techniques. Likewise, one might synthesize multiple variables into principal component (PCA) scores. One could then readily partition the ordination space or principal components into discrete binary domains. For example, PCA scores could be rendered binary by assigning all positive scores a 1 and all others, 0. An ROC curve could then be generated by systematically changing this threshold.

Hierarchical ROCs. Arranging the process of data reclassification into a hierarchical format (Figure 10.3) allows one to evaluate the change in model adequacy with changes in the specificity of the decision criterion. Figure 10.3 illustrates a hypothetical example for a model whose adequacy (Q/S) is high when the criterion is general (i.e., "forest" versus "nonforest") but declines as the criterion becomes more specific (i.e., "Norway maple" versus "other maples"). The successive ROCs may indicate that adequacy declines to that of a random model when predictions fail to distinguish between "maples" and "other deciduous types" (level c, Figure 10.3). A number of interesting research questions emerge from this example, including the determination of the statistical power associated with this procedure and possible relationships with classification and regression tree (CART) methods (Chipman et al. 1998, Vayssières et al. 2000).

Interdependent Dimensions. A different situation exists when we wish to use ROC curves to assess the adequacy of model predictions in both the spatial and temporal dimensions. Examples of such possibilities abound, including the movement of an ecotone with climate change, the spread of urbanization into surrounding suburban and agricultural zones, or ecological succession due to disturbances within heterogeneous landscapes. The essential question in these cases is a conditional one: Did the system change state and, if so, over what interval of time? The "data" to test such a model would be a series of spatially distributed sites where temporal observations of each ecological transitions are recorded. The interdependency of time and space becomes clear when one attempts to form an ROC: a lower threshold for change results in a quicker ecosystem response. Thus, the variation of a single criterion results in changes in both the spatial and temporal dimensions. The resulting ROC would be a three-dimensional dome, allowing an assessment of model adequacy across the range of the combined (and perhaps interactive) criteria.

Conclusions and Prospectus

Many thoughtful and insightful articles have been written concerning model validation and testing since the earliest days of systems analysis. These articles have provided extensive discussions of the practical, philosophical, and semantic difficulties involved in verifying the usefulness and reliability of ecological models (e.g., Oreskes 1994; Rykiel 1996). Although a broad spectrum of methods has been suggested to meet this challenge (e.g., Table 10.2), a general approach for model–data testing has yet to emerge. We have illustrated in this chapter a potentially general approach that tests model performance across a spectrum of possible evaluation criteria. The resulting curves (referred to as receiver-operator characteristics or ROC curves) provide insight into model performance across a series of model–data comparisons. Although our examples have been centered on predictions of habitat type, the performance of any model may be evaluated if sufficient data are available and empirically defined evaluation criteria (or "decision rules") can be estimated. Nevertheless, the challenge remains to fully generalize these methods to all models—especially complex ecosystem simulators for which some element of "tuning" (calibration) is often unavoidable; yet the implications of tunings remain unexplored.

Generalizations of model–data comparisons are desirable because they would provide a "single currency" against which results could be compared. However, we should emphasize that extensions or generalizations need not take the exact form of an ROC curve. For example, Costanza and Maxwell (1994) proposed a scheme for characterizing model goodness-of-fit as a function of the spatial resolution of model–data comparisons. In this case, the tuning variable was *scale* with the analysis assisting the model user in choosing an appropriate spatial resolution for model applications. From this perspective, a challenge in extending a similar approach to complex ecosystem models is that model behavior (and the data as well) must be collapsed in a meaningful way into simple metrics that lend themselves to an ROC-like summary.

The extensions we propose are new but many familiar themes can be identified if one considers earlier procedures for model evaluation. Recalling the large body of model-testing approaches previously developed within a regression framework (Table 10.2), one of the most appealing features of regressions is that they can provide information not only on *whether* the model misses the data but also on *how* the model misses the data. This is possible because regression-based indices can emphasize spread, bias, or other components of goodness-of-fit, giving the user a basis to make decisions about how to interpret model applications. In this same spirit, the ROC approach provides additional information beyond how well the model predicts the data. This additional information allows informed decisions when model bias might be appropriate for a given application. As we suggested previously, predictions about *potential* habitat for an endangered species might admit lower specificity, while predictions of *optimal* habitat would demand high selectivity; in either application maximizing (Se + Sp) might not be the best choice! In the sense of the Mankin et al. (1975) scheme, this is equivalent to claiming that for a given application, adequacy and reliability are not equivalent measures of model validity. Ecologists familiar with contrasting type I versus type II statistical errors accept the notion of choosing an acceptable or even preferable bias for a model. In general, we might attempt to find a more general framework and approach for evaluating models in terms of their inherent biases and predictive behavior over a wide range of conditions. Then, the model could be used more intelligently within its known domain of applicability.

Over the past decade, formal research on the statistics of model validation has lagged far behind the pace of new developments in modeling itself. More powerful computers, new programming languages and algorithms, and the increasing availability of massive spatiotemporal data sets acquired via remote-sensing technologies have resulted in new classes of simulators. Many of these models do not lend themselves easily to regression-based model–data comparisons, yet we have been slow to develop new approaches for evaluating these models.

There is a pressing need for new research on the theory and statistical methods of model evaluation. High-priority needs include (1) the development of consistent standards for model–data comparisons that can be applied over a wide variety of kinds of models; and (2) a more rigorous understanding of how the various model-testing methods are related statistically—much in the way that we are now developing a better appreciation for how general linear models, generalized linear models, and generalized additive models are related as a family of regression methods (Dobson 1990; Yee and Mitchell 1991). The goal of this new research should be broader than model–data comparisons in the sense of simple data-matching; rather, the goal should be on acquiring a thorough and rigorous understanding of *how* and *why* a model behaves as it does. This understanding can then inform model applications by those who develop models as well as client end-users who are the ultimate users of models.

This research on modeling will require a substantial investment by ecologists and their funding agencies. At present, there seems to be a general consensus among modelers that funding opportunities for research on the process of model identification, development, and evaluation is rather limited: modeling is funded mostly when proposed in conjunction with field measurements or experiments. While we wholeheartedly agree that models should be linked closely to field studies, the need for research on modeling itself is also crucial and this research can (indeed, perhaps *should*) stand on its own. We see this research as fundamental to the task of returning modeling to its rightful place as a central tool of ecosystem science.

Acknowledgments. Support for RHG was provided by the EPA STAR program as part of the Multiscale Experimental Ecosystem Research Center at the University of Maryland Center for Environmental Science, and for DLU from National Science Foundation grants IBN-9652656 and SBR-9817755 and USGS/BRD grants in support of the Sierra Nevada Global Change Research Program.

References

Aber, J.D. 1997. Why don't we believe the models? *Bulletin of the Ecological Society of America* 78: 232–233.

Burnham, K.P., and D.R. Anderson. 1998. *Model Selection and Inference: A Practical Information-Theoretic Approach.* New York: Springer.

Burrough, P.A., R. van Rijn, and M. Rikken. 1996. Spatial data quality and error analysis issues: GIS functions and environmental modeling. Pp, 29–34 in M.F. Goodchild, L. Steyaert, B. Parks, C. Johnston, D. Maidment, M. Crane, and S. Glendinning, editors. *GIS and Environmental Modeling Progress and Research Issues.* Fort Collins: GIS World Books.

Campbell, J.B. 1996. *Introduction to Remote Sensing.* 2d ed. New York: The Guilford Press.

Caswell, H. 1976. The validation problem. Pp. 313–325 in B.C. Patten, editor. *Systems Analysis and Simulation in Ecology.* New York: Academic Press.

———. 1988. Theory and models in ecology: A different perspective. *Ecological Modelling* 43: 33–44.

Caswell, H., and M.C. Trevisan. 1994. Sensitivity analysis of periodic matrix models. *Ecology* 75: 1299–1303.

Cherrill, A., and C. McClean. 1995. An investigation of uncertainty in-field habitat mapping and the implications for detecting land-cover change. *Landscape Ecology* 10: 5–21.

Child, G.K., and H.H. Shugart. 1972. Frequency response analysis of magnesium cycling in a tropical forest ecosystem. Pp. 102–135 in B.C. Patten, editor. *Systems Analysis and Simulation in Ecology.* New York: Academic Press.

Chipman, H.A., E.I. George, and R.E. McCulloch. 1998. Bayesian CART model search. *Journal of the American Statistical Association* 93: 935–948.

Costanza, R., and T. Maxwell. 1994. Resolution and predictability: An approach to the scaling problem. *Landscape Ecology* 9: 47–57.

Dale, V.H., H.I. Jager, R.H. Gardner, and A.E. Rosen. 1988. Using sensitivity and uncertainty analysis to improve predictions of broad-scale forest development. *Ecological Modelling* 42: 165–178.

Deutsch, C.V., and A.G. Journel. 1998. *GSLIB: Geostatistical Software Library and User's Guide.* New York: Oxford University Press.

Dobson, A.J. 1990. *An Introduction to Generalized Linear Models.* Boca Raton: Chapman and Hall.

Funderlic, R.E., and M.T. Heath. 1971. *Linear compartmental analysis of ecosystems.* IBP Report 7104. Oak Ridge, TN: Oak Ridge National Laboratory.

Gardner, R.H., R.V. O'Neill, J.B. Mankin, and J.H. Carney. 1981. A comparison of sensitivity analysis and error analysis based on a stream ecosystem model. *Ecological Modelling* 12: 177–194.

Gelman, A., J.B. Carlin, H.S. Stern, and D.B. Rubin. 1995. *Bayesian Data Analysis.* London: Chapman and Hall.

Giavelli, G., O. Rossi, and E. Siri. 1988. Stability of natural communities: Loop analysis and computer-simulation approach. *Ecological Modelling* 40: 131–143.

Goodall, D.W. 1972. Building and testing ecosystem models. Pp. 173–194 in J.N.R. Jeffers, editor. *Mathematical Models in Ecology.* Oxford: Blackwell.

Haefner, J.W. 1996. *Modeling Biological Systems: Principles and Applications.* New York: Chapman and Hall.

Håkanson, L. 1996. A general method to define confidence limits for model predictions based on validations. *Ecological Modelling* 91: 153–168.

Hall, C.A.S., H. Tian, Y. Qi, G. Pontius, and J. Cornell. 1995. Modelling spatial and temporal patterns of tropical land use change. *Journal of Biogeography* 22: 753–757.

Henderson-Sellers, A., M.F. Wilson, and G. Thomas. 1985. The effect of spatial resolution on archives of land cover type. *Climate Change* 7: 391–402.

Henebry, G.M. 1995. Spatial model error analysis using autocorrelation indices. *Ecological Modelling* 82: 75–91.

Hilborn, R., and M. Mangel. 1997. *The Ecological Detective: Confronting Models with Data.* Monographs in Population Biology 28. Princeton: Princeton University Press.

HPS (High Performance Systems, Inc.) 2000. *Stella.* Hanover, NH.

Hwang, D.M., H.A. Karimi, and D.W. Byun. 1998. Uncertainty analysis of environmental models within GIS environments. *Computers and Geosciences* 24: 119–130.

Janssen, P.H.M., and P.S.C. Heuberger. 1995. Calibration of process-oriented models. *Ecological Modelling* 83: 55–66.

Klepper, O. 1997. Multivariate aspects of model uncertainty analysis: Tools for sensitivity analysis and calibration. *Ecological Modelling* 101: 1–13.

Kemp, W.M., J.E. Petersen, and R.H. Gardner. 2001. Scale-dependence and the problem of extrapolation: Implications for experimental and natural coastal ecosystems. Pp. 3–57 in R.H. Gardner, W.M. Kemp, V.S. Kennedy, and J.E. Petersen, editors. *Scaling Relations in Experimental Ecology.* New York: Columbia University Press.

Lane, P. 1986. Symmetry, change, perturbation, and observation mode in natural communities. *Ecology* 67: 223–239.

Leggett, R.W., and L.R. Williams. 1981. A reliability index for models. *Ecological Modelling* 13: 303–312.

Levins, R. 1966. The strategy of model building in population biology. *American Scientist* 54: 421–431.

Lynn, H., C.L. Mohler, S.D. DeGloria, and C.E. McCulloch. 1995. Error assessment in decision-tree models applied to vegetation analysis. *Landscape Ecology* 10: 323–335.

Mankin, J.B., R.V. O'Neill, H.H. Shugart, and B.W. Rust. 1975. The importance of validation in ecosystem analysis. Pp. 63–71 in G.S. Innis, editor. *New Directions in the Analysis of Ecological Systems*. Part 1. Simulation Councils Proceedings Series. LaJolla, CA: Society for Computer Simulation (Simulation Councils).

Miller, D.R., G. Butler, and L. Bramall. 1976. Validation of ecological system models. *Journal of Environmental Management* 4: 383–401.

Monserud, R.A., and R. Leemans. 1992. Comparing global vegetation maps with the kappa-statistic. *Ecological Modelling* 62: 275–293.

Monte, L., L. Håkanson, U. Bergstrom, J. Brittain, and R. Heling. 1996. Uncertainty analysis and validation of environmental models: The empirically based uncertainty analysis. *Ecological Modelling* 91: 139–152.

Olson, J.S. 1963. Analog computer model for movement of nuclides through ecosystems. Pp. 121–125 in V. Schultz and A. Klement, editors. *Radioecology*. New York: Reinhold.

Oreskes, N., K. Shrader-Frechette, and K. Belitz. 1994. Verification, validation, and confirmation of numerical models in the earth sciences. *Science* 263: 641–646.

Overton, S. 1977. A strategy of model construction. Pp. 49–73 in C. Hall and J. Day, editors. *Ecosystem Modeling in Theory and Practice: An Introduction with Case Histories*. New York: John Wiley and Sons.

Pearce, J., and S. Ferrier. 2000. Evaluating the predictive performance of habitat models developed using logistic regression. *Ecological Modelling* 133: 225–245.

Pontius, R.G. 2000. Quantification error versus location error in comparison of categorical maps. *Photogrammetric Engineering and Remote Sensing* 66: 1011–1016.

Pontius, R.G., and L.C. Schneider. 2001. Land-cover change model validation by an ROC method for the Ipswich watershed, Massachusetts, USA. *Agriculture Ecosystems and Environment* 85: 239–248.

Power, M. 1993. The predictive validation of ecological and environmental models. *Ecological Modelling* 68: 33–50.

Rose, K.A., and E.P. Smith. 1998. Statistical assessment of model goodness-of-fit using permutation tests. *Ecological Modelling* 106: 129–139.

Rose, K.A., and G.L. Swartzman. 1981. *A Review of Parameter Sensitivity Methods Applicable to Ecosystem Models*. NUREG/CR-2016. Washington, DC: U.S. Nuclear Regulatory Commission.

Rossi, R.E., P.W. Borth, and J.J. Tollefson. 1993. Stochastic simulation for characterizing ecological spatial patterns and appraising risk. *Ecological Applications* 3: 719–735.

Rykiel, E.J. 1996. Testing ecological models: The meaning of validation. *Ecological Modelling* 90: 229–244.

Smith, E.P., and K.A. Rose. 1995. Model goodness-of-fit analysis using regression and related techniques. *Ecological Modelling* 77: 49–64.

Tomovic, R., and W.J. Karplus. 1963. *Sensitivity Analysis of Dynamic Systems.* New York: McGraw-Hill.

Turner, M.G., R.V. O'Neill, R.H. Gardner, and B.T. Milne. 1989. Effects of changing spatial scale on the analysis of landscape pattern. *Landscape Ecology* 3: 153–162.

Urban, D.L., C. Miller, P.N. Halpin, and N.L. Stephenson. 2000. Forest gradient response in Sierran landscapes: The physical template. *Landscape Ecology* 15: 603–620.

Vayssières, M.P., R.E. Plant, and B.H. Allen-Diaz. 2000. Classification trees: An alternative non-parametric approach for predicting species distributions. *Journal of Vegetation Science* 11: 679–694.

Wiegert, R.G. 1975. Simulation modeling of the algal-fly components of a thermal ecosystem: Effects of spatial heterogeneity, time delays, and model condensation. Pp. 157–181 in B.C. Patten, editor. *Systems Analysis and Simulation in Ecology.* New York: Academic Press.

Wilder, J.W. 1999. A predictive model for gypsy moth population dynamics with model validation. *Ecological Modelling* 116: 165–181.

Yee, T.W., and N.D. Mitchell. 1991. Generalized additive models in plant ecology. *Journal of Vegetation Science* 2: 587–602.

11

Standards of Practice for Review and Publication of Models: Summary of Discussion

John D. Aber, Emily S. Bernhardt, Feike A. Dijkstra,
Robert H. Gardner, Kate H. Macneale, William J.
Parton, Steward T.A. Pickett, Dean L. Urban, and
Kathleen C. Weathers

Introduction

Papers presenting ecological data are expected to follow a prescribed pattern. There is no similar, standard method for writing or presenting ecological models. The structure and content of modeling papers varies widely, frequently failing to adequately present the structure of the model and the parameters required to run it. Readers of modeling papers often cannot reproduce the results presented. Because it is unclear what should be included in modeling papers, or reviews of modeling papers, there are no standards as to what constitutes a rigorous evaluation of this type of scientific effort. The result is a wide range in the quality and completeness of modeling papers, and a reduction of the impact of modeling as a tool for research in ecology and ecosystem studies.

Models can be the most accessible and transferable form of ecological information. The actual code that produces the dynamics of the model is a quantitative and unambiguous expression of how the results were obtained. Any researcher running the model with the same inputs should obtain the same outputs. However, this form of communication is open only if the actual code of a model is available to other research groups. There is a wide range of practice in terms of availability of code for published models.

A Straw-Man Standard

A discussion group met during the conference to consider the process of publication of papers that presented ecosystem models. The group's discussions began with the proposition that a lack of standards by which papers presenting ecological models are reviewed leads to a lack of rigor in the field and reduces the impact of modeling in general. A recent commentary in the *Bulletin of the Ecological Society* proposing a specific set of standards for the review of modeling papers (Aber 1997) served as a starting point for discussions of standards and goals. That commentary suggested that all modeling papers include the following sections:

Model Structure

This section of a modeling paper should present the entire model structure, both diagrammatically and in equation form. The goal is to allow a complete reconstruction of the model by other users, just as the description of methods in a data paper should allow a reconstruction of the experiment. This section of the paper also should review previous field and modeling work on the processes described, thus ensuring that the author is aware of previous relevant work, and also indicate what is unique about the approach presented.

Parameterization

All of the parameters used in the model should be listed (with units), and all values for those parameters given, along with references to the sources of those parameters. Parameters derived by calibration should be clearly indicated, the calibration method described, and the calibrated values given. For models that are presented in several steps, or in papers that use an existing model without modification, there should be a clear reference trail back to the original papers that describe the model. Web-based documentation may constitute a valid means for presenting material that is too extensive, dense, and detailed for inclusion in journals.

Testing (Validation)

The term used to describe this step may vary but its meaning is clear. Any model that passes beyond the theoretical or conceptual stage needs to be tested against independent field data. Independence means that the data against which model output are compared were not used in any way to determine the structure or parameters of the model.

Sensitivity Analysis

Sensitivity analysis is the systematic evaluation of model response to perturbations of individual model parameters (Chapter 10). This step should be included in modeling papers to give readers and reviewers some idea of model responsiveness to these changes and a quantitative evaluation of the relative importance of correctly specifying each input value. A greater degree of uncertainty can be tolerated in parameters to which the model is relatively insensitive. A second type of sensitivity analysis might be called the "null model" approach, in which a process model is compared against a simple statistical model to indicate the increase in predictive accuracy achieved by including knowledge of the processes in the system. It should be remembered that statistical models are likely to be more accurate than process models within the domain of observations but may not extrapolate well into new domains, such as a change in environmental conditions.

Projection

Only after the above standards have been addressed can the model be used to project the effects of policy alternatives or changing conditions. Perhaps the greatest disservice ecologists can provide comes from allowing poorly described and unvalidated models to be used to test policy scenarios (Chapter 7). This would be equivalent to basing policy decisions on data in which we have no confidence whatsoever. The use of untested models also fosters the false impression that we know more than we do about the systems under study.

Revisions to the Straw Man

The discussion group quickly suggested that the criteria listed above were most appropriate for models used for projection and scenario testing or for application to policy questions. Models are produced for other purposes, as well, and may need to be judged by different standards.

Three types of models were described (Table 11.1).

(1) Conceptual models are often the first step in the description of a new research endeavor and can be valuable in organizing a program and developing hypotheses. Other chapters (e.g., Chapter 12) have suggested that modeling should be the first step in the development of a new research plan rather than the last, as is often the case. Conceptual models provide a framework for a quantitative expression or summary of the hypothesized interactions among the processes of interest. Models at this stage could be publishable as part of literature review or synthesis papers that lay out the current state and proposed extension of an area of knowledge. No actual runs of the model need be made at this point.

Table 11.1. Different types of models used for different purposes.

Model Type	Purpose	Should Include	Can't Do
Conceptual	Synthesis Hypothesis generation	Structure	Sensitivity Projection
Realized	Organize research Data repository	Parameterization Sensitivity analysis	Projection
Applied	Projection	Testing (validation) Sensitivity analysis	

(2) Realized models have actual code and may use initial parameters gathered from the literature. They may become a means of organizing and summarizing data from different individual research projects into a coherent whole; they can act as data repositories. Sensitivity analyses can be used to determine the relative precision with which each parameter needs to be specified or which parameters need more study. Models published at this stage should include a complete presentation of model structure and parameterization as described above.

(3) Finally, if models are to be applied to policy questions, then they must be first tested against independent data in order to establish some basis for the validity of the projections presented.

Different Models, Different Goals

The group also discussed other dimensions of the modeling process, including the continuum from theoretical to empirical (or statistical) and from simple to complex, and the trade-offs among these in realizing three fundamental goals of models: precision (and accuracy), generality, and realism. These three ideas were placed in a conceptual space (Figure 11.1). In general, the group, and the larger conference, recognized that theoretical, or process, models only differed from empirical models in degree of complexity of the structure overlain on the data and that almost all process models were built on empirical relationships at lower levels.

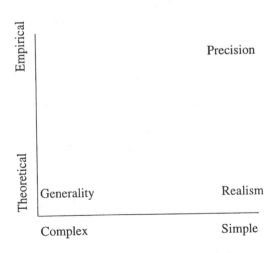

Figure 11.1. Generalized interaction among model characteristics

From this, Figure 11.1 suggests that the models which will emulate any system most precisely will be those which contain only a few major processes described by site-specific, empirical relationships (short-term predictions of weather patterns or tree growth were given as examples). In contrast, the most complex and theoretically complete models provide a more comprehensive or generalized picture of ecosystem function—and perhaps sacrifice precision for completeness (Chapter 2). Simpler models containing relationships well founded in theory may provide a less complete but more realistic view of system function.

Validation and Calibration

The group discussed the need for model testing, or validation, and the process of calibration to obtain parameter values. The article by Aber (1997), used as the straw man, cited the need for clarity in understanding the difference between these two. The heart of the distinction is that calibration is a process by which data describing model outputs are used to determine the input values, while model testing, evaluation, or validation compares model outputs against independent data that were *not* used in deriving the input parameters (Chapter 10).

One example might be the use of climate and physiological data to predict water yield from a forested catchment. Using the same model structure and the same data sets, the calibrated model would use one of a variety of fitting procedures to produce a set of parameters that minimize the disagreement between measured and modeled stream flow. Model predictions compared against these same data provide only a measure of goodness of fit and do not reveal the adequacy of model structure and parameters for predicting other sets of observations. It is possible that parameter values that result from calibration will lay well beyond the range of reasonable values derived by measurement.

There is an important relationship between model complexity and calibration. In models having many parameters calibrated to limited sets of measurements, there are many combinations of input parameters that will result in an equally good fit to the output data. The alternate extreme is to enter in this same model only values that result from direct measurement. Driving this model with measured climate values yields an estimate of water yield that can be compared against the measured values because they were not used in the derivation of the model parameters.

It should be said that almost all ecosystem models in use today, especially those used in the global change context, are calibrated. It should also be said that a calibrated model can be tested against independent data. Using the example above, the calibrated model could be compared against future measurements of water yield, or against measurements from other watersheds for which input parameters have been respecified based on direct measurements (not a recalibration).

There was considerable discussion in both the discussion group and the conference about Bayesian methods (Chapter 9). It was clear that the range of understanding of these methods among the participants varied widely. It was unclear to most whether this approach provided a method for continuous updating and calibration of parameters to achieve a more precise and empirical model or a Monte Carlo method for producing estimates of uncertainty around model predictions. The approach appeared to be powerful but remained mysterious to most.

An Index of Validity

It is extremely difficult to compare objectively different models (or progressive modifications of a single model). Therefore, we attempted to conceptualize and combine, and possibly even quantify, the characteristics that increase user confidence in a particular model. We began with the concept of predictive power presented by Håkanson (Chapter 8). In this calculation, the actual predictive precision (the mean R^2 value), the consistency of this precision (the CV of the R^2 values) and the mean bias (slope of linear regressions) were incorporated into the metric.

In an effort only partly whimsical, our discussion group expanded upon this concept and discussed a possible "validity index" (Figure 11.2B). The first term in our index is the inverse of the mean standard error of the estimate (SEE) of the predicted outputs. This value is then multiplied by 1 minus the number of calibrated input parameters (c) divided by the number of independently predicted outputs (p). This term has the advantage of driving the entire index negative if $c > p$. Finally, the third term increases the value of the index as the number of predicted outputs increases, giving credit to models that can predict several processes (e.g. photosynthesis, NPP, water yield, species diversity) simultaneously. Derivation and meaning of the half-validation constant (V) gave rise to an all-too-brief and rather creative discussion.

One additional concept presented relative to model testing was that of comparisons of process models against simpler statistical models. The idea was that the statistical models represent the null case or null model in which all process-level understanding has been removed. A comparison between the tested and the null model based on the mean error (predicted – observed) for all predicted variables gives a quantitative estimate of how much has been added to the precision of our understanding of the modeled process by the addition of information from lower levels of organization. It was recognized, however, that this is really a comparison between the upper-right corner of and the lower-right corner of Figure 11.1. As such, better precision might be expected from the null

A.
  Predictive Power

$$\frac{R^2}{(1-\alpha)\,CV(R^2)} \quad (OR) \quad \frac{Precision}{Bias \;\; CV\,(Precision)}$$

B.
  Validity Index

$$\frac{p}{\sum_{i=1}^{p}(SEE_p)} * \left(1 \cdot \frac{c}{p}\right) * \frac{p}{p+V}$$

  p= # of predicted outputs
  c = # of calibrated inputs
  V= The "half-validation constant"

Figure 11.2. Two metrics for determining the value of different models in projections and scenario testing: (A) the predictive power metric of Håkanson et al. (Chapter 8) and (B) a possible "validity index." Both are based on the precision of models in recreating independently measured data. The second metric adds parameters related to the number of variables predicted to the number of parameters calibrated.

model, but a more realistic response to altered environmental conditions from the process model.

Standardized Data Sets for Model Testing

At the current time there are no common standards used in the evaluation of ecological modeling papers. Models are typically tested using data sets not available to other modeling groups, thus making it difficult to evaluate the relative advantage or disadvantage of a particular model. We support the establishment of common guidelines for evaluating ecological modeling papers and the availability of standard data sets that can be used to test different types of ecological models. These standard data sets will allow us to evaluate the relative merits of different modeling approaches. There have been a number of model comparison activities (Chapter 12) during the last ten years that highlight a variety of approaches for comparing models and provide the data sets that could be used as standard data sets for evaluating model performance. Unfortunately, there are not any current standard approaches used by the modeling community for evaluating different models, or standard guidelines for reviewing modeling papers.

References

Aber, J.D. 1997. Why don't we believe the models? *Bulletin of the Ecological Society of America* 78: 232–233.

12

The Collision of Hypotheses: What Can Be Learned from Comparisons of Ecosystem Models?

Edward B. Rastetter

I am myself an empyric in natural philosophy, suffering my faith to go no further than my facts. I am pleased, however, to see the efforts of hypothetical speculation, because by the collisions of different hypotheses, truth may be elicited, and science advanced in the end.—Thomas Jefferson[1]

Summary

Model comparisons in ecology are typically conducted in one of two contexts. First, models are compared to data that have already been collected, although comparisons would be more profitable before data collection because gaps could be avoided in the data needed to differentiate among alternate models. In the second context, models are compared for situations where data cannot be obtained (e.g., long-term responses to climate change). These comparisons can build confidence in the model predictions, but only if the models differ in their underlying structure. It is demonstrated that model dynamics are controlled largely by the overall model structure and not by the details of how individual processes are represented. Models structured from the perspective of several different resources might all yield similar predictions for slowly changing or steady-state ecosystems because resource cycles tend to synchronize as ecosystems develop. This synchrony is disrupted by perturbations; therefore, models can be more readily differentiated based on their responses to particular disturbances. Time-series techniques can be used to compare the dynamic behavior of models. These techniques can yield useful information on how the model structure might be improved. Finally, modeling is put into the broader perspective of the process of ecological science. It is argued that a division of labor is needed that includes expertise in modeling, tool making, and data collection/experimentation.

Introduction

The comparison of models lies at the very heart of the scientific method. In the context of this essay, models are formal, quantitative representations of scientific hypotheses or theories. By contrasting models of this type, quantitative tests can be devised to assess the superiority of one of the embedded theories or hypotheses over another. This central role of model comparisons is reflected in most philosophical works on the scientific method. For example, in Popper's *The Logic of Scientific Discovery* (1959) prior to the attempt at "falsification" of a theory, the predictions of a new theory are compared to current theory to determine "which are not derivable from current theory, and more especially. . . which the current theory contradicts." Similarly, Platt's method of strong inference (1964) relies strongly upon a comparison of models. Advancement by means of strong inference proceeds by

(1) Devising alternative hypotheses

(2) Devising a crucial experiment . . . [to] exclude one or more of the hypotheses

(3) Carrying out the experiment so as to get a clean result

(4) Recycling the procedure . . . to refine the possibilities that remain

The design of the crucial experiment in step 2 requires a comprehensive comparison of models, not only to reveal differences among models, but also to identify among those differences the ones most likely to be differentiated by means of an empirical test.

Comparisons of ecosystem models are now commonly done (e.g., VEMAP 1995; Ryan et al. 1996 a and b; Pan et al. 1998; Cramer and Field 1999; Kicklighter et al. 1999; Moorhead et al. 1999; Clein et al. 2000; Jenkins et al. 2000; Schimel et al. 2000; McGuire et al. 2001). However, these comparisons are rarely, if ever, done to identify the crucial experiment that would eliminate one or more models from the group, as is required in both Popper's and Platt's methods. Instead, these comparisons either are against data that already exists or are for situations where data are unlikely to be acquired (e.g., for a broader geographical range than could be sampled directly or for conditions in the far future or past). In the former case, the models provide an a posteriori synthesis of the data that can yield insights into the system being modeled but lack the virtual guarantee of advancement provided by Platt's method. In the latter case, the motivation for the models is to compensate for the lack of data, and the model comparison provides the only available means to assess the predictions.

Model Comparisons Based on Preexisting Data

Examples of model comparison using preexisting data are the study by Moorhead et al (1999) using litter-bag data from the Long-Term Inter-Site Decomposition Experiment Team (LIDET 1995) and the study by Ryan et al (1996 a and b) using data from the Swedish Coniferous Biome (SWECON) study (Persson 1980) and the Australian Biology of Forest Growth (BFG) study (Benson et al. 1992). The conclusions from these model comparisons are typical of similar studies. First, all the models generally fit the data. This result suggests that there was a successful preselection of model structures based on the original data used in their derivation. However, it also suggests that the experimental design was not focused enough to be able to differentiate among model structures and thereby unambiguously reject or falsify one or more of them. Second, there is a call for more or better data falling in one of three categories.

(1) In clarity of hindsight, there are always other important factors that should have been measured concurrently. These factors are usually needed to either drive or initialize the models. Thus, there is uncertainty about whether the deviations between the models and data are the result of a flaw in model structure or of the assumptions about these unmeasured factors.

(2) There is often a need for higher temporal resolution in data to discern transient dynamics. This problem arises because of the emphasis on statistics, rather than dynamics, in experimental design. The dynamic behavior of a model is more revealing of its structure than its steady-state behavior and therefore provides a better test of that structure (see below).

(3) There is very often a need for higher resolution data along environmental gradients to discern a response surface. This problem arises from the prevalence of an analysis-of-variance (ANOVA) experimental design rather than a design intended to test mechanistically meaningful equations. Thus, the data reveal that a factor is important but do not reveal the details of how the system responds to that factor.

These conclusions about data needs would clearly aid in the design of the crucial experiment specified in step 2 of Platt's method of strong inference (1964). However, they fall short in two important ways. First, they are not specific enough to identify the crucial experiment and crucial measurements needed to differentiate one model from another. Second, the experiment is rarely, if ever, repeated with the benefit of the insights gained in the model comparison. It would be almost impossible to get funding to rerun an experiment like LIDET, SWECON, or BFG.

Despite these shortcomings, this type of model comparison sometimes yields very useful insights into the structure of the models. For example, in the Ryan et al. analysis (1996a), seven models were compared to the data from the SWECON and BFG studies. Among these models, the BIOMASS and

HYBRID models fit the data particularly well. These two models used foliar N as an input to the models, whereas the remaining five models predicted N allocation to the canopy. This result implied a particular importance of N-allocation patterns in the model structures and suggests that the next iteration of model comparisons and empirical test might productively focus on the controls on N allocation. The results of the Ryan et al. (1996a) analysis are nonetheless fortuitous in that the experiments were not designed specifically to test this difference among the models.

Model Comparisons Where Data Are Not Directly Available

Many recent model comparisons have focused on issues related to the global C budget and to long-term regional or global responses to changes in climate and atmospheric CO_2 (e.g., VEMAP 1995; Cramer and Field 1999). Because of the broad spatial extent or the long time horizon of these studies, direct measurement of the systems being modeled is impossible. Nevertheless, some means of evaluating the models is needed to build confidence in their projections. The model comparison provides this confidence in an indirect, two-step process.

First, the individual models are developed and tested based on fine-scale data from a variety of sources. These data will typically be from specific sites and of short duration relative to the model projections, and they might be observational or experimental. The models themselves might be developed from a variety of different perspectives (e.g., productivity models based on water use, absorbed radiation, or C-N interactions) and therefore use very different types of data in their development and initial testing.

Second, to the extent that the models agree during the comparison itself, they transfer at least some of the confidence developed during the initial development and testing to the other models in the comparison (Rastetter 1996). This transfer of confidence is particularly significant if the models differ greatly in perspective. For example, if a productivity model based upon C-water relations and a model based on C-N interactions are consistent in their predicted responses to elevated CO_2 and altered climate, then the C-water model has not conflicted with the constraints imposed by C-N interactions and the C-N model has not conflicted with the constraints imposed by C-water interactions. Because the model perspectives differ, it is unlikely that one model could have been tested against the type of data used in the development and initial testing of the other. Each model thereby serves to translate these data into a form that can be compared to the other model.

The emphasis of these model comparisons where data are not directly available is generally on an assessment of how well the models can answer questions such as How, When, Where, etc. For example, the three biogeochemical models in the VEMAP assessment agreed within 25% on the current C sink associated with elevated CO_2 and climate variation in the conterminous United States

(Schimel et al 2000). This degree of consensus is remarkable given that one of the models was developed from a water-use perspective, the second from a C-N perspective with a heavy emphasis on soil processes, and the third from a C-N perspective with more of an emphasis on plant processes. However, the models diverged in their long-term projections. Predictably, these divergences can be attributed to the different perspectives on hydrologic versus N feedbacks on production in these models (Pan et al. 1998). Ryan et al. (1996b) came to a similar conclusion in his model comparison where productivity in the C-water models responded negatively to increased temperature because of higher water stress and the C-nutrient models responded positively to increased temperature because of higher nutrient mineralization rates.

These comparisons have been valuable in that they have assessed the degree of consensus among models and allowed assessment of confidence limits around the suite of projections. They have also highlighted the consequences of different perspectives and model structures and have thereby identified general areas of needed future research. However, these data-independent comparisons can never be the basis for rejecting one model structure over another (Rastetter 1996). Nor do these comparisons provide a clear next step for improving the models, because it is not clear which of the models is better than the others.

Comparative Analysis of Model Structure

A recommendation of the VEMAP (1995) and several similar studies is to "modularize" the models so they can exchange modules. Thus, the models would be reformulated in such a way that the photosynthesis, respiration, mineralization, etc. modules could be easily removed and replaced with the analogous modules from other models. By exchanging modules, the processes responsible for differences in model predictions can be identified and isolated for further study.

Although modular model structures are good practice from the perspective of code development, I have two strong objections to a modular structure designed specifically to facilitate the exchange of modules among models. To illustrate these two objections, I have developed three model structures, all intended to represent the same hypothetical system (Figure 12.1, Table 12.1). I have deliberately not placed ecological meaning on any of the variables or components so as not to confound the heuristic value of the example with preconceived notions of what the process formulations *should* be.

Objection 1. If the models are restructured so they can exchange modules, they tend to converge on a similar structure (convergent evolution). For one module to be exchangeable with another, then

    (1) the two modules must perform the same function in the model (e.g., a gross production module could not be exchanged with a net production module) and,

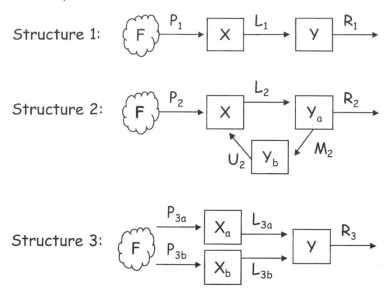

Figure 12.1. Three model structures that all represent the same hypothetical system. To be compatible with a particular model structure, the processes must refer only to variables in the model. This compatibility criterion makes it difficult to exchange process modules among models without restructuring the model.

(2) the module must be compatible with the model structure in that it can only refer to variables already in the model (e.g., a production module based in part on N availability could only be inserted in a model that includes N availability).

Based on these criteria, model structures 1 and 2 can easily exchange modules for process P, at least as they are formulated in Table 12.1. However, model 2 could also incorporate modules for P that make reference to variables Y_a and Y_b, model 1 cannot. To incorporate such modules, model 1 would have to be restructured to split Y into Y_a and Y_b, which would make its structure converge on that of model 2. A similar restructuring would be required to exchange modules for L between the two models because process L in model 1 represents the net of process L minus U in model 2, and process U makes reference to variable Y_b, which is not in model 1. Similar compatibility problems arise for process R because of its reference to Y in model 1 and Y_a in model 2. Restructuring would also be required to allow exchange of modules with model 3. Model 3 is similar in structure to model 1 except that variable X has been divided in two. This division allows for modules based on competitive interactions between X_a and X_b. For these competition modules to be incorporated into either model 1 or 2, the models would have to be restructured by dividing X in two, thereby making their structure converge on that of model 3.

Table 12.1. Equations for the three models depicted in Figure 12.1.

Structure 1

$$P_1 = \alpha X^{\eta} \frac{F}{\beta + F} \qquad L_1 = a_1 X \qquad R_1 = b_1 Y$$

$$\frac{dX}{dt} = P_1 - L_1 \qquad \frac{dY}{dt} = L_1 - R_1$$

$\alpha = 1$, $\beta = 1$, $\eta = 0.5$, $a_1 = 0.05$, $b_1 = 0.05$, $F(0) = 1$, $X(0) = 100$, and $Y(0) = 100$

Structure 2

$$P_2 = \phi\left(1 - e^{-\lambda X}\right) F^{\varepsilon} \qquad L_2 = a_2 X \qquad R_2 = b_2 Y_a$$

$$M_2 = c_2 Y_a \qquad U_2 = g_2 Y_b$$

$$\frac{dX}{dt} = P_2 + U_2 - L_2 \qquad \frac{dY_a}{dt} = L_2 - M_2 - R_2 \qquad \frac{dY_b}{dt} = M_2 - U_2$$

$\phi = 7.91$, $\lambda = 0.01$, $\varepsilon = 0.42$, $a_2 = 0.1$, $b_2 = 0.1$, $c_2 = 0.1$, $g_2 = 0.1$, $F(0) = 1$, $X(0) = 100$, $Y_a(0) = 50$, and $Y_b(0) = 50$

Structure 3

$$P_{3a} = \alpha_a \left(\frac{X_a}{1 + X_b}\right)^{\eta_a} \frac{F}{\beta_a + F} \qquad L_{3a} = a_{3a} X_a$$

$$P_{3b} = \alpha_b \left(\frac{X_b}{1 + X_a}\right)^{\eta_b} \frac{F}{\beta_b + F} \qquad L_{3b} = a_{3b} X_b \qquad R_3 = b_3 Y$$

$$\frac{dX_a}{dt} = P_{3a} - L_{3a} \qquad \frac{dX_b}{dt} = P_{3b} - L_{3b}$$

$$\frac{dY}{dt} = L_{3a} + L_{3b} - R_3$$

$\alpha_a = 4.42$, $\alpha_b = 4.36$, $\beta_a = \beta_b = 1$, $\eta_a = \eta_b = 0.5$, $a_{3a} = a_{3b} = 0.05$, $b_3 = 0.05$, $F(0) = 1$, $X_a(0) = 75$, $X_b(0) = 25$, and $Y(0) = 100$

While the agreement among several models in a model comparison increases confidence in the model predictions, that increase in confidence is derived from the diversity of perspectives represented in the models. It would not be surprising for models with similar structures to agree in such a comparison. If the structures of the models converge, that diversity decreases and the amassed confidence also decreases. Convergence of model structures should be acceptable after testing the models; that is simply progression along the path of strong inference. However, it should not be acceptable as a means to test or compare the models.

Objection 2. My second objection to a structure designed to facilitate the exchange of modules among models is that the exchange is not likely to reveal anything meaningful. If different formulations of a process result in wildly different response surfaces, then those formulations should be isolated from the models and compared directly to data, not modularized and exchanged among models. On the other hand, several formulations of a process might have reasonable response surfaces that are all consistent with the data. The differences among these formulations will not be the cause of major differences in model dynamics; the differences in model dynamics will instead result from the way the processes are linked to one another. That is, it is the structure of the model itself that imparts the important dynamics, not the details of how individual processes are represented in the model modules.

The three models discussed above can be used to illustrate the relative importance of process formulations versus model structure on model dynamics. The representations of process P in the three models are clearly different. However, when these formulations are implemented, they will have very similar shapes because their parameters will have to be estimated based on similar data. If the data are of very high quality, then it should be possible to reject one or more formulations in favor of the others. However, with the level of uncertainty typical of most ecological data, it is far more likely that no one formulation will provide a significantly better fit than the others will. Indeed, there are an infinite number of formulations with about the same shape that will fit the data and are compatible with a particular model structure (e.g., any polynomial above some minimum order). Selection of reasonable formulations from among these should be based on criteria other than goodness of fit to the data (e.g., parsimony, some theoretical foundation). When any of these formulations are used in a model, the dynamics of the model will be about the same. To illustrate this similarity, I estimated parameters for the three formulations of process P from models 1, 2, and 3 using some hypothetical data for P, X, and F. I ran the models with the value of F increasing steadily from 1 to 2 beginning at time 10 and ending at time 100, then exchanged the modules for process P between models 1 and 2 and reran the simulations (Figure 12.2). The structure of model 3 does not allow for the module P to be exchanged with the other two models. The dynamics of the individual models differed greatly from one another, but exchanging the P modules had very little effect on those dynamics. Again, it is the structure that is the main determinant of the model dynamics; the details of the process formulations are far less important.

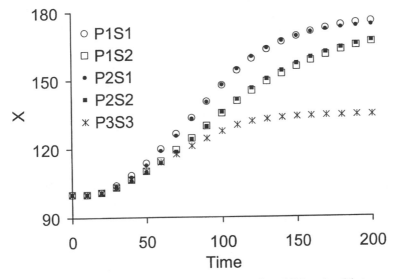

Figure 12.2. Simulated response of variable X to a ramping of F from 1 to 2 between time 10 and 100 using the three model structures presented in Figure 12.1 and the three formulations of process P presented in Table 12.1. Processes P_1 and P_2 can be used with either structure 1 (S1) or 2 (S2). Process P_3 can only be used with structure 3 (S3). Structure 1 (circles) yields similar results regardless of whether P_1 or P_2 is used. Similarly, structure 2 (squares) yields similar results regardless of whether P_1 or P_2 is used.

Convergent Evolution, Functional Convergence, and Resource Optimization

I have argued that confidence in model predictions should increase if those predictions are consistent among several models that differ widely in underlying perspective. Thus, predictions based on the agreement among models built from a water-use, nutrient-use, and light-use perspective should be more credible than predictions from any one model alone. But, why should it be expected that predictions from these different perspectives should ever agree?

The agreement arises from a combination of evolutionary, ecological, and physiological processes in nature. Convergent evolution tends to produce species with similar characteristics to fill similar niches. These species interact on ecological time scales such that "ecosystems tend toward combinations of species with functional properties that vary consistently with climate and resources" (functional convergence; Field et al. 1995). Physiological adjustments within these species serve to optimize resource acquisition. "Generally, resources [within an organism] are allocated most efficiently when growth is equally limited by all resources [in the environment]" (Chapin et al. 1987, 56).

In combination, these processes tend toward a synchronization of all the re-source cycles in an ecosystem.

The consequence of this synchronization is that models that differ markedly in approach and complexity will yield comparable estimates of ecosystem be-havior near a steady state (Field et al. 1995). However, these differences in ap-proach should yield very different predictions following a perturbation when the species are reoptimizing resource acquisition and community structure is reconverging on a configuration consistent with the new conditions. Under such conditions, the synchronization of resource cycles should collapse, and one resource cycle should dominate the ecosystem dynamics until the other cycles fall in step. Therefore, the crucial criteria to distinguish among alternate models of ecosystem function are likely to be the transient responses to sudden pertur-bations, and not those occurring during slowly changing or near steady-state conditions. The most appropriate modeling perspective is likely to differ for different types of perturbation. To assess differences in modeling perspective in this setting will require a dynamic or time series approach.

Time Series Analysis

Time series analysis is less formally defined in ecology (e.g., Powell and Steele 1995) than in engineering (e.g., Box and Jenkins 1976, Young 1984). The less formal definition in an ecological context is, at least in part, due to the difficulty in obtaining the long records of regularly spaced time series data needed for most engineering analyses. Nevertheless, when such data can be obtained for ecological systems, these techniques can be used to good effect (e.g., Cosby and Hornberger 1984; Cosby et al. 1984; Rastetter 1986). Of particular perti-nence to the issue of assessing model structures is the test of model adequacy devised by Cosby and Hornberger (1984) based on the Extended Kalman Filter (EKF).

The EKF is a model-based control algorithm that might be used, for exam-ple, to regulate inputs to an industrial chemostat. Control decisions are made based on model projections one time step into the future. Measurements are then made after the time step has elapsed and are compared to the model predic-tions. This comparison is used to update both the parameter estimates and a matrix of parameter and measurement variance and covariance. The magnitude of the update correction on each parameter is weighted based on the current estimate of variance for that parameter. Thus, the discrepancy between meas-urements and model predictions is partitioned among the parameters and an assumed measurement error based on how stable the parameter estimates have been in the past. By monitoring past estimates, the EKF can be used to generate time series of the deviations between measurement and model predictions and of the parameter estimates.

The test of model adequacy used by Cosby et al. (1984) is as follows: The model embedded in the EKF is adequate if (1) the time series of model deviations from the measurements are random white noise about a zero mean (i.e., no trend and no autocorrelation), (2) the time series of parameter estimates are random white noise about a fixed mean, and (3) there is no cross-correlation among parameters or between parameters and any of the model variables. This test is very sensitive to system dynamics and should therefore be useful for testing the adequacy of model structures. More importantly, the way that a model fails this test can be very revealing about how the model structure needs to be changed. For example, a correlation between a parameter time series and a variable suggests a missing link between that variable and the process to which the parameter belongs and also reveals a good deal of information about the form of that linkage. A nonrandom variation in a parameter estimate or in the deviations might suggest a missing variable in the model and again reveal a great deal about its characteristics.

Cosby et al. (1984) used the EKF test on eight formulations of the photosynthesis-irradiance (PI) relationship in a model of O_2 concentrations in a macrophyte-dominated stream. Unfortunately, the use of these eight formulations of the PI relationship constitute an exchange of process modules rather than a change in model structure (see above). Because the model structure remained constant, all eight PI modules failed in basically the same way; the time series for the parameter controlling the initial slope of the PI relationship had a strong diel cycle, yet again illustrating that it is the structure of the model, not the details of the process representations, that controls the dynamics of the system. What all the formulations failed to capture was the hysteresis between morning and afternoon PI relationships. This failure suggests a model restructuring to incorporate a feedback on the PI relationship that would depress afternoon photosynthesis. For example, the O_2-based model might be linked to a CO_2-based model if there is an afternoon depletion of CO_2 in the water or a limitation on photosynthesis if carbohydrate sinks become saturated. Obviously, the exact nature of the appropriate new structure will require a deeper investigation of the system, but the EKF results reveal some useful clues about where to look.

As suggested above, implementation of the EKF and most other time series techniques requires long records of regularly spaced measurements. This type of time series is very rare in ecology (e.g., eddy flux data, well maintained autosampling of stream chemistry). The EKF is therefore not likely to be of much use for comparing models directly to data. However, it might be a very useful tool for comparing model structures to one another. For example, a time series of data might be generated using a model with a strong water-use perspective on production. The EKF could then be used to analyze how well models with, for example, a nutrient-use or light-use perspective fit the time series. Such an analysis would reveal which dynamics generated with the water-use perspective are not captured from the other perspectives and also reveal where and perhaps how those dynamics might be introduced into these other models.

Models and the Scientific Method

Model comparisons are an essential component of the scientific method. However, full and efficient use of models and model comparisons in ecology will require better coordination between modeling and empirical efforts. Progress derived from this coordination will necessarily be iterative in nature. The major role of the modeler in this iteration is to formalize and explore the ramifications of alternate hypotheses. The goal of this exploration should be to identify the crucial tests that will differentiate among the alternate hypotheses. Real progress on ecological issues will require a team effort with a clear division of labor reflecting needed expertise in three areas.

(1) Modelers are needed to formalize and explore the ramifications of alternate hypotheses. If this analysis is to serve as a means to design experiments, it must obviously be done before the data needed to test the embedded theories and hypotheses are collected.

(2) Instrument makers are needed to devise new methods and equipment needed to test the hypotheses. Frequently the characteristics identified in a modeling analysis as crucial for differentiating among alternate hypotheses cannot be measured with current technology. Rather than allow new technologies to filter passively into ecology (e.g., eddy flux, isotope tracers, remote sensing), ecologists should be preemptive and solicit solutions to measurement problems from other fields of research. Of particular use for dynamic models is automation to improve the collection of concurrent time series data.

(3) Empiricists/experimentalists are and will always be the main drivers behind the science; the intuition derived from field studies is the main source of new hypotheses. However, their results would have more influence and would be more useful to modeling efforts if the experimental design were motivated more by the dynamics of ecosystem processes and less by statistical design.

Acknowledgments. This chapter is derived from work sponsored by the National Science Foundation (DEB-0108960, OPP-9732281), the National Aeronautics and Space Administration (NASA NAG5-5160), and the United States Environmental Protection Agency (EPA QT-RT-00-001667).

Note

1. Llewellyn, R., and D. Day. 1982. *The Academic Village. Thomas Jefferson's University*. Charlottesville: Thomasson-Grant Publishing.

References

Benson, M.L., J.J. Lansberg, and C.J. Borough. 1992. The biology of forest growth experiment: An introduction. *Forest Ecology and Management* 52: 1–16.

Box, G.E.P., and G.M. Jenkins. 1976. *Time Series Analysis: Forecasting and Control.* San Francisco: Holden-Day.

Chapin, F.S. III, A.J. Bloom, C.B. Field, and R.H. Waring. 1987. Plant responses to multiple environmental factors. *BioScience* 37: 49–57.

Clein, J.S., B.L. Kwiatkowski, A.D. McGuire, J.E. Hobbie, E.B. Rastetter, J.M. Melillo, and D.W. Kicklighter. 2000. Modeling carbon responses of moist tundra ecosystems to historical and projected climate: A comparison of fine- and coarse-scale ecosystem models for identification of process-based uncertainties. *Global Change Biology* 6(1): 127–140.

Cosby, B.J., and G.M. Hornberger. 1984. Identification of photosynthesis-light models for aquatic systems. I. Theory and simulations. *Ecological Modelling* 23: 1–24.

Cosby, B.J., G.M. Hornberger, and M.G. Kelly. 1984. Identification of photosynthesis-light models for aquatic systems. II. Application to a macrophyte-dominated stream. *Ecological Modelling* 23: 25–51.

Cramer, W., and C.B. Field, editors. 1999. The Potsdam NPP Model Intercomparison. *Global Change Biology* 5(1).

Field, C.B., J.T. Randerson, and C.M. Malström. 1995. Global net primary production. *Remote Sensing of Environment* 51: 74–88.

Jenkins, J.C., D.W. Kicklighter, and J.A. Aber. 2000. Regional impacts of increased CO2 and climate change on forest productivity. Pp. 383–423 in R.H. Mickler, R.A. Birdsey, and J. Horn, editors. *Responses of Northern U.S. Forests to Environmental Change.* New York: Springer-Verlag.

Kicklighter, D.W., M. Bruno, S. Dönges, G. Esser, M. Heimann, J. Helfrich, F. Ift, F. Joos, J. Kaduk, G.H. Kohlmaier, A.D. McGuire, J.M. Melillo, R. Meyer, B. Moore III, A. Nadler, I.C. Prentice, W. Sauf, A.L. Schloss, S. Sitch, U. Wittenberg, and G. Würth. 1999. A first-order analysis of the potential role of CO_2 fertilization to affect the global carbon budget: A comparison study of four terrestrial biosphere models. *Tellus* 51B: 343–366.

LIDET (Long-Term Intersite Decomposition Experiment Team). 1995. *Meeting the Challenge of Long-Term, Broad-Scale Ecological Experiments.* Publ. No. 19 Seattle: Long-term Ecological Research Office.

McGuire, A.D., S. Sitch, J.S. Clein, R. Dargaville, G. Esser, J. Foley, M. Heimann, F. Joos, J. Kaplan, D.W. Kicklighter, R.A. Meier, J.M. Melillo, B. Moore III, I.C. Prentice, N. Ramankutty, T. Reichenau, A. Schloss, H. Tian, L.J. Williams, and U. Wittenberg. 2001. Carbon balance of the terrestrial biosphere in the twentieth century: Analyses of CO_2, climate and land-use effects with four process-based ecosystem models. *Global Biogeochemical Cycles* 15: 183–206.

Moorhead, D.L., W.S. Currie, E.B. Rastetter, W.J. Parton, and M.E. Harmon. 1999. Climate and litter quality controls on decomposition: An analysis of modeling approaches. *Global Biogeochemical Cycles* 13: 575–589.

Pan, Y., J.M. Melillo, A.D. McGuire, D.W. Kicklighter, L.F. Pitelka, K. Hibbard, L.L. Pierce, S.W. Running, D.S. Ojima, W.J. Parton, D.S. Schimel, and other VEMAP Members. 1998. Modeled responses of terrestrial ecosystems to elevated atmospheric CO_2: A comparison of simulations by the biogeochemistry models of the Vegetation/Ecosystem Modeling and Analysis Project (VEMAP). *Oecologia* 114: 389–404.

Persson, T., editor. 1980. Structure and Function of Northern Coniferous Forests-an Ecosystem Study. *Ecological Bulletines* (Stockholm) 32.

Platt, J.R. 1964. Strong inference. *Science* 146: 347–353.

Popper, K.R. 1959. *The Logic of Scientific Discovery.* New York: Harper and Row.

Powell, T.M., and J.H. Steele, editors. 1995. *Ecological Time Series.* New York: Chapman and Hall.

Rastetter, E.B. 1986. Analysis of community interactions using linear transfer function models. *Ecological Modelling* 36: 101–117.

———. 1996. Validating models of ecosystem response to global change. *BioScience* 46: 190–198.

Ryan, M.G., E.R. Hunt Jr., R.E. McMurtrie, G.I. Ågren, J.D. Aber, A.D. Friend, E.B. Rastetter, W. Pulliam, J. Raison, and S. Lindner. 1996a. Comparing models of ecosystem function for temperate conifer forests. I. Model description and validation. Pp. 313–362 in A. Breymeyer, D.O. Hall, J. M. Melillo, and G.I. Ågren, editors. *Global Change: Effects on Coniferous Forests and Grasslands.* SCOPE (Scientific Committee on Problems of the Environment), no. 56. New York: John Wiley and Sons.

Ryan, M.G., R.E. McMurtrie, G.I. Ågren, E.R. Hunt Jr., J.D. Aber, A.D. Friend, E.B. Rastetter, and W.M. Pulliam. 1996b. Comparing models of ecosystem function for temperate conifer forests. II. Simulations of the effect of climatic change. Pp. 363–387 in A. Breymeyer, D.O. Hall, J.M. Melillo, and G.I. Ågren, editors. *Global Change: Effects on Coniferous Forests and Grasslands.* SCOPE (Scientific Committee on Problems of the Environment), no. 56. New York: John Wiley and Sons.

Schimel, D., J. Melillo, H. Tian, A.D. McGuire, D. Kicklighter, T. Kittel, N. Rosenbloom, S. Running, P. Thornton, D. Ojima, W. Parton, R. Kelly, M. Sykes, R. Neilson, and B. Rizzo. 2000. Contribution of increasing CO_2 and climate to carbon storage by ecosystems in the United States. *Science* 287: 2004–2006.

VEMAP Members. 1995. Vegetation/ecosystem modeling and analysis project: Comparing biogeography and biogeochemistry models in a continental-scale study of terrestrial ecosystem responses to climate change and CO_2 doubling. *Global Biogeochemical Cycles* 9: 407–437.

Young, P.C. 1984. *Recursive Estimation and Time-Series Analysis.* Berlin: Springer-Verlag.

13

Evaluating and Testing Models of Terrestrial Biogeochemistry: The Role of Temperature in Controlling Decomposition

Ingrid C. Burke, Jason P. Kaye, Suzanne P. Bird,
Sonia A. Hall, Rebecca L. McCulley, and
Gericke L. Sommerville

Summary

Simulation models have played an important role in the development of terrestrial biogeochemistry. One contemporary application of biogeochemical models is simulating interactions between global climate change and terrestrial carbon balance. The largest global pool of terrestrial carbon is detrital (nonliving) soil organic matter, and one ongoing debate is whether warmer temperatures will increase the amount of soil C released to the atmosphere via microbial decomposition (oxidation to CO_2). While much of the literature suggests that decomposition rates increase with temperature, several recent papers cast doubt on this general conclusion. Given the difficulty of directly estimating field rates of total organic matter decomposition, models are playing an important role in assessing how ecosystem carbon balance will respond to global change. We evaluated a suite of models to ask three main questions: (1) What are the nature and origin of the equations used to simulate organic matter decomposition? (2) Is there a consensus understanding of the role of temperature in controlling decomposition? and (3) How well do these models serve as resources for the scientific community?

Our review resulted in several important conclusions. First, current models of decomposition are based on very few empirical studies of the process. Instead, soil organic matter decomposition is simulated using data from soil respiration, short-term laboratory studies, or decomposition of recently senesced foliage. Second, while most models represent decomposition as a process that increases with temperature, the shape of the temperature-decomposition curve, and their interactions with soil moisture varied among the models. Ultimately, the development of realistic, mechanistically based models of organic matter decomposition is limited by field data. There is a strong need for long-term experiments with estimates of detrital inputs and detrital pools to test the understanding of decomposition currently incorporated into simulation models. Finally, our evaluation of the models was limited by incomplete documentation of the source of the relationships used in the model and by the evolution of the models through time.

Introduction

Models provide a forum for synthesis (Parton et al. 1987; Aber et al. 1991) and a tool for extrapolating our understanding to longer time scales and broader spatial extents than we can measure (Running 1986; Burke et al. 1991; Aber et al. 1993). Simulation models also enable us to evaluate complex interactions among element cycles and the processes that drive these cycles. The potential interaction between global climate change and terrestrial carbon (C) balance is one of the most important unknowns of the day (Pastor and Post 1988; Rastetter et al. 1991; Aber et al. 1995; Parton et al. 1995; and many others). Increases in atmospheric CO_2 and potential changes in climate may influence and be influenced by terrestrial C storage. For example, the largest global pool of terrestrial C is detrital soil organic matter (OM), which is decomposed by microorganisms, releasing CO_2 to the atmosphere. If warmer temperatures increase decomposition rates, the resulting increase in atmospheric CO_2 could induce warming that would increase decomposition. The role of temperature in controlling decomposition is an important debate in contemporary global-change science. While much of the literature suggests that decomposition rates increase predictably with temperature (Meentemeyer 1984; Kirschbaum 1995; 2000; Townsend et al. 1997), some recent analyses (Liski et al. 1999; Giardina and Ryan 2000; Epstein et al. 2002) suggest that decomposition may be less responsive to temperature than previously thought.

The process of decomposition is an especially appropriate one to model because it is a conceptual construct, representing an amalgamation of multiple soil processes rather than a real biological process. Ecologists generally conceptualize decomposition as the transformation of organic materials into inorganic materials by heterotrophs, usually focusing on C. Decomposition includes both intracellular and extracellular enzyme-mediated breakdown of multiple classes of molecules and particular chemical bonds, with subsequent intracellular oxidation to CO_2. Some soil scientists favor a second, counterintuitive definition of decomposition as the transformation of plant litter into soil humic materials

(Paul and Clark 1996). We use the former definition because it is most closely related to the issue of terrestrial C storage.

Initially, biogeochemists abstracted the complex intra- and extracellular processes into a mass-specific litter (recently senesced foliage) decomposition rate, or k, assuming first- (or larger) order decay kinetics (Olson 1963). However, most detrital soil OM has been microbially, chemically, and physically altered (called humus) and decomposes more slowly than litter. In order to model decomposition of the entire soil OM pool (fresh litter plus humus), some biogeochemical models represented detrital OM as a series of pools with different decomposition rates. Organic matter pools were delineated by the type of material being decomposed; soluble plant and microbial materials were given more rapid decomposition constants than plant structural tissues or stabilized soil humic materials (Jenkinson and Rayner 1977; van Veen and Paul 1981). Many current models use this multiple-pool construct to simulate decomposition (Parton et al. 1994; King et al. 1997) and interactions among decomposition rates, atmospheric CO_2, and global warming.

Given the widespread use of simulation models to address global-change questions related to decomposition, it is important to evaluate how and how well biogeochemical models simulate the process. Evaluating how models simulate decomposition reveals whether there is consensus among biogeochemists regarding mechanistic controls on CO_2 release from soils. Model comparisons also allow us to evaluate how specific differences in model structure affect predicted decomposition rates. This type of comparison also generates a range of decomposition scenarios that can be used to evaluate our confidence in global-change predictions (VEMAP Members 1995, Chapter 12).

In this chapter, we use the relationship between temperature and OM decomposition as a case study for evaluating terrestrial biogeochemical models. Our approach is to compare extant models to determine if there is consensus about the effects of temperature on decomposition. We first evaluate the source of the empirical data used to develop the temperature-decomposition relationships in several terrestrial biogeochemistry models. Second, we evaluate the different model structures used to represent the process of decomposition and its control by temperature. Third, we compare the net relationship of temperature to decomposition across models; in doing so, we evaluate whether the models have sufficient congruence to conclude that terrestrial biogeochemists understand the role of temperature in controlling decomposition sufficiently well to provide this kind of information to policy makers. Finally, we suggest some alternative modeling approaches and comment on the ability of the extant models to serve as resources for the scientific community.

Empirical Data Sources

All simulation models require empirical data for development and validation. Several different sources of information have served as the foundation for temperature-decomposition relationships in current biogeochemical models. Throughout the chapter, "litter" always refers to recently senesced plant foliage, and "soil OM" includes all detrital soil OM (litter plus humus).

Litter Decomposition

Numerous experiments have evaluated litter decomposition rates across environmental gradients (Meentemeyer 1978, 1984, Meentemeyer and Berg 1986). These experiments consistently show a strong temperature response of litter decay; from their data Vitousek et al. (1994) and Gholz et al. (2000) calculated Q_{10}'s from 2 to 11. As with all geographic analyses, these studies are limited by the complex environmental gradients used to mimic climate variability. Factors that co-vary with temperature (precipitation, soil parent material, elevation, etc.) confound simple temperature-decomposition relationships. A second limitation is that these studies are confined to the litter component of decomposition; total soil OM decomposition is not measured.

Soil Organic C Pools

A second potential source of empirical data is geographic analysis of soil organic-C distributions along regional- or continental-scale temperature gradients. Trends in decomposition rates of the total OM pool can be inferred or calculated from knowledge or assumptions of the rates of OM production. These studies have strength in that they evaluate the long-term effect of temperature on total soil OM; however, they require assumptions or measurements of total net primary production. To date, such studies have not consistently shown increases in decomposition rates with increasing temperature. While global analyses (Post et al. 1982; Meentemeyer et al. 1985) and those from grasslands and Hawaii (Burke et al. 1989; Townsend et al. 1995; 1997; Epstein et al. 2002) indicated decreasing soil C with increasing temperature, two forest studies did not. Homann et al. (1995) analyzed soil organic-C patterns in the coniferous region of the Pacific Northwest U.S. and did not find the decreases in soil C that they expected with increases in mean annual temperature. Grigal and Ohmann (1992) found that temperature and AET (Actual Evapotranspiration) played relatively small and inconsistent roles in determining soil C. In addition, Epstein et al. (2002) found that the effect of temperature on decomposition was smaller than anticipated after eliminating the interaction between precipitation and temperature in the Great Plains region.

Soil Respiration

Field experiments have been conducted that analyze patterns in soil respiration across temperature gradients in space or time (Reiners 1968; Kicklighter et al. 1994; and many others reviewed by Singh and Gupta 1977; Schlesinger 1977; Raich and Schlesinger 1992; Lloyd and Taylor 1994) or that analyze the consequences of experimental warming for soil respiration (Billings et al. 1982; Peterjohn et al. 1993; McKane et al. 1997; Hobbie and Chapin 1998; Rustad and

Fernandez 1998; Bridgham et al. 1999; Saleska et al. 1999; and others reviewed in Rustad et al. 2001). Warming studies have strength in that they are controlled experimental manipulations; however, soil or ecosystem warming does not exclusively change temperature. In most systems, warming increases evaporation, decreases soil moisture, and potentially increases aeration, any of which may be proximal controls on decomposition (Shaver et al. 1992; Harte et al. 1995; Bridgham et al. 1999). These results be interpreted as a strength of an integrated field experiment approximating global warming, but it does not provide the kind of simple response to temperature that might be useful in developing or parameterizing simulation models.

Most warming and climate-gradient experiments show that soil respiration increases with temperature, and it is generally assumed that this is at least in part due to increases in decomposition (Schlesinger and Andrews 2000). However, decomposition rates cannot be directly inferred from soil respiration data because the fluxes include both plant-root respiration and heterotrophic respiration. A few studies have separated root respiration from OM decomposition by estimating soil respiration in plots with roots excluded. These studies suggest that Q_{10}'s for soil OM decomposition fall in the narrow range of 2 to 2.5 (Brumme 1995; Nakane et al. 1996; Boone et al. 1998).

Laboratory Incubations

Many laboratory incubations have been conducted in which temperature is varied and either litter decay or soil CO_2 evolution is measured in the absence of plants (reviewed by Kirschbaum 1995; Ågren et al. 1996; and Katterer et al. 1998). These experiments generally show increasing laboratory rates of decomposition with increased temperatures; Q_{10} values range between 2 and 5, with lower Q_{10} values at higher temperatures. As we will describe later, some of the early incubation experiments (e.g. Drobnik 1962, cited in Hunt 1977; Sorenson 1981) were used to develop the temperature-response curves of current OM simulation models.

While short incubations almost always show increases in decomposition at higher temperatures, longer incubations highlight interactions between substrate availability and temperature in controlling decomposition rates. Holland et al. (2000) incubated a variety of tropical soils and found consistent exponential increases (mean $Q_{10} = 2.37$) in heterotrophic respiration with temperature (up to 55°C) only during the first few days of the incubation. After the first week, the optimum temperature for decomposition shifted from 55°C to 45°C and then to 35°C between 10 and 24 weeks of incubation. Changes in the temperature sensitivity of decomposition presumably resulted from declines in labile C availability as the incubation progressed; decomposition of slower turnover C substrates (late in the incubation) may be less sensitive to temperature than more labile substrates. A synthesis of forest-soil incubation studies yielded similar results; decomposition measured in year-long incubations varied only slightly (and negatively) with site mean annual temperature (Giardina and Ryan 2000).

   While the relationship between OM decomposition and temperature is clearly isolated in laboratory experiments, soils are typically disturbed such that decomposition is not realistically limited by spatial C substrate heterogeneity or physicochemical protection of OM. In addition, experiments that focus on temperature responses are generally conducted under ideal moisture conditions, so that the many combinations of climatic limitations that might occur in the field are not tested. Finally, even simplified laboratory experiments show complex responses to temperature with time, prompting Daubenmire and Prusso (1963, 591) to state "It appears that outside the natural environment of the forest floor these ratings are of limited scientific value, that the speed of decomposition is to a remarkable degree determined by the temperature levels under which the saprobic communities develop and operate, and that higher temperatures do not always result in greater net decomposition over periods of many weeks."

Isotopic Methods

Carbon-14 dating and bomb ^{14}C tracer studies (a pulse of atmospheric ^{14}C derived from thermonuclear bomb testing in the 1960s) have both been used to justify simulating soil OM decomposition as a series of pools with different turnover times (Jenkinson and Rayner 1977; van Veen and Paul 1981; Trumbore et al. 1997). While recalcitrant soil C decomposition (turnover time = from centuries to millennia by ^{14}C dating) is apparently insensitive to regional temperature gradients (Paul et al. 1997), decomposition of actively cycling soil C (turnover time = from years to decades by ^{14}C bomb tracer) showed a Q_{10} of 3 to 3.8 in one cross-site comparison (Trumbore et al. 1997).

   The stable isotope ^{13}C can also be used to estimate decomposition rates when land-use change is accompanied by a vegetation shift from C_3 to C_4 (or vice versa) photosynthetic pathways. Giardina and Ryan (2000) collected data from 44 such land-use change studies and found that decomposition rates were not correlated with mean annual temperature at the sites.

Brief Introduction to the Models

We selected several terrestrial biogeochemical models that have been validated against field data and are being broadly used to represent ecosystem processes (Table 13.1). For space reasons, we have not included every biogeochemical model for description, but have included what we think are the key varieties of the models (e.g. we did not include MBL-GEM [Rastetter et al. 1991], which is similar in many ways to RothC, CENTURY, and PnET). Three of the models have been compared with one another in a simulation of global responses to climate change (VEMAP Members 1995); several were compared in terms of their representation of litter decomposition (Moorhead et al. 1999) or coniferous forest function (Ryan et al. 1996). Except for the litter decomposition comparison, these analyses revealed large differences in simulated decomposition rates that in part result from differences in the ways that the temperature-decomposition relationship is treated. Below, we describe each of the selected models, including the empirical data sources, validation against temperature changes, the structure of the decomposition component, and the functional relationship of temperature to decomposition.

Table 13.1. Biogeochemical models evaluated in this chapter.

Model	Decomposition Equation	Terms	C Pool Structure
TEM (Raiche et al. 1991)	$k = k_Q W_s e^{0.0693T}$	k = Decomposition rate (mo-1) k_Q = Site-specific litter quality constant (mo-1) W_s = Soil moisture/texture scalar T = Mean monthly air temperature	Detrital C
Forest-BGC (Running and Gower 1991)	$k_L = k_{max}[(T_s+W_s)/2]$ $k_s = 0.03k_L$	k_L = Leaf and root decomposition rate (yr^{-1}) k_{max} = Fixed maximum decomp. rate (0.5 yr^{-1}) T_s = Soil temperature scalar W_s = Soil moisture scalar k_s = Soil C decomposition rate (yr^{-1})	Leaf- and root-litter C Other detrital soil C
Biome-BGC (Hunt et al. 1996)	$k_L = k_Q T_s W_s$ $k_s = k_C T_s W_s$	k_L = leaf and root decomposition rate (d^{-1}) k_Q = site-specific litter quality constant (d^{-1}) T_s = soil temperature scalar W_s = soil moisture scalar	Leaf- and root-litter C Other detrital soil C

Table 13.1. *Continued*

Model	Decomposition Equation	Terms	C Pool Structure
		k_s = soil C decomposition rate (d^{-1}) k_c = fixed decomposition rate from CENTURY (d^{-1})	
CENTURY (Parton et al. 1994)	$k_1 = k_{max}T_sW_sC_s$ $k_2 = k_{max}T_sW_sQ_s$ $k_3 = k_{max}T_sW_s$	k_1 = Soil microbial decomposition rate (yr^{-1}) k_2 = Structural plant decomposition rate (yr^{-1}) k_3 = All other pools decomposition rate (yr^{-1}) k_{max} = Fixed maximum decomp. rate (yr^{-1}) T_s = Air temperature scalar W_s = Soil moisture scalar C_s = Soil texture scalar Q_s = Litter quality (lignin) scalar	Structural plant C Metabolic plant C Surface microbial C Soil microbial C Slow soil C Passive soil C
FAEWE (Van der Peijl and Verhoeven 1999)	$k = k_{max}(T_{as}/T_{ms})$	k = decomposition rate (wk^{-1}) k_{max} = maximum decomposition rate (wk^{-1}) T_{as} = actual soil temperature scalar T_{ms} = mean annual soil temperature scalar	Detrital soil C
PnET-II (Aber et al. 1997)	$R = 27.46e^{0.0684T}$	R = soil respiration (g $m^{-2} mo^{-1}$) T = mean monthly temperature	No detrital C pools

Table 13.1. *Continued*

Model	Decomposition Equation	Terms	C Pool Structure
Linkages (Pastor and Post 1986)	$k_L = -\ln\{1 - [0.98 + 0.09\text{AET} + (0.5 - 0.002\text{AET})(\text{L:N})]/100\}$ $k_t = 0.2$ $k_{sw} = 0.1$ $k_{lw} = 0.03$ $k_{dw} = 0.05$ $ks = H\{-0.0004(\text{N:C})/[-0.03 + (\text{N:C})]\}/N$	k_L = Root and leaf decomposition rate (yr^{-1}) AET = Actual evapotranspiration L:N = Litter lignin to nitrogen ratio k_t = Twig decomposition (yr^{-1}) k_{sw} = Small wood decomposition (yr^{-1}) k_{lw} = Large wood decomposition (yr^{-1}) k_{dw} = Decayed wood decomposition (yr^{-1}) k_s = Soil humus decomposition (yr^{-1}) H = Humus mass (Mg/ha) N = Total humus N (Mg/ha) C = Total humus C (Mg/ha)	Leaf + root litter C Soil humus C Twig C Small wood C Large wood C Decayed wood C
RothC (Coleman and Jenkinson 1999, user guide)	$k = 1 - e^{(-T_s W_s S_s k_{max}/12)}$	k = Decomposition rate for each pool (mo^{-1}) k_{max} = Maximum decomposition rate (yr^{-1}) T_s = Air temperature scalar W_s = Soil moisture scalar S_s = Soil cover scalar (typically 0.6)	Metabolic plant C Structural plant C Microbial biomass C Humic organic matter

Note: The left hand portions of decomposition equations are mass specific decomposition rates (heterotrophic respiration/carbon pool size) except PnET. The subscript s denotes a unitless scalar function. Constants in the Linkages model were rounded considerably. Temperature is in Celsius for all models.

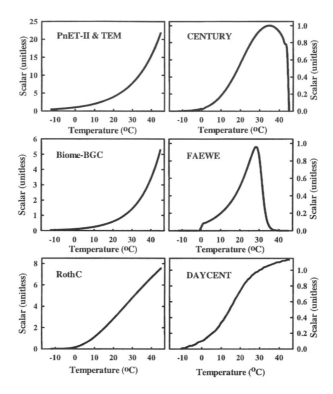

Figure 13.1. The temperature scalars used by models evaluated in this chapter. These scalars are multiplied by fixed decomposition constants to simulate the temperature effect on organic matter decomposition.

Rothamsted (RothC)

The Rothamsted model was the first simulation model published that represented soil OM decomposition as a multipool process (Jenkinson and Rayner 1977; Coleman and Jenkinson 1999). RothC models four detrital OM pools on a monthly time step using the same basic equation (Table 13.1); preset maximum decomposition rates (k_{max} in yr^{-1}) are modified by temperature, moisture, and plant cover scalars. Pools include structural plant material ($k_{max} = 0.3$), metabolic plant material ($k_{max} = 10$), microbial biomass ($k_{max} = 0.66$), and humified organic matter ($k_{max} = 0.02$). The pool structure was based on laboratory incubations of ^{14}C-labeled materials, radiocarbon dating of soil OM, and long-term measurements of soil OM content. Fixed k_{max} values were based on model runs tuned to field data from Rothamsted (Coleman and Jenkinson 1999). The temperature scalar (T_s; Figure 13.1a) is:

$$T_s = \frac{47.9}{1 + e^{(\frac{106}{T+18.3})}} \qquad (13.1)$$

where T is the average monthly air temperature (°C). This relationship was developed from decomposition of ^{14}C-labeled litter (Jenkinson and Ayanaba 1977; Ayanaba and Jenkinson 1990) but is used for all OM pools. The model was used to simulate changes in decomposition following global climate change (King et al. 1997), and results suggested that decomposition would increase but would be nearly balanced by increases in net primary production producing no net change in terrestrial C storage.

CENTURY

The CENTURY model was originally developed for grasslands and semi-arid agroecosystems (Parton et al. 1983, 1987, 1994); it has been modified to represent numerous ecosystem types (Schimel et al. 1996). Current versions exist with either monthly (Parton et al. 1994) or daily (Kelly et al. 2000) time steps. Similar to RothC, CENTURY represents OM decomposition as several pools with preset k_{max} values (mo^{-1}) modified by site-specific scalars (Table 13.1). The pools include aboveground plant structural material ($k_{max} = 3.9$), belowground plant structural material ($k_{max} = 4.9$), aboveground metabolic plant components ($k_{max} = 14.8$), belowground plant metabolic plant material ($k_{max} = 18.5$), aboveground microorganisms ($k_{max} = 6.0$), soil microorganisms (a.k.a. "active" soil OM, $k_{max} = 7.3$), slow-turnover soil OM ($k_{max} = 0.2$), leached OM (does not decompose), and passively turning over soil OM ($k_{max} = 0.0045$). The source of the initial k values for these pools is not described in the literature. CENTURY modifies k_{max} using temperature, moisture, and for some pools, soil-texture and litter-quality scalars (Table 13.1). The temperature scalar (T_s) is a generalized Poisson function (Parton et al. 1987; 1994):

$$T_s = (t_{max} - t)/(t_{max} - t_{opt})^{0.2} * \exp[(0.2/2.63)(1 - (t_{max} - t/t_{max} - t_{opt})^{2.63})] \qquad (13.2)$$

where t_{max} is 45°C and t_{opt} is 35 °C.

The temperature scalar is multiplied by a logistic moisture scalar to derive an abiotic scalar (DEFAC [Decomposition factor based on temperature and moisture]; Kelly et al. 2000). The original form of T_s (Parton et al. 1987) was determined by fitting data from an incubation experiment in which cellulose was labeled and decomposed in the laboratory at three different temperatures (Sorenson 1981). Monthly versions of CENTURY use this original T_s, and the daily version (DAYCENT) uses an arctangent function (R. Kelly, pers. comm.; Figure 13.1). The effect of temperature on maximum decomposition is identical for each pool.

CENTURY has been tested against field data in several ways. Simulated and measured soil OM values were compared across the Great Plains of the United States (Burke et al. 1989); this "validation" did not directly test the relationship of temperature to decomposition but provided support for the integrated-model representation of all processes influencing soil OM. Litter decomposition data were compared directly to the model (Vitousek et al. 1994), resulting in some modification of the litter decomposition model (which did not influence the temperature relationship). More recently, Gholz et al. (2000) found a strong correlation between DEFAC and litter decomposition rates from a continental scale field experiment, LIDET. Short-term estimates of CO_2 flux were compared to DAYCENT simulations of decomposition; this comparison is somewhat limited because soil CO_2 fluxes represent both heterotrophic and autotrophic respiration (Kelly et al. 2000).

When CENTURY was used to predict the responses of ecosystem C to global change, the simulations predicted that simple climatic warming reduced soil C globally, but that combinations of CO_2 increases with global climate change generally resulted in net C storage (Parton et al. 1995, VEMAP Members 1995, Schimel et al. 2000). With this model, simulated land-use change has more impact on stored C than does climate change or CO_2 increases.

TEM

The Terrestrial Ecosystem Model (TEM) model (Raich et al. 1991; Melillo et al. 1993; McGuire et al. 1995; 1997) was developed to simulate continental-to-global scale C and N balance. TEM contains only one detrital soil C pool, and decomposition is the only C loss from this pool (Raich et al. 1991; McGuire et al. 1995). Rather than scaling a fixed k_{max}, as in CENTURY and RothC, TEM determines a site-specific decomposition constant (here called k_Q) by comparing litter quality at the site to litter quality at one of the TEM calibration sites. This k_Q is then modified by temperature and moisture scalars on a monthly time step to simulate decomposition rates (Table 13.1). The temperature scalar is a simple exponential such that decomposition has a Q_{10} of 2.0 over all temperatures (Table 13.1). This fixed Q_{10} is based on a literature review of soil respiration from temperate forest soils (Kicklighter et al. 1994). The sensitivity of the model results to temperature have been explored extensively and compared with total C values (McGuire et al. 1995). Simulations of global climate change with TEM showed decreases in soil C and ecosystem C as a consequence of the decomposition sensitivity to temperature; however, simultaneous increases in CO_2 offset C losses by increasing NPP (Melillo et al. 1993; McGuire et al. 1995; 1997; VEMAP members 1995).

PnET-II

The PnET model (Aber et al. 1995) was designed to simulate C and water balance in northeastern U.S. temperate forests. PnET-II does not contain a complete C budget, in that it does not represent biomass production–decomposition feedbacks, track total soil C content, or allocate soil C into various turnover pools. Rather, it combines heterotrophic and live-root respiration into a simple logarithmic equation for soil respiration with mean monthly temperature as the only parameter (Table 13.1). Like TEM, the soil respiration-temperature relationship (Q_{10} = 2 for all temperatures) was developed for temperate zone forests by Kicklighter et al. (1994). PnET has been validated against field data on total net ecosystem CO_2 exchange and biomass production data. It is not used for climate change assessments since it lacks a soil C pool and feedbacks between temperature-induced decomposition increases and net primary production. An alternate version of the model (PnET-CN) contains one soil C pool similar in turnover to the active pool in CENTURY (Aber et al. 1997).

Linkages

Pastor and Post (1986) generated an individual-based model of tree growth with a link between productivity and decomposition through N availability. The purpose of the model was to simulate the interactions between plant community structure and ecosystem processes. The model differs from others that we evaluated in that cohorts of litter are modeled separately; each year's litter is tracked as a separate pool in the model. The simulations are for plots of 1/12 ha, considered the average gap size created by a dominant tree (Pastor and Post 1986). Linkages contains six detrital OM pools: leaf plus root litter; twigs; wood less than 10 cm DBH (diameter at breast height); wood greater than 10 cm DBH; well decayed wood; and humus. Leaf and root decomposition are modeled using field litterbag results from Meentemeyer (1978) and Melillo et al. (1982), which suggest that litter decomposition depends on AET, litter lignin concentration, and litter nitrogen concentration (Table 13.1).

Every time step, L:N is modified for a given cohort of litter (L:N ranges from 5 to 70) following relationships found by Aber and Melillo (1980) and Berg et al. (1985) between the fraction of OM remaining and N and lignin concentrations. Once the litter reaches a species-specific critical N concentration, it starts to mineralize N and is transferred to the soil humus pool. The humus pool is then decomposed according to the following equation derived from field net N mineralization (N_m) data in Wisconsin:

$$N_m = H\{-0.000379(N:C)/[-0.02984 + (N:C)]\} \tag{13.3}$$

where H is humus mass and N:C is the elemental ratio of litter forming the humus. Decomposition rates are determined by assuming that C is released from the humus pool in the same proportion as N_m is released from the humus pool (Table 13.1). Twig, small wood, large wood, and well-decayed wood pools have fixed decomposition constants (Table 13.1; Pastor and Post 1986), thus, only leaf and root litter decomposition are affected by temperature.

The model has been validated against primary production, biomass, nitrogen cycling, and plant species composition data for sites in Wisconsin, Michigan,

New Hampshire, and Minnesota (Pastor and Post 1986), but apparently not for soil OM content. Two climate-change sensitivity analyses (Pastor and Post 1988, Post and Pastor 1996) demonstrated significant simulated interactions among temperature, N mineralization, species composition, and net primary productivity.

Forest-BGC and Biome-BGC

Forest-BGC was developed to simulate C, water, and N cycles in forested eco-systems (Running and Coughlan 1988). The litterfall and decomposition elements of the model have an annual time step, while water balance and canopy gas exchange are modeled on a daily basis. The model includes two detrital soil OM pools: leaf plus root litter, and all other soil OM. Inputs into both pools come from leaves and roots, and the fraction of litterfall allocated to soil OM is determined by lignin content. Decomposition of large, woody components is not defined.

Forest-BGC initially used Meentemeyer's (1978) multiple regression to incorporate environmental and litter quality controls on decomposition of leaves and roots (Running and Coughlan 1988). A more recent version (Running and Gower 1991) calculates root and leaf decomposition based on climatic variables alone (Table 13.1); the source of these relationships is not clear from descriptive literature. Maximum litter decomposition is assumed to be 0.5 (yr^{-1}) and, like RothC and CENTURY, this maximum rate is then modified by temperature and moisture scalars. The temperature scalar (T_s) is:

$$T_s = [\Sigma(T_d/365)]/T_{opt} \tag{13.4}$$

where T_d is daily soil temperature and T_{opt} is optimum soil temperature set at $50°C$ (Running 1994). The rate of soil OM decomposition is a fixed proportion of litter decomposition rates. Running and Gower (1991) and Running (1994) use a fractional constant of 0.03 (i.e., the decomposition rate of the soil OM pool is 3% of the litter pool decomposition rate).

Biome-BGC (Hunt et al. 1996), a recent version of Forest-BGC uses an entirely different decomposition equation but still contains one litter and one soil OM pool (Table 13.1). In Biome-BGC, litter decomposition is determined similarly to TEM: as a site-specific decomposition rate based on litter quality, modified by soil moisture and temperature scalars. Soil OM decomposition is modeled by modifying a fixed k value (0.00035 d^{-1}; actually a combination of slow and active k's from CENTURY) by the same moisture and temperature scalars used for litter. The temperature scalar (T_s) yields a Q_{10} of 2.4 (Figure 13.1) and is based on soil respiration data (Raich and Schlesinger 1992):

$$T_s = e^{[0.08755(T-26)]} \tag{13.5}$$

where T is soil temperature.

Forest-BGC has been tested for a range of sites across a climatic gradient in Oregon (Running 1994), focusing on aboveground net primary production, stem biomass, and leaf nitrogen concentration. Biome-BGC showed more sensitivity to combined climate and CO_2 change scenarios than CENTURY or TEM in a recent model comparison (VEMAP Members 1995); warming caused losses in soil organic C that led to total ecosystem C losses. An entirely new decomposition subroutine exists in the current, unpublished, Biome-BGC code (© 2000. Peter Thornton. Biome-BGC Version 4.1.1. Numerical Terradynamics Simulation Group, School of Forestry, University of Montana, Missoula, MT). This new version models decomposition almost identically to CENTURY, with seven soil OM pools decomposed by modifying fixed k_{max} values with (again, new) moisture and temperature scalars.

FAEWE

We selected one wetland simulation model (Van der Peijl and Verhoeven 1999) as an example of decomposition modeled for systems with rare moisture limitation. The model was developed as part of the Functional Analysis of European Wetland Ecosystems (FAEWE) project and simulates C, N, and P dynamics in freshwater wetlands on a weekly time-step. The C submodel simulates three detrital OM pools: above- and belowground plant litter, and all other soil OM. Only the soil OM pool produces CO_2; the plant litter pools are inputs into the soil organic pool, but they do not respire. The model has been calibrated and run for only one site, a riverine grassland in southwestern England.

Like CENTURY, Forest-BGC, and RothC, FAEWE models decomposition by modifying a maximum decomposition rate (k_{max}) by temperature and moisture. However, in the wetland model, k_{max} is not fixed; rather it is a function of the redox potential of the soil, the level of the groundwater, and the oxygen content of the soil atmosphere ($k_{max} = 7.7 \times 10^{-4}$ in anaerobic conditions and 7.7×10^{-5} in aerobic conditions). Temperature modifies k_{max} through a ratio of temperature scalars (Table 13.1):

$$\text{Temperature effect} = T_{as}/T_{ms} \tag{13.6}$$

where T_{as} is based on actual soil temperature and T_{ms} is based on mean annual soil temperature via the following equation:

$$T_{as} \text{ or } T_{ms} = [0.003T*10^{(10.93686-3259.18/T)}]/[1+10^{(-632.649+172713.1/T)}+10^{(113.5406-34516.4/T)}] \tag{13.7}$$

where T is either actual or mean annual temperature in degrees Kelvin. These 0–1 scalar functions are described by an optimum temperature curve (Figure 13.1) that is based on absolute reaction-rate theory (Schoolfield et al. 1981). The maximum process rate is reached at approximately 30°C, and the left hand side of the curve is comparable to an exponential function with a Q_{10} near 2. This model has not been used to predict changes in soil C loss following global climate change. A sensitivity analysis showed that the model is most responsive to changes in the growth rates of plants (Van der Peijl and Verhoeven 1999).

Model Comparison I: Which Data Were Used?

One of the key questions we asked in evaluating models with respect to the temperature control over decomposition was: How similar were the data used to develop the models? RothC, CENTURY, and FAEWE base the temperature response on laboratory incubations across a range of temperatures under "optimum" moisture conditions. The incubations were conducted on litter, soils, and cellulose, and the models were originally developed for nonwoody ecosystems (grasslands, agroecosystems, and wetlands). These temperature relationships, generated from one substrate, are used to simulate all pools of OM. In addition, Roth-C and CENTURY are currently applied to all types of ecosystems, including forests (VEMAP Members 1995, King et al. 1997). Early Forest-BGC and Linkages both represent temperature-decomposition relationships using the forest-litter decomposition experiments from Meentemeyer (1978, 1984), Meentemeyer and Berg (1986) and Melillo et al. (1982). In more recent versions of Forest-BGC, the litter decay characteristics do not depend on litter quality, and soil OM decomposition is a constant percentage (3%) of litter decomposition. Finally, the temperature-decomposition relationships in TEM, Biome-BGC, and PnET-II were developed primarily from field data on soil respiration (warming experiments and interannual variability in undisturbed systems). The data represent both autotrophic and heterotrophic respiration but are being applied to heterotrophic respiration. Few data points exist above 22°C, and none below 0°C, though TEM and Biome-BGC are applied globally (Hunt et al. 1996; McGuire et al. 1997). As Daubenmire and Prusso (1963) and Niklińska et al. (1999) suggest, it is very likely that optimum temperatures for decomposition vary with the climate and evolutionary history of individual locations.

This review suggests that the temperature-decomposition relationships in modern biogeochemical models (assuming the models we reviewed are representative of the discipline) are based on just a few, imperfect data sources: environmental gradients in litter decomposition or soil respiration, or laboratory decomposition experiments. On one hand, these shortcomings are not surprising; there have been few, if any experiments that are completely appropriate for the development of relationships between OM decomposition and temperature (see Empirical Data Sources section), and model builders worked with the best data available. On the other hand, it is clear that these very restricted datasets (representing specific OM substrates, ecosystems, and climatic condition) have been used to generate models that are broadly applied outside the range of those datasets. We feel that there is a very strong need for long-term experimental data on the relationship of temperature to decomposition. Many such experiments have been initiated (Rustad et al. 2001); the challenge now is to estimate decomposition rates (as opposed to CO_2 flux) from those experiments.

Model Comparison II: Consensus Understanding of Temperature Controls on Decomposition?

The models evaluated in this chapter contain fundamentally different structures based on a few imperfect data sources. To test how variation in model structure affects simulated decomposition rates, we conducted a sensitivity analysis of decomposition to changing air (and soil) temperature.   All models were parameterized for a single site, the Konza Prairie Long-Term Ecological Research Site in Kansas (Appendix A includes details regarding parameterization data [http://www.ecostudies.org/cary9/appendicics.html]). In most cases, only the decomposition equations (Table 13.1) were parameterized, but for models with complex C-pool structure (CENTURY and RothC), models were run to steady state to obtain pool sizes for the model-specific C fractions. We could not parameterize Linkages for this grassland site because species- and site-specific relationships determining critical N concentrations and net N mineralization rates were not available. We did not simulate interactions between temperature and C inputs; rather, we focused solely on decomposition.

There was great variability in the sensitivity of simulated decomposition rates to large changes in air temperature (Figure 13.2). At the mean annual air temperature for Konza (~ 13°C), decomposition rates varied by an order of magnitude (from 0.02 for TEM to 0.18 for CENTURY) among the models. For reference, field data for total net primary production (225 g C m^{-2}yr^{-1} both above and below ground; John Blair, Kansas State University, pers. comm.) and total soil C (5000 g C m^{-2} in top 15 to 20 cm) suggest that actual decomposition rates at Konza are about 0.09 yr^{-1}. However, soil depth varies greatly (from 10 to 200 cm) at Konza, so soil C values could be higher, and thus k values lower, in deeper soil profiles. Similarly, C inputs vary greatly as a result of wildfire, and lower inputs could decrease our calculated reference k value.

The shape of the response of k to changing temperature ranged from linear to exponential to unimodal. In general, the shape of the temperature scalar (Figure 13.1) could be used to predict the shape of the k response. Biome-BGC and PnET-II had exponential temperature scalars, and both showed exponential responses to changing temperature, although k values differed greatly between these two models. RothC had a linear temperature scalar over most of the range we analyzed, and k responded linearly to changes in temperature. Models with optimal temperature scalars (CENTURY and FAEWE, Figure 13.1) showed optimal responses to changing temperature. However, the peaks in k (between 18 and 20°C) did not coincide with maxima in the temperature scalars for these models (30 or 35°C). Two factors likely caused this shift in maximum decomposition. In both models, decomposition during some months (CENTURY) or

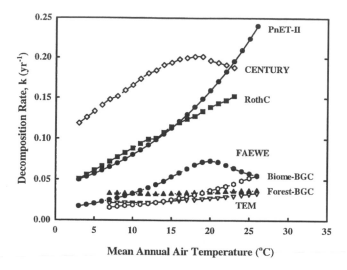

Figure 13.2. The relationship between temperature and mass specific decomposition at the Konza Prairie Long-Term Ecological Research site as simulated by seven biogeochemical models.

weeks (FAEWE) exceeded the optimum temperature, causing a decrease in annual k values at higher temperatures, and in CENTURY, the soil moisture scalar decreased k at higher temperatures.

In one model, the temperature scalar and k response curves did not have similar shapes. TEM has an exponential temperature scalar, but k changed only slightly with temperature and in our sensitivity analysis. The lack of temperature response in TEM resulted from a strong moisture limitation to decomposition. When we fixed W_s in TEM to reflect optimal soil moisture, k increased exponentially (from 0.02 to 0.09 yr^{-1}) with temperature.

Finally, in Forest-BGC the temperature scalar and k had similar shapes (linear), but both were insensitive to temperature changes. This model contains a site-specific temperature scalar that increased linearly from 0.31 to 0.44 at Konza. Similarly, k increased linearly from 0.033 yr^{-1} near mean annual temperature to 0.038 yr^{-1} at 15°C above mean temperature. When we optimized W_s (fixed soil moisture at field capacity) in Forest-BGC, the k-temperature relationship still varied only slightly with temperature, increasing from 0.034 to 0.049 yr^{-1} over the entire range.

Many of the other models included moisture scalars in the decomposition equation. In these cases, moisture limitations dampened linear or exponential increases in k with temperature (Biome-BGC, RothC) or shifted the temperature at which k was greatest (CENTURY and FAEWE; see above). In PnET, soil moisture has no direct effect on mass-specific soil respiration (here decomposition), so increases in temperature result in a simple Q_{10} response. Soil moisture would also affect actual (as opposed to mass-specific) decomposition in most of the models by affecting C inputs and thus pool sizes. In this temperature sensitivity analysis we avoided moisture effects on substrate availability; however, in a complete climate-change simulation, variation in moisture or any other factor that alters primary production might alter k if simulated C inputs change at a different rate than the C losses we are focusing on here.

Most of the models simulate decomposition by modifying a base or maximum k (e.g., k_{max} and k_Q in Table 13.1) by temperature and moisture scalars. Variations in these base or maximum k values lead to variation in the magnitude of k without affecting the shape of the k-temperature relationship. Direct connections between field data and these fixed k_{max} or k_Q values were rarely available in model documentation. In some models, baseline k's are calculated using litter quality (TEM and Biome-BGC). Thus, we assume that litterbag studies were important sources. In other cases, very recalcitrant OM pools are included, so we assume that ^{14}C dating (RothC and CENTURY) was an important source.

In a final analysis of model consensus regarding temperature controls on decomposition, we calculated changes in the Q_{10} of the temperature scalars for each model over a range of temperatures (Fig. 13.3). The Q_{10} for a given temperature was calculated by dividing the scalar value at 5 degrees above the temperature of interest by the scalar value at 5 degrees below the temperature of interest [(scalar at $T + 5°C$)/(scalar at $T - 5°C$)]. This analysis assumes that temperature is the only factor affecting decomposition rates and that the Q_{10} of the scalar is equal to the Q_{10} of decomposition (which is true for the multiplicative scalars).

All of the models suggest that OM decomposition should have a Q_{10} between 1 and 3 over the temperature range of 17 to 22°C (Fig. 13.3). PnET, TEM, and Biome-BGC had constant Q_{10} values over the entire temperature range. All of the other models predict that decomposition is highly sensitive to temperature at low temperatures and less sensitive to temperature at higher temperatures. CENTURY and RothC show Q_{10} values greater than 4 for temperatures less than 10°C. Similarly, DAYCENT and FAEWE show large increases in Q_{10}'s for temperatures less than 5°C. At temperatures greater than 25°C, CENTURY, RothC, and DAYCENT suggest a Q_{10} near 1, and with FAEWE, it drops to zero.

Our review suggests three points of consensus in the way simulation models treat the temperature-decomposition relationship. First, there appears to be a consensus that a practical mathematical representation of decomposition is a fixed maximum or baseline rate modified over time by a series of temperature, moisture, and litter-quality scalars. Second, all models agree that the temperature scalar should increase with increasing temperature (at least up to 30°C) and that a Q_{10} of 1 to 3 is likely for temperatures between 17 and 22°C. Finally, the

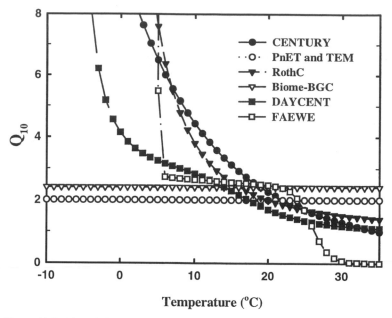

Figure 13.3. The relationship between temperature and the Q_{10} of temperature scalars used to calculate decomposition in the models evaluated in this chapter.

models agree that mass-specific decomposition is not negatively affected by increasing temperatures.

Beyond these (perhaps trivial) points of agreement, the models varied greatly. Carbon-pool structure ranged from a single detrital C pool to seven pools (Table 13.1). Temperature scalars were linear, exponential, optimal, or mixed (Figure 13.1). In some models moisture had no effect on k, but in others the moisture scalars caused k to remain constant with increasing temperature (Figure 13.2). Differences in these three factors (C-pool structure, temperature scalars, and moisture interactions) caused simulated k values to differ by an order of magnitude among the models. We conclude that there is little consensus among models in the response of k to temperature change and that this lack of consensus constrains our ability to predict ecosystem responses to global change. For example, in the VEMAP (1995) comparison, predicted terrestrial C storage ranged from -39 to +32% of current C storage depending on which biogeochemical model was used. The best way to reduce variability in simulated decomposition is not to make all models arbitrarily similar, but rather, to increase the amount of real OM decomposition data available to build and validate the models.

Some Alternative Modeling Approaches

Our sense is that modeling is such an important activity for our discipline (Lauenroth et al. 1998 and this volume) that it is very important that a diversity of models be developed and tested. A diverse array of models provides alternate hypotheses that may stimulate further empirical investigations and augment our understanding of ecosystem behavior (Chapter 12). We suggest that two under-used modeling approaches could yield insight into decomposition dynamics under climate-change scenarios. First, all of the models we evaluated represent decomposition as a series of discrete pools with different decomposition rates. However, another interpretation of decomposition is that OM decays through a continuum of stages starting with fresh litter and continuing through recalcitrant humus (Bossatta and Ågren 1991). A continuous, rather than discrete, interpretation of OM decomposition can be modeled by following individual cohorts of litter and calculating OM quality as a function of time, mass loss, or N concentration (Pastor and Post 1988; Bossatta and Ågren 1991). An important advantage of these models is that they could simultaneously predict changes in plant species composition and biogeochemistry under altered climates (Pastor and Post 1988). Currently, species changes are predicted from biogeography models, and biogeochemical changes are predicted in separate biogeochemical models such as those evaluated here (VEMAP Members 1995).

A second underused approach is to model discrete soil C pools with differential responses to changing temperature. The models we evaluated apply one temperature scalar to all detrital C pools. For example, in CENTURY passive OM decomposition uses the same temperature scalar as active OM decomposition. The assumption that all pools of OM respond similarly to temperature is contradicted by both theory and data (Trumbore et al. 1997; Ågren 2000; Giardina and Ryan 2000; Holland et al. 2000). In at least one case, simulating recalcitrant OM turnover with a weaker response to temperature improved the model agreement with a regional OM gradient (Liski et al. 1999), although this result may reflect the structure of the model more than a mechanistic explanation of the climate gradient (Ågren 2000). The models we evaluated also assumed that the temperature-decomposition relationship was constant across ecosystem types and climatic locations. A recent laboratory incubation suggests that this assumption is not always valid: decomposition of OM beneath Scots pine in northern Europe was more sensitive to changes in temperature than OM beneath Scots pine in southern Europe (Niklińska et al. 1999).

Models as a Community Resource

Model builders have the unique responsibility of synthesizing theory and data from the greater scientific community into a cohesive representation of the state of knowledge in biogeochemistry. Model users (we acknowledge a false dichotomy here because many are both builders and users of models), in turn, have the responsibility of applying these models to test and develop biogeochemical theory and discover their implications for global change. Thus, we view models as resources, by which our collective knowledge is synthesized into working theories that are tested and retested through various applications. One goal of this model comparison was to evaluate certain models as tools for the scientific community.

It seems clear that model builders agree that models are a community resource; most models used for this review were readily available at websites with detailed documentation and instructions for use. It was relatively simple to run the models and dissect model structure to isolate the decomposition equations and parameters that controlled decomposition. It was much more difficult to determine the source of the empirical data used to develop the relationship that defined how temperature controls decomposition. In some cases we noticed significant changes in this functional relationship among papers that represented different versions of the models, but contained no descriptions of the data used to adjust those relationships. In other papers, we found detailed descriptions of the models and the functional influence of temperature on decomposition without citations of how the relationship was developed. Details were also lacking on the origin of k_{max} and k_Q values. This lack of documentation placed constraints on our ability to evaluate the models and to test them on independent datasets (since we do not know what is independent).

Synthesis

Perhaps the most important conclusion in our evaluation of decomposition-temperature relationships in current biogeochemical models is that they are based on a few, imperfect data sources. Litter decomposition, soil respiration, and laboratory incubations were the main sources of data used to develop temperature-decomposition relationships in the models we evaluated. Well-controlled field measurements of OM decomposition rates are exceedingly rare but are required to build more realistic decomposition algorithms.

The lack of field data was not the only cause of variability among models; similarities and differences in model structures were also important. Most models fix a maximum or baseline decomposition rate that is modified by temperature and moisture scalars. The temperature scalars all suggest that decomposition should increase with temperature; however, there is little agreement among models regarding the shape of the temperature-decomposition relationship. Scalar shape was a particularly important difference among the models because it usually predicted the shape of the k-temperature relationship. Finally, the origins of the temperature scalars and of maximum and baseline k values are poorly documented, making model comparison and validation difficult.

While there was some agreement among the models, our analysis suggests that there are sufficient differences to cast doubt on the solidity of our understanding of temperature-decomposition relationships. Clearly, temperature increases the decomposition rates of some soil OM, but we lack knowledge about the range of the response for parts of the globe that are very cool or very warm, and for recalcitrant pools of OM, which comprise most of the terrestrial C storage. Thus, modeled estimates of global changes in C due to warming will have a very high degree of variability and depend strongly upon the model used. Consequently, we suggest prudence in providing detailed simulation results on global warming to policy makers (Chapter 7). We also suggest that there is an exciting area of new research into the mechanistic control of decomposition by temperature.

References

Aber, J.D., C. Driscoll, C.A. Federer, R. Lathrop, G. Lovett, J.M. Melillo, P. Steudler, and J. Vogelmann. 1993. A strategy for the regional analysis of the effects of physical and chemical climate change on biogeochemical cycles in northeastern (U.S.) forests. *Ecological Modelling* 67: 37–47.

Aber, J.D., and J. M. Melillo. 1980. Litter decomposition: Measuring relative contributions of organic matter and nitrogen to forest soils. *Canadian Journal of Botany* 58: 416–421.

Aber, J.D., J.M. Melillo, K.J. Nadelhoffer, J. Pastor, and R.D. Boone. 1991. Factors controlling nitrogen cycling and nitrogen saturation in northern temperate forest ecosystems. *Ecological Applications* 1: 303–315.

Aber, J.D., S.W. Ollinger, and C.T. Driscoll. 1997. Modeling nitrogen saturation in forest ecosystems in response to land use and atmospheric deposition. *Ecological Modelling* 101: 61–78.

Aber, J.D., S.V. Ollinger, C.A. Federer, P.B. Reich, M.L. Goulden, D.W. Kicklighter, J.M. Melillo, and R.G. Lathrop. 1995. Predicting the effects of climate change on water yield and forest production in the northeastern United States. *Climate Research* 5: 207–222.

Ågren, G. 2000. Temperature dependence of old soil organic matter. *Ambio* 29: 55.

Ågren, G.I., M.U.F. Kirschbaum, D.W. Johnson, and E. Bossatta. 1996. Ecosystem physiology: Soil organic matter. Pp. 207–228 in A.I. Breymeyer, D.O. Hall, J.M. Melillo, and G.I. Ågren, editors. *Global Change: Effects on Coniferous Forests and Grasslands*. New York: John Wiley and Sons.

Ayanaba, A., and D.S. Jenkinson. 1990. Decomposition of C-14-labeled ryegrass and maize under tropical conditions. *Soil Science Society of America Journal* 54: 112–115.

Berg, B., G. Ekbohm, and C.A. McClaugherty. 1985. Lignin and hemicellulose relations during long-term decomposition of some forest litters. *Canadian Journal of Botany* 62: 2540–2550.

Billings, W.D., J.O. Luken, D.A. Mortensen, and K.M. Peterson. 1982. Arctic tundra: A source or sink for atmospheric carbon dioxide in a changing environment? *Oecologia* 53: 7–11.

Boone, R.D., K.J. Nadelhoffer, J.D. Canary, and J.P. Kaye. 1998. Roots exert a strong influence on the temperature sensitivity of soil respiration. *Nature* 396: 570–572.

Bossatta, E., and G.I. Ågren. 1991. Dynamics of carbon and nitrogen in the organic matter of the soil: A generic theory. *American Naturalist* 138: 227–245.

Bridgham, S.D., J. Pastor, K. Updegraff, T.J. Malterer, K. Johnson, C. Harth, and J. Chen. 1999. Ecosystem control over temperature and energy flux in northern peatlands. *Ecological Applications* 9: 1345–1358.

Brumme, R. 1995. Mechanisms of carbon and nutrient release and retention in beech forest gaps. III. Environmental regulation of soil respiration and nitrous oxide emissions along a microclimatic gradient. *Plant and Soil* 168/169: 593–600.

Burke, I.C., T.G.F. Kittel, W.K. Lauenroth, P. Snook, C.M. Yonker, W.J. Parton. 1991. Regional analysis of the central great plains, sensitivity to climate variability. *BioScience* 41: 685–692.

Burke, I.C., C.M. Yonker, W.J. Parton, C.V. Cole, K. Flach, and D.S. Schimel. 1989. Texture, climate, and cultivation effects on soil organic matter content in U.S. grassland soils. *Soil Science Society of America Journal* 53: 800–805.

Coleman, K., and D.S. Jenkinson. 1999. *RothC-26.3: A model for the turnover of carbon in soil.* Model description and Windows users guide. November 1999. Harpenden, UK: Lawes Agricultural Trust.

Daubenmire, R., and D.C. Prusso. 1963. Studies of decomposition rates of tree litter. *Ecology* 44: 589–592.

Drobnik, J. 1962. The effect of temperature on soil respiration. *Folia Microbiology* 7: 132–140.

Epstein, H.E., I.C. Burke, and W.K. Lauenroth. 2002. Regional patterns of decomposition and primary production rates in the U.S. Great Plains. *Ecology* 83: 320327.

Gholz, H.L., D.A. Wedin, S.M. Smitherman, M.E. Harmon, and W.J. Parton. 2000. Long-term dynamics of pine and hardwood litter in contrasting environments: Toward a global model of decomposition. *Global Change Biology* 6: 751765.

Giardina, C.P., and M.G. Ryan. 2000. Evidence that decomposition rates of organic carbon in mineral soil do not vary with temperature. *Nature* 404: 858–861.

Grigal, D.F., and L.F. Ohmann. 1992. Carbon storage in upland forests of the Lake States. *Soil Science Society of America Journal* 56: 935–943.

Harte, J., M.S. Torn, F.-R. Change, B. Feifarek, A.P. Kinzig, R. Shaw, and K. Shen. 1995. Global warming and soil microclimate: results from a meadow-warming experiment. *Ecological Applications* 5: 132–150.

Hobbie, S.E., and F.S. Chapin. 1998. Response of tundra plant biomass, aboveground production, nitrogen, and CO_2 flux to experimental warming. *Ecology* 79: 1526–1544.

Holland, E.A., J.C. Neff, A.R. Townsend, and B. McKeown. 2000. Uncertainties in the temperature sensitivity of decomposition in tropical and subtropical ecosystems: Implications for models. *Global Biogeochemical Cycles* 14: 1137–1151.

Homann, P.S., P. Sollins, H.N. Chappell, and A.G. Stangenberger. 1995. Soil organic carbon in a mountainous, forested region: Relation to site characteristics. *Soil Science Society of America Journal* 59: 1468–1475.

Hunt, H.W. 1977. A simulation model for decomposition in grasslands. *Ecology* 58: 469–484.

Hunt, E.R. Jr., S.C. Piper, R. Nemani, C.D. Keeling, R.D. Otto, and S.W. Running. 1996. Global net carbon exchange and intra-annual atmospheric CO_2 concentrations predicted by an ecosystem process model and three-dimensional atmospheric transport model. *Global Biogeochemical Cycles* 10: 431–456.

Jenkinson, D.S., and A. Ayanaba. 1977. Decomposition of carbon-14-labeled plant material under tropical conditions. *Soil Science Society of America Journal* 41: 12–915.

Jenkinson, D.S., and J.H. Rayner. 1977. The turnover of soil organic matter in some of the Rothamsted classical experiments. *Soil Science* 123: 298–305.

Katterer, T., M. Reichstein, O. Andren, and A. Lomander. 1998. Temperature dependence of organic matter decomposition: A critical review using literature data analyzed with different models. *Biology and Fertility of Soils* 27: 258–262.

Kelly, R.H., W.J. Parton, M.D. Hartman, L.K. Stretch, D.S. Ojima, and D.S. Schimel. 2000. Intra-annual and interannual variability of ecosystem processes in shortgrass steppe. *Journal of Geophysical Research* 105 (D15): 20093–20100.

Kicklighter, D.W., J.M. Melillo, W.T. Peterjohn, E.B. Rastetter, and A.D. McGuire. 1994. Aspects of spatial and temporal aggregation in estimating regional carbon dioxide fluxes from temperate forest soils. *Journal of Geophysical Research* 99: 1303–1315.

King, A.W., W.M. Post, S.D. Wullschleger. 1997. The potential response of terrestrial carbon storage to changes in climate and atmospheric CO_2. *Climatic Change* 35: 199–227.

Kirschbaum, M.U.F. 1995. The temperature dependence of soil organic matter decomposition, and the effect of global warming on soil organic C storage. *Soil Biology and Biochemistry* 27: 753–760.

———. 2000. Will changes in soil organic carbon act as a positive or negative feedback on global warming? *Biogeochemistry* 48: 21–51.

Lauenroth, W.K, C.D. Canham, A.P. Kinzig, K.A. Poiani, W.M. Kemp, and S.W. Running. 1998. Simulation modeling in ecosystem science. Pp. 404–415 in M.L. Pace and P.M. Groffman, editors. *Successes, Limitations, and Frontiers in Ecosystem Science*. New York: Springer-Verlag.

Liski, J., H. Ilvesniemi, A. Makela, and C.J. Westman. 1999. CO_2 emissions from soil in response to climatic warming are overestimated: The decomposition of old soil organic matter is tolerant of temperature. *Ambio* 28: 171–174.

Lloyd, J., and J. Taylor. 1994. On the temperature dependence of soil respiration. *Functional Ecology* 8: 315–323.

McGuire, A.D., J.M. Melillo, D.W. Kicklighter, and L.A. Joyce. 1995. Equilibrium responses of soil carbon to climate change: Empirical and process-based estimates. *Journal of Biogeography* 22: 785–796.

McGuire, A.D., J.M. Melillo, Y. Pan, X. Xiao, J. Jelfrich, B.I. Moore, C.J. Vörösmarty, and A.L. Schloss. 1997. Equilibrium responses of global net primary production and carbon storage to doubled atmospheric carbon dioxide: Sensitivity to changes in vegetation nitrogen concentration. *Global Biogeochemical Cycles* 11: 173–189.

McKane, R.B., E.B. Rastetter, G.R. Shaver, K.J. Nadelhoffer, A.E. Giblin, J.A. Laundre, and F.S. Chapin. 1997. Climatic effects on tundra carbon storage inferred from experimental data and a model. *Ecology* 78: 1170–1187.

Meentemeyer, V. 1978. Macroclimate and lignin control of litter decomposition rates. *Ecology* 59: 465–472.

———. 1984. The geography of organic decomposition rates. *Annals of the Association of American Geographers* 74: 551–560.

Meentemeyer, V., and B. Berg. 1986. Regional variation in rate of mass loss of *Pinus sylvestris* needle litter in Swedish pine forests as influenced by climate and litter quality. *Scandinavian Journal of Forest Research.* 1: 167–180.

Meentemeyer, V., J. Gardner, and E.O. Box. 1985. World patterns and amounts of detrital soil carbon. *Earth Surface Processes and Landforms* 10: 557–567.

Melillo, J.M., J.D. Aber, and J.F. Muratore. 1982. Nitrogen and lignin control of hardwood leaf litter decomposition dynamics. *Ecology* 63: 621–626.

Melillo, J.M., A.D. McGuire, D.W. Kicklighter, B. Moore III, C.J. Vörösmarty, and A.L. Schloss. 1993. Global climate change and terrestrial net primary production. *Nature* 363: 234–240.

Moorhead, D.L., W.S. Currie, E.B. Rastetter, W.J. Parton, and M.E. Harmon. 1999. Climate and litter quality controls on decomposition: An analysis of modeling approaches. *Global Biogeochemical Cycles* 13: 575–589.

Nakane, K., T. Kohno, and T. Horikoshi. 1996. Root respiration rate before and just after clear-felling in a mature, deciduous, broad-leaved forest. *Ecological Research* 11: 111–119.

Niklińska, M., M. Maryański, and R. Laskowski. 1999. Effect of temperature on humus respiration rate and nitrogen mineralization: Implications for global climate change. *Biogeochemistry* 44: 239–257.

Olson, J.S. 1963. Energy storage and the balance of producers and decomposers in ecological systems. *Ecology* 44: 322–331.

Parton, W.J., D.S. Ojima, C.V. Cole, and D.S. Schimel. 1994. A general model for soil organic matter dynamics: Sensitivity to litter chemistry, texture, and management. *Quantitative Modeling of Soil Forming Processes, SSSA Special Publication* 39: 147–167.

Parton, W.J., D.W. Anderson, C.V. Cole, and J.W.B. Stewart. 1983. Simulation of soil organic matter formations and mineralization in semiarid agroecosystems. Pp. 533–550 in R.R. Lowrance, R.L. Todd, L.E. Asmussen, and R.A. Leonard, editors. *Nutrient Cycling in Agricultural Ecosystems.* Special Publication No. 23. Univ. of Georgia. College of Agricultural Experiment Stations. Athens, GA.

Parton, W.J., D.S. Schimel, C.V. Cole, and D.S. Ojima. 1987. Analysis of factors controlling soil organic matter levels in great plains grasslands. *Soil Science Society of America Journal* 51: 1173–1179.

Parton, W.J., J.M.O. Scurlock, D.S. Ojima, D.S. Schimel, D.O. Hall, and SCOPEGRAM members. 1995. Impact of climate change on grassland production and soil carbon worldwide. *Global Change Biology* 1: 13–22.

Pastor, J., and W.M. Post. 1986. Influence of climate, soil moisture, and succession on forest carbon and nitrogen cycles. *Biogeochemistry* 2: 3–27.

———. 1988. Response of northern forests to CO_2-induced climate change. *Nature* 334: 55–58.

Paul, E.A., and F. Clark. 1996. *Soil Microbiology and Biochemistry.* New York: Academic Press.

Paul, E.A., R.F. Follett, S.W. Leavitt, A. Halvorson, G.A. Peterson, and D.J. Lyon. 1997. Radiocarbon dating for determination of soil organic matter pool sizes and dynamics. *Soil Science Society of America Journal* 61: 1058–1067.

Peterjohn, W.T., J.M. Melillo, F.P. Bowles, and P.A. Steudler. 1993. Soil warming and trace gas fluxes: Experimental design and preliminary flux results. *Oecologia* 93: 18–24.

Post, W.M., W.R. Emanuel, P.J. Zinke, and A.G. Stangenberger. 1982. Soil carbon pools and world life zones. *Nature* 298: 156–159.

Post, W.M., and J. Pastor. 1996. Linkages: An individual-based forest ecosystems model. *Climatic Change* 34: 253–261.

Raich, J.W., E.B. Rastetter, J.M. Melillo, D.W. Kicklighter, P.A. Steudler, B.J. Peterson, A.L. Grace, B.I. Moore, and C.J. Vörösmarty. 1991. Potential net primary productivity in South America: Application of a global model. *Ecological Applications* 1: 399–429.

Raich, J.W., and W.S. Schlesinger. 1992. The global carbon dioxide flux in soil respiration and its relationship to vegetation and climate. *Tellus* 44B: 81–99.

Rastetter, E.B., M.G. Ryan, G.R. Shaver, J.M. Melillo, K.J. Nadelhoffer, J.E. Hobbie, and J.D. Aber. 1991. A general biogeochemical model describing the responses of the C and N cycles in terrestrial ecosystems to changes in CO_2, climate, and N deposition. *Tree Physiology* 9: 101–126.

Reiners, W.A. 1968. Carbon dioxide evolution from the floor of three Minnesota forests. *Ecology* 49: 471–483.

Running, S.W. 1986. Global primary production from terrestrial vegetation: Estimates integrating satellite remote sensing and computer simulation technology. *Science of the Total Environment* 56: 233–242.

————. 1994. Testing Forest-BGC ecosystem process simulations across a climatic gradient in Oregon. *Ecological Applications* 4: 238–247.

Running, S.W., and J.C. Coughlan. 1988. A general model of forest ecosystem process for regional applications. I. Hydrologic balance, canopy gas exchange and primary production processes. *Ecological Applications* 42: 125–54.

Running, S.W., and S.T. Gower. 1991. Forest-BGC, a general model of forest ecosystem processes for regional applications. II. Dynamic carbon allocation and nitrogen budgets. *Tree Physiology* 9: 147–160.

Rustad, L.E., J.L. Campbell, G.M. Marion, R.J. Norby, M.J. Mitchell, A.E. Hartley, J.H.C. Cornelissen, and J. Gurevitch. 2001. A meta-analysis of the response of soil respiration, net nitrogen mineralization, and aboveground plant growth to experimental ecosystem warming. *Oecologia* 126: 543–562.

Rustad, L.E., and I.J. Fernandez. 1998. Experimental soil warming effects on CO_2 and CH_4 flux from a low elevation spruce-fir forest soil in Maine, USA. *Global Change Biology* 4: 597–605.

Ryan, M.G., E.R. Hunt Jr., R.E. McMurtrie, G.I. Ågren, J.D. Aber, A.D. Friend, E.B. Rastetter, W.M. Pulliam, R.J. Raison, and S. Linder. Comparing models of ecosystem function for temperate conifer forests. I. Model description and validation. Pp. 313–362 in A.I. Breymeyer, D.O. Hall, J.M. Melillo, and G.I. Ågren, editors. *Global Change: Effects on Coniferous Forests and Grasslands.* New York: John Wiley and Sons.

Saleska, S.R., J. Harte, and M.S. Torn. 1999. The effect of experimental ecosystem warming on CO_2 fluxes in a montane meadow. *Global Change Biology* 52: 125–141.

Schimel, D.S., J. Melillo, H.Q. Tian, A.D. McGuire, D. Kicklighter, T. Kittel, N. Rosenbloom, S. Running, P. Thornton, D. Ojima, W. Parton, R. Kelly, M. Sykes, R. Neilson, and B. Rizzo. 2000. Contribution of increasing CO_2 and climate to carbon storage by ecosystems in the United States. *Science* 287: 2004–2006.

Schlesinger, W.H. 1977. Carbon balance in terrestrial detritus. *Annual Review of Ecology and Systematics* 8: 51–81.

Schlesinger, W.H., and J.A. Andrews. 2000. Soil respiration and the global carbon cycle. *Biogeochemistry* 48: 7–20.

Schoolfield, R.M., P.J.H. Sharpe, and C.E. Magnuson. 1981. Non-linear regression of biological temperature-dependent rate models based on absolute reaction-rate theory. *Journal of Theoretical Biology* 88: 719–731.

Shaver, G.R., W.D. Billings, F.S. Chaplin III, A.E. Giblin, K.J. Nadelhoffer, W.C. Oechel, and E.B. Rastetter. 1992. Global change and the carbon balance of arctic ecosystems. *BioScience* 42(6): 433–441.

Singh, J.S., and S.R. Gupta. 1977. Plant decomposition and soil respiration in terrestrial ecosystems. *Botanical Review* 43: 449–528.

Sorenson, C.H. 1981. Carbon-nitrogen relationships during the humification of cellulose in soils containing different amounts of clay. *Soil Biology and Biochemistry* 13: 313–321.

Thornthwaite, C.W., and J.R. Mather. 1955. The water balance. *Publications in Climatology* 8: 1–104.

Townsend, A.R., P.M. Vitousek, D.J. Desmarais, and A. Tharpe. 1997. Soil carbon pool structure and temperature sensitivity inferred using CO_2 and $^{13}CO_2$ incubation fluxes from five Hawaiian soils. *Biogeochemistry* 38: 1–17.

Townsend, A.R., P.M. Vitousek, and S.E. Trumbore. 1995. Soil organic matter dynamics along gradients in temperature and land use on the island of Hawaii. *Ecology* 76(3): 721–733.

Trumbore, S.E., O.A. Chadwick, and R. Amundsen. 1997. Rapid exchange between soil carbon and atmospheric carbon dioxide driven by temperature change. *Science* 272: 393–396.

Van der Peijl, M.J., and J.T.A. Verhoeven. 1999. A model of carbon, nitrogen and phosphorus dynamics and their interactions in river marginal wetlands. *Ecological Modelling* 118: 95–130.

Van Veen J.A., and E.A. Paul. 1981. Organic carbon dynamics in grasslands soils. 1. Background information and computer simulation. *Canadian Journal of Soil Science* 61: 185–201.

VEMAP Members. 1995. Vegetation/ecosystem modeling and analysis project: Comparing biogeography and biogeochemistry models in a continental-scale study of terrestrial ecosystem responses to climate change and CO_2 doubling. *Global Biogeochemical Cycles* 9: 407–437.

Vitousek P.M., D.R. Turner, W.J. Parton, and R.L. Sanford. 1994. Litter decomposition on the Mauna Loa environmental matrix, Hawaii patterns, mechanisms, and models. *Ecology* 75: 418–429.

14

Representing Biogeochemical Diversity and Size Spectra in Ecosystem Models of the Ocean Carbon Cycle

Robert A. Armstrong

Summary

On time scales of millennia and longer, ocean biogeochemistry determines the atmospheric concentration of carbon dioxide. Understanding exchanges of carbon and other elements between the deep ocean and the surface ocean is therefore critical to predicting climate change. Pelagic ecosystem models for studying the ocean carbon cycle must emphasize aspects of surface ocean biology that determine production and export of sinking particles, while aspects not directly related to these processes may be omitted. Representing biogeochemical diversity of phytoplankton is critical to this effort, because biogenic minerals produced by certain phytoplankton taxa add ballast that promotes particle sinking while protecting organic carbon from remineralization on its way to the sea floor. Size structure is also critical, because large phytoplankton tend to be grazed by large, fecal-pellet-producing zooplankton, and because size-structured predation and nutrient acquisition define the ecological context within which biogeochemically diverse communities are assembled. Here I review the scientific motivation for including biogeochemical diversity and size structure in ecosystem models and highlight promising modeling approaches for incorporating this structure.

Introduction

On time scales of thousands of years, ocean biogeochemical processes determine the concentration of carbon dioxide in the atmosphere. Carbon fixed by photosynthesis in the surface ocean is exported to depth by sinking particles, and the nutrients liberated at depth are returned to the surface ocean through a combination of regional upwelling and deep winter mixing. Carbon dioxide is exchanged between the atmosphere and the ocean surface by physical processes. In seawater, dissolved carbon dioxide reacts with carbonate ions to form bicarbonate according to the equation

$$H_2O + CO_2 + CO_3^{2-} \leftrightarrow 2HCO_3^- \tag{14.1}$$

Carbon dioxide is more soluble in cold water than in warm water, and colder water is heavier and so tends to sink. Cold, CO_2-rich water sinks in two regions: the North Atlantic off Iceland and Greenland, and the waters surrounding Antarctica (the Southern Ocean). Water that sinks in the North Atlantic travels southward as North Atlantic Deep Water until, several hundred years later, it is upwelled in the Southern Ocean, where it is cooled even further and distributed to all the major ocean basins (Broecker and Peng 1982). This circulation forms a giant "conveyer belt" that distributes heat, water, and dissolved materials throughout the world ocean (Broecker 1991).

Three processes—production, export, and remineralization—define the role of ocean biology in determining the atmospheric concentration of carbon dioxide. In production, phytoplankton "fix" carbon dioxide into organic material in the sunlit photic zone. In organic form, carbon can no longer exchange with the atmosphere, so that the production of organic matter lowers the partial pressure of carbon dioxide (pCO_2) in the surface ocean, driving a net influx of CO_2 from the atmosphere. Most of the material produced by photosynthesis is remineralized (returned to inorganic [mineral] form) within the photic zone; a small part is exported to the deep ocean, where it is either remineralized or buried in sediments. The export of carbon and associated nutrients to the deep ocean sequesters carbon from the surface for centuries to millennia, setting the critical timescales of ocean-atmosphere interaction (Broecker and Peng 1982; Sarmiento and Toggweiler 1984; Siegenthaler and Sarmiento 1993).

Carbon that is exported from surface waters to the deep ocean must either sink as particles or be mixed downward as dissolved organic matter. Since organic matter has approximately the same density as seawater, dense mineral matter is necessary for efficient sinking of particulate organic carbon to ocean depths (Honjo 1996). The minerals with which organic matter may be associated are either biogenic (silica from diatom frustules and radiolarian tests; carbonate minerals from coccolithophorids, foraminifera, and pteropods) or lithogenic (dust). Recent results (Armstrong et al. 2002) suggest a relatively fixed ratio of organic carbon to mineral materials at depths greater than 2000 m. In addition, Armstrong and Jahnke (2001) have pointed out that respiratory oxygen demand in sediments (which is closely tied to organic matter flux to the sediments) does not have a close relationship to surface organic matter production immediately above a given location but is instead higher near coasts, where inputs of mineral matter may help ballast organic matter to the sea floor. Taken together, these results suggest that export of particulate organic matter to the deep ocean may be more tightly linked to surface production of mineral ballasts than to organic matter production. It follows that the ability to predict abundances of silicifying and carbonate-producing plankton should be a key goal of pelagic ecosystem modeling.

In addition to the ballasting effect, removal of calcium carbonate from surface seawater affects air-sea exchange of carbon dioxide by lowering the net cationic charge balance, or "alkalinity," of seawater (Broecker and Peng 1982). This can be seen by writing the equilibrium equation that describes the carbonate system reaction (Eq. 14.1) as

$$K' = \frac{[HCO_3^-]^2}{[H_2O][CO_2][CO_3^{2-}]} \qquad (14.2)$$

where K' is the equilibrium constant. From Equation 14.2, it is clear that if carbonate ion is removed as calcium carbonate (as in the formation of coccoliths), the reaction in Equation 14.1 will be driven to the left; additional carbonate and carbon dioxide are then produced from bicarbonate, raising the pCO_2 and causing net outgassing of CO_2 from the ocean to the atmosphere.

Thus, the fact that some organisms make silica frustules and others precipitate carbonate minerals are not just biological curiosities but play a fundamental part in the ocean carbon system. Within the realm of ocean biogeochemical modeling, the question is not whether we need to represent diversity in models but rather how to accomplish this aim reliably, reproducibly, and with maximum computational efficiency (Evans 1988; Hofmann and Lascara 1998; Armstrong 1999; Doney 1999).

Models with Single Phytoplankton and Zooplankton Taxa

The impact of biology on the flux of organic carbon from the atmosphere to the ocean has been studied and modeled in considerable detail. The vehicles of choice for such investigations are "coupled physical-biogeochemical models" (Doney 1999). The physical part of these models is usually supplied by a General Circulation Model (GCM) of the ocean, while the biogeochemical side is supplied by a variety of models of varying complexity, known collectively as "ecosystem models." The physical model determines temperatures and currents; more important, it also determines delivery of limiting nutrients to the surface layer from depth. The ecosystem model allows phytoplankton to grow in response to light and nutrients, then redistributes both the fixed carbon and the associated nutrients throughout the water column.

In the simplest ecosystem models (e.g., Fasham et al. 1990, 1993; Sarmiento et al. 1993; McGillicuddy et al. 1995; Chai et al. 1996; Doney et al. 1996; Loukos et al. 1997; Friedrichs and Hofmann 2001), primary production is modeled as a function of processes occurring in a single phytoplankton "box" (state variable). Grazing on the phytoplankton component (and usually on bacterial and detrital components as well) is represented by a single "zooplankton" state variable. Zooplankton produce fecal pellets that sink at a predetermined speed, without regard to ballasting. Deep-water remineralization processes are also given very little attention in these models, being modeled as a simple power-law function of depth (Martin et al. 1987), again independent of ballast considerations.

The simplicity of these models is dictated in large part by the computational demands of large-scale ocean circulation models, coupled with the recognition that physics dominates nutrient supply and that nutrient supply dominates production (Dugdale and Goering 1967; Eppley and Peterson 1979). Therefore, biology is intentionally kept as simple as possible. Only when models fail to incorporate processes of obvious biogeochemical importance are biogeochemistry modelers willing to invest computational effort into resolving biological complexity.

HNLC (High Nutrient–Low Chlorophyll) Areas: The Model of Pitchford and Brindley

The total amount of carbon that can be fixed by phytoplankton is ultimately controlled by the total amount of mineral nutrients available in the photic zone. In most of the world ocean, the macronutrients nitrate and phosphate, which are required in substantial amounts in living cells, limit total production. However, in certain High Nutrient–Low Chlorophyll (HNLC) regions, the micronutrient iron limits total production, while the concentrations of nitrate and phosphate remain high. In these regions, cell sizes are typically small, presumably because the larger surface-to-volume ratio of small cells allows them to outcompete larger cells for scarce iron (Hudson and Morel 1990; Morel, Hudson, and Price 1991; Morel, Reuter, and Price 1991; Price et al. 1994; Landry et al. 1997). Size, acting through phytoplankton physiology, appears to be a powerful structuring force in plankton communities.

But size acts not only through nutrient uptake; it also acts through size-specific predation. This fact was illustrated most clearly by the IronEx II experiment (Coale et al. 1996). In this study, iron was added to a large patch (72 km^2) of water in the equatorial Pacific Ocean. The resulting phytoplankton bloom increased total chlorophyll levels from initial values of 0.15–0.20 µg chl/liter to values approaching 4 µg chl/liter on day 9. More interesting, the majority of the increase was in the large (greater than 10 µm) size fraction (Coale et al. 1996; Cavender-Bares et al. 1999; Landry, Ondrusek, et al. 2000). In particular, large pennate diatoms (*Nitzschia* spp.) accounted for most of the chlorophyll buildup, while many other groups, including *Prochlorococcus, Synechococcus,* prymnesiophytes, and small centric diatoms, did not bloom (Coale et al. 1996; Landry, Ondrusek, et al. 2000), presumably because their population densities were kept under control by small grazers (Landry, Constantinou, et al. 2000).

These observations support neither pure "top down" nor pure "bottom up" control of phytoplankton biomass (Hairston et al. 1960; Hunter and Price 1992). Instead, the picture that emerges is one of dual controls: bottom-up control of total phytoplankton biomass, which allows larger size classes to exist only when sufficient nutrients (in this case iron) are present, coupled to top-down control of individual size classes. The idea of dual controls is implicit in the "ecumenical" hypothesis of Morel, Reuter, and Price (1991), which is "ecumenical" in the sense that it incorporates two (potentially) competing explanations of HNLC areas—bottom-up control of total phytoplankton growth by iron limitation, and top-down control of individual size classes by zooplankton grazing—within the same hypothesis. This hypothesis was explored more fully by Armstrong (1994) and by Price et al. (1994), and its implications for the subarctic Pacific were explored by Armstrong et al. (1994).

To capture the dual nature of control in the IronEx experiment—the release from predation of large pennate diatoms, and the control of smaller phytoplankton species by their grazers—Pitchford and Brindley (1999) proposed a model based on "excitable medium" theory (Murray 1989; Truscott and Brindley 1994). Essentially, the diatom population is modeled as a compressed spring or wound-up rubber band, an explosion just waiting to happen. When the right stimulus is applied long enough and hard enough, the spring releases, and a bloom ensues.

The food web structure of the Pitchford-Brindley model is shown in Figure 14.1. The defining equations are:

$$\frac{dp_1}{dt} = p_1(\mu_1 - g_1 z_1) \tag{14.3}$$

$$\frac{dz_1}{dt} = z_1\left[c_1 g_1 p_1 - g_2 z_2 \frac{\phi}{\phi z_1 + (1-\phi)p_2 + k} - \varepsilon_1\right] \tag{14.4}$$

$$\frac{dp_2}{dt} = p_2\left[\mu_2 - g_2 z_2 \frac{(1-\phi)}{\phi z_1 + (1-\phi)p_2 + k}\right] \tag{14.5}$$

$$\frac{dz_2}{dt} = z_2\left[c_2 g_2 \frac{\phi z_1 + (1-\phi)p_2}{\phi z_1 + (1-\phi)p_2 + k} - \varepsilon_2\right] \tag{14.6}$$

In this model, both phytoplankton species (p_1 and p_2) would grow exponentially (at rates μ_1 and μ_2, respectively) in the absence of grazing by zooplankton species z_1 and z_2. Zooplankton species z_1 represents a microzooplankton that grazes only on the smaller phytoplankton species p_1, while the larger zooplankton species z_2 consumes both the microzooplankton and the larger phytoplankton species p_2. The larger predator switches between its prey species according to their

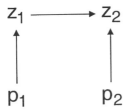

Figure 14.1. The food web studied by Pitchford and Brindley (1999). The p's denote phytoplankton size classes, and the z's zooplankton. See text for further details.

relative densities, favoring the prey that is currently in greater supply; this switching is described by the function

$$\phi = \frac{\psi_{z_1} z_1}{\psi_{z_1} z_1 + \psi_{p_2} p_2} \tag{14.7}$$

where the ψ's are "intrinsic" preferences. The terms g_1 and g_2 are maximum grazing rates; c_1 and c_2 are metabolic conversion efficiencies; and k is a half-saturation constant for grazing. The terms ε_1 and ε_2 represent respiration and other unavoidable losses.

In this model, in the absence of enough iron, large zooplankton subsist on a combination of microzooplankton and large phytoplankton. Large phytoplankton are maintained in the model, even with a small iron-limited growth rate, because when they become rare, large zooplankton switch their feeding almost entirely to small zooplankton; this switching behavior is critical to model behavior. If, at a certain time, a large amount of iron is put into the system, the growth rate μ_2 of the large phytoplankton species suddenly increases, and large phytoplankton start to grow much more rapidly. For a bloom to occur, large phytoplankton must grow fast enough that they are not overtaken by the growth and prey-switching of large zooplankton during this critical period; if this condition is not met, the nascent bloom will be squelched. Evans and Parslow (1985) earlier explored the same sort of model to explain differences between the spring blooms in the North Atlantic and North Pacific; but their model, which contained many more details, was not so clear as the Pitchford-Brindley model.

The Pitchford-Brindley model is complex enough to exhibit behaviors thought to be essential for explaining the IronEx II results. However, in order to capture the fact that several small phytoplankton taxa were kept under control while a single large taxon increased, their model must be expanded to include multiple phytoplankton and zooplankton taxa. Pitchford and Brindley suggested that the multiple-size-class approach of Moloney and Field (1991) might be a useful starting point for generalizing their results.

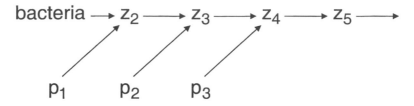

Figure 14.2. The food web model of Moloney and Field (1991). The p's denote phyto-plankton size classes, and the z's zooplankton. See text for further details.

Multiple Size Classes: The Model of Moloney and Field

Moloney and Field (1991) and Moloney et al. (1991) developed a generalized framework for modeling multiple size classes (Figure 14.2). In this approach, the state variables are cell densities or limiting nutrients in fixed size classes. Successive size classes are assumed to differ by a fixed multiple of equivalent spherical diameter. Moloney and Field (1991) used intervals that differed by a factor of 10: they defined phytoplankton size classes of 0.2–2.0, 2.0–20, and 20–200 μm equivalent spherical diameter. In contrast, Moloney et al. (1991) used size classes that differed by a factor of 5.

In the complete version of their model, zooplankton could prey on a range of size classes smaller than themselves; in published simulations, however, zoo-plankton in each size class were allowed to prey only on phytoplankton and zooplankton of the next smaller size class (Moloney and Field 1991; Moloney et al. 1991). With this simplifying assumption, the defining equations of their model become

$$\frac{dp_i}{dt} = p_i \mu_{\max,i} \frac{N}{N + k_{N,i}} - z_{i+1} I_{\max,i+1} \frac{p_i - ref_{p,i}}{k_{z,i+1} + (p_i - ref_{p,i}) + (z_i - ref_{z,i})} - R_{p,i} p_i \quad (14.8)$$

$$\frac{dz_{i+1}}{dt} = z_{i+1} c_{i+1} I_{\max,i+1} \frac{(p_i - ref_{p,i}) + (z_i - ref_{z,i})}{k_{z,i+1} + (p_i - ref_{p,i}) + (z_i - ref_{z,i})}$$

$$- z_{i+2} I_{\max,i+2} \frac{z_{i+1} - ref_{z,i+1}}{k_{z,i+2} + (p_{i+1} - ref_{p,i+1}) + (z_{i+1} - ref_{z,i+1})} - R_{z,i+1} z_{i+1} \quad (14.9)$$

for $i = 1, \ldots, i_{\max}$. Here p_i and z_i are phytoplankton and zooplankton biomasses in the ith size class; $\mu_{\max,i}$ is the maximum growth rate, and $k_{N,i}$ the half-saturation constant, for the growth of phytoplankton species i on nutrient N; $I_{\max,i}$ is a maximum ingestion rate, and c_i a growth efficiency, for zooplankton in size class i; R_i is a respiration rate; and ref_i is a refuge size below which predation ceases (Frost 1975). All parameters are indexed by size class i, and their values vary allometrically (as power laws) with size (Moloney and Field 1989, 1991). In addition, there is a final equation for bacterial growth, which is of the same form as the phytoplankton equations but is a function of organic carbon and nitrogen; this state variable would take the position of z_1 in the notation used here (cf. Figure 14. 2).

The resulting model was used to simulate conditions on the Agulhas bank (coastal stratified conditions), the Benguela upwelling region, and the oligotrophic southeast Atlantic (Moloney et al. 1991). Carr (1998) used a variant of this model to explore food chain dynamics in three upwelling regions: the Benguela Current, coastal Peru, and the Canary Current. Finally, Gin et al. (1998) used another variant of the model to simulate conditions at the Bermuda Atlantic Time-series Station (BATS).

Dynamically, the original Moloney-Field model was found to exhibit severe swings in the abundances of individual size classes; in some of the cases considered by Moloney et al. (1991), variations of 100-fold occurred at intervals of less than a day. (These problems were recognized early on by Moloney and Field [1991]; see p. 1031 of their paper.) Armstrong (1999) analyzed the reasons for these oscillations (some of which were apparently chaotic); he concluded that they arose (1) because all phytoplankton species share a single resource, while each has a different predator species (cf. Armstrong 1983), and (2) because the Holling zooplankton type-II feeding curves (Holling 1959) in the model are inherently destabilizing, at least without feeding thresholds (Frost 1975). In most of the simulations, refuge sizes were taken to be zero and could not stabilize the system. However, the oscillations persisted even when sensitivity tests were run with nonzero refuge sizes (Moloney and Field 1991).

In contrast, Gin et al. (1998) apparently had little problem with inherent oscillatory behaviors. Their model differed from the original Moloney-Field model in that (1) the connections between successive zooplankton species were severed and (2) stabilizing mortality terms were added. These mortality ("closure") terms were of the form suggested by Steele and Henderson (1981, 1992), who recommended adding a loss term in (biomass)^2 to mimic the effects of (unmodeled) higher trophic levels. While such terms do not invariably lend stability (Edwards and Yool 2000; see also Evans 1977), the studies of McGillicuddy et al. (1995), Gin et al. (1998), and Armstrong (1999) show that they can often serve an important stabilizing function. The severing of connections between successive zooplankton size classes is more problematical because it precludes representation of switching, as in the Pitchford-Brindley model. A simple multi-phytoplankton model that allows switching behaviors, while not suffering the stability problems inherent in simpler versions of the Moloney-Field approach, is presented in the following section.

A Size-Spectrum Model of Distributed Grazing

A common observation in aquatic systems is that when numbers of organisms in successive (logarithmic) size classes are plotted against size on a log-log scale, they tend to fall on a straight line (Sheldon et al. 1972; Sprules 1988). More recently, and in the context of phytoplankton community structure, Cavender-Bares et al. (2001) used a plot of Pr ($V \geq v$)—the probability that a cell of volume V will be larger than or equal to a reference volume v—versus volume to explore patterns of community composition on a transect across the Sargasso Sea (Figure 14.3). They found that counts of bacteria, *Prochlorococcus, Synechococcus*, and eukaryotic phytoplankton were different at various points in the transect but that the cumulative curve over all species was relatively smooth and that its slope changed systematically (but not dramatically) from south to north. A complementary observation (Raimbault et al. 1988; Chisholm 1992) is that as total chlorophyll in a system increases, this increase tends to occur through the addition of larger phytoplankton size classes rather than by the expansion of smaller size classes.

These observations suggest that mechanism-based plankton models should from the outset be constructed to reproduce a desired phytoplankton biomass spectrum; this strategy is in direct contrast to constructing a model without this criterion in mind and then *hoping* that it will yield a reasonable spectrum. While this strategy has been tried before in the context of both energetics (Platt and Denman 1977; Silvert and Platt 1978; Kerr and Dickie 2001) and simplified predator-prey relationships (Kiefer and Berwald 1992), an additional goal here is to offer a framework that might be expanded to include biogeochemical diversity.

Following Armstrong (1999), let $n(x)$ represent the biomass density of a phytoplankton-bacteria spectrum, where x is the logarithm of size. The currency of $n(x)$ could be the density of some limiting nutrient (e.g., nitrogen), the density of a biochemical constituent, such as carbon or chlorophyll, or cell volume or number. For example, if $n(x)$ is number density, the connection between $n(x)$ and $\Pr(X \geq x)$ (Cavender-Bares 2001; see Figure 14.3) is that

$$\Pr(X \geq x) = \int_x^\infty n(x')dx' / \int_{-\infty}^\infty n(x')dx' \tag{14.10}$$

Let total cell density of phytoplankton (or bacterial) type i be denoted p_i, $i=1$ ... k, and assume that it is distributed across size according to some function $f_i(x)$, which is normalized such that $\int f_i(x)dx = 1$. The biomass spectrum of species i is then $p_i f_i(x)$, and the spectrum of total algal biomass is $n(x) = \sum_i p_i f_i(x)$. We can choose values for p_i that best match any prespecified density spectrum $\tilde{n}(x)$ in the sense of least squares by minimizing the quantity

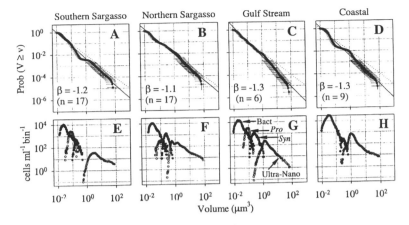

Figure 14.3. Phytoplankton size spectra along a transect from Bermuda (Sargasso Sea) to Woods Hole, MA. (A–D) Plots of $\Pr(V \geq v)$, the probability that volume V exceeds some threshold value v, versus v (see Equation 14.10). The slope of β of each spectrum is indicated, as is the number of stations averaged in a given plot (n). (A) the southern Sargasso Sea; (B) the northern Sargasso; (C) the Gulf Stream; (D) coastal Massachusetts. The dotted line is the 1:1 line. (E-H) Representative plots of cell concentration (per logarithmic counting "bin") versus cell volume for each of the four regions in (A-D). Bact: bacteria; *Pro*: *Prochlorococcus*; *Syn*: *Synechococcus*; Ultra-Nano: larger size classes. Note the uniform spacing between successive size classes and that bacteria lie on the same spectrum as phytoplankton, suggesting a taxon-independent origin (predation) for the size-spectral pattern. (Adapted from Figure 5 of Cavender-Bares et al. 2001.)

$$\int\limits_{-\infty}^{\infty} \left[\tilde{n}(x) - \sum_i p_i f_i(x) \right]^2 dx \tag{14.11}$$

which will happen when the p_i are chosen to satisfy

$$0 = \int\limits_{-\infty}^{\infty} \tilde{n}(x) f_i(x) dx - \sum_{j=1}^{\infty} p_j \int\limits_{-\infty}^{\infty} f_i(x) f_j(x) dx \tag{14.12}$$

for $i = 1, \ldots, k$. (See MacArthur 1970, 1972 for a similar approach to avian community structure.)

To incorporate these observations into a pelagic ecosystem model, Armstrong (1999) proposed a model of "distributed grazing" on phytoplankton. The premise was that while the internal workings of a zooplankton community might be impossibly difficult to model explicitly, it might nevertheless be pos-

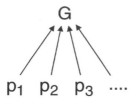

Figure 14.4. The food web model of Armstrong (1999). The p's denote phytoplankton size classes, and G represents the total herbivorous grazing capacity of the zooplankton community. This grazing is distributed across phytoplankton taxa in proportion to their abundances. See text for further details.

sible to represent implicitly the grazing impact of the zooplankton community as a whole on multiple phytoplankton species. The proposed model is

$$\mathrm{d}p_i / \mathrm{d}t = p_i[\mu_i(N) - bw_i G] \tag{14.13}$$

$$\mathrm{d}w_i / \mathrm{d}t = \alpha_i w_i[1 - w_i / w_i^*] \tag{14.14}$$

$$\mathrm{d}G / \mathrm{d}t = f(G, w, p) \tag{14.15}$$

(Figure 14.4). Here G is the total predation pressure exerted by the grazer community on the phytoplankton community, and the w_i are weights ("distribution fractions") of predation pressure. The latter are defined such that the per-capita predation pressure on phytoplankton species i is proportional to $w_i G$. These weights are assumed to decay logistically (Eq. 14.14) towards target weights

$$w_i^* = y_i^* / \sum_j y_j^* \tag{14.16}$$

where

$$y_i^* = \sum_j a_{ij} p_j \tag{14.17}$$

is the predation pressure that could be sustained at steady state by current phytoplankton densities, and where the "competition coefficients" a_{ij} are defined by the inner products $\int f_i(x) f_j(x) \, \mathrm{d}x$ in Equation 14.12 (Armstrong 1999).

To complete the model, low-density growth rates $\mu_i(N)$ must be specified. A functional form that is directly useful in constructing size spectra was proposed by Aksnes and Egge (1991):

$$\mu_i(N) = \mu_{\max} N / [N + K_N(0) \exp(\beta x_i)] \tag{14.18}$$

Here $\mu_{\max}$ is a maximum growth rate, $K_N(0)$ is a half-saturation constant for growth on nutrient N at $x_i = 0$, and β is the (allometric) rate at which K_N increases with size.

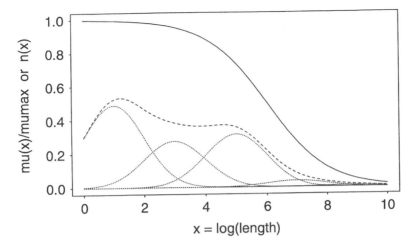

Figure 14.5. An example of how the model of distributed grazing (Eqs. 14.13–14.17) can be used to generate size spectra. The solid line is a plot of normalized growth rate μ/μ_{max} defined by Eq. 14.18. The dotted lines are size spectra $p_i f_i(x)$ of individual species; the dashed line is the sum $n(x) = \sum_i p_i f_i(x)$ of the dotted lines. (Adapted from Figure 2 of Armstrong 1999.)

Specifying an allometric equation for $\mu_i(N)$ leads directly to a compact representation of size spectra (Armstrong 1999; see Figure 14.5). The key to this behavior is that at steady state,

$$\mu_i(N) = bw_i G \propto \sum_j a_{ij} p_j \qquad (14.19)$$

so that total (weighted) algal abundance is proportional to $\mu_i(N)$, which is itself allometrically related to size. The representation of growth could easily be modified to include taxonomic information by specifying different values of μ_{max} and $K_N(0)$ for different phytoplankton taxa.

General Discussion

Many technical issues arise in modeling the coexistence of phytoplankton species. In order to permit steady-state coexistence of n species, there must be at least n limiting factors sensu Levin (1970): n descriptors of population regulation that are not linearly dependent on one another (Armstrong and McGehee 1980). Predator-prey interactions in which zooplankton feeding responses depend on the density of their phytoplankton prey, either explicitly (as in the model of Moloney and Field 1989) or implicitly (e.g., Steele-Henderson terms), provide one common type of limiting factor. Limiting factors can also be generated by the switching of one or more predators onto different prey species, as in the models of Pitchford and Brindley (1999; see Eq. 14.7) or Armstrong (1999; see Eq. 14.14). Additional coexistence could potentially be provided by nonlinear interactions in nonequilibrium settings (Hutchinson 1961; Armstrong and McGehee 1980; Huisman and Weissing 1999, 2000; Lundberg et al. 2000); but to capture the size-spectrum regularity seen in Figure 14.3, it would seem prudent to include in models enough limiting factors to insure that coexistence can occur near steady state.

This desideratum exposes a potential weakness of the Moloney-Field approach, which employs equally spaced size classes. Size-spectrum data often suggest that coexisting phytoplankton species are packed along a gradient of size (equivalent spherical diameter) with centers of adjacent distributions that allow from 5 to 7 species to be packed into a 10-fold increase in length (Figure 14.3; see also Figure 7 of Havlicek and Carpenter 2001). If a Moloney-Field model is to allow such tight packing, and if zooplankton grazing is supplying the needed limiting factors, then the distance between adjacent boxes must have this spacing or less; it cannot have the spacings of 5 or 10 that are typically employed. On the other hand, factors of 5 or 10 are chosen to represent typical spacings between the sizes of adjacent zooplankton trophic levels (Kiefer and Berwald 1994; Kerr and Dickey 2001), so that tighter packing of phytoplankton would require a much more complicated food web structure than that depicted in Figure 14.2 (cf., e.g., the ridiculously complicated structures considered by Armstrong et al. 1994).

In contrast, the model of distributed grazing allows competing species (including those from biogeochemically distinct taxa) to be located at arbitrarily spaced points along a size axis; two competing species could, for example, be located very close to each other on this axis. For example, Gin (1996) observed that in eutrophic environments, bacteria increase in abundance at the expense of small prochlorophyte phytoplankton. This observation could result from predation on bacteria and on small phytoplankton of roughly the same size by grazers common to both taxa; this mechanism could be efficiently represented in a model of distributed grazing, unifying the description of the microbial food web.

A serious challenge to any simple size-based approach is that while small phytoplankton species tend to be spherical, larger phytoplankton species can play "tricks" that render them "large" with respect to predation and "small" with respect to nutrient uptake. For example, diatoms may accomplish this feat either by being very elongate (e.g., pennate diatoms) or by forming chains; such tricks require special attention in models that rely on size structure (e.g., Carr 1998). In the context of the models discussed here, a dual size categorization may be required for nutrient uptake and predation (cf. May 1975).

A further consideration is that zooplankton may contribute substantially to the rain of ballast minerals (Honjo 1996; Takahashi et al. 1999), so that zooplankton diversity must be represented in addition to phytoplankton diversity. These groups (radiolarians, foraminifera, and pteropods) may need to be represented in feeding-size spectra, which might require the explicit representation of zooplankton taxa. In particular, Steele (1998) has suggested that food web considerations may be paramount in constructing a general theory of export from the photic zone, both because large zooplankton such as copepods "package" their food into fecal pellets that are large enough to sink and because intermediate microzooplankton may add ballast to the copepods' food.

Finally, the models I have discussed were all developed for describing production in the surface ocean. However, patterns of remineralization in deeper waters (Armstrong and Jahnke 2001; Armstrong et al. 2002) suggest that extension of these models to regions below the photic zone may soon become essential. In these regions, there are no phytoplankton, but bacteria and other decomposers will take their place in models. Our models must eventually extend into the "twilight zone" between 100 m and 1000 m, where empirical knowledge is thin. Hopefully, a size-spectrum approach will prove useful in this region as well.

Acknowledgments. I thank Kent Cavender-Bares for redrawing Figure 14.3 for use in this review. I also thank Jon Cole and Cindy Lee for helpful comments on the manuscript. This work was supported by award OCE-0049009 from the U.S. National Science Foundation; it is contribution #809 from the U.S. JGOFS program, and #1249 from the Marine Sciences Research Center at Stony Brook University.

References

Aksnes, D.L., and J.K. Egge. 1991. A theoretical model for nutrient uptake in phytoplankton. *Marine Ecology Progress Series* 70: 65–72.

Armstrong, R.A. 1983. The role of symmetric food web models in explicating the stability/diversity connection. Pp. 95–99 in D.L. DeAngelis, W.M. Post, and G. Sugihara, editors. *Current Trends in Food Web Theory*. Springfield, VA: National Technical Information Service.

―――. 1994. Grazing limitation and nutrient limitation in marine ecosystems: Steady-state solutions of an ecosystem model with multiple food chains. *Limnology and Oceanography* 39: 597–608.

―――. 1999. Stable model structures for representing biogeochemical diversity and size spectra in plankton communities. *Journal of Plankton Research* 21: 445–464.

Armstrong, R., S. Bollens, B. Frost, M. Landry, M. Landsteiner, and J. Moisan. 1994. Food webs. Pp. 25–35 in C.S. Davis and J.H. Steele, editors. *Biological/Physical Modeling of Upper Ocean Processes*. Technical Report WHOI-94-32. Woods Hole, MA: Woods Hole Oceanographic Institution.

Armstrong, R.A., and R.A. Jahnke. 2001. Decoupling surface production from deep remineralization and benthic deposition: The role of mineral ballasts. *U.S. Joint Global Ocean Flux Study News* 11(2): 1–2.

Armstrong, R.A., C. Lee, J.I. Hedges, S. Honjo, and S.G. Wakeham. 2002. A new, mechanistic model for organic carbon fluxes in the ocean based on the quantitative association of POC with ballast minerals. *Deep-Sea Research II* 49: 219–236.

Armstrong, R.A., and R. McGehee. 1980. Competitive exclusion. *American Naturalist* 115: 151–170.

Broecker, W.S. 1991. The great ocean conveyer. *Oceanography* 4(2): 79–89.

Broecker, W.S., and T.-H. Peng. 1982. *Tracers in the Sea.* Palisades, NY: Lamont-Doherty Earth Observatory.

Carr, M.-E. 1998. A numerical study of the effect of periodic nutrient supply on pathways of carbon in a coastal upwelling regime. *Journal of Plankton Research* 20: 491–516.

Cavender-Bares, K.K., E.L. Mann, S.W. Chisholm, M.E. Ondrusek, and R.R. Bidigare. 1999. Differential response of equatorial Pacific phytoplankton to iron fertilization. *Limnology and Oceanography* 44: 237–246.

Cavender-Bares, K.K., A. Rinaldo, and S.W. Chisholm. 2001. Microbial size spectra from natural and nutrient enriched ecosystems. *Limnology and Oceanography* 46: 778–789.

Chai, F., S.T. Lindley, and R.T. Barber. 1996. Origin and maintenance of a high nitrate condition in the equatorial Pacific. *Deep-Sea Research II* 43: 1031–1064.

Chisholm, S.W. 1992. Phytoplankton size. Pp. 213–237 in P.G. Falkowski and A.D. Woodhead, editors. *Primary Productivity and Biogeochemical Cycles in the Sea.* New York: Plenum Press.

Coale, K.H., et al. 1996. A massive phytoplankton bloom induced by an ecosystem-scale iron fertilization experiment in the equatorial Pacific Ocean. *Nature* 383: 495–501.

Doney, S.C. 1999. Major challenges confronting marine biogeochemical modeling. *Global Biogeochemical Cycles* 13: 705–714.

Doney, S.C., D.M. Glover, and R.G. Najjar. 1996. A new coupled, one-dimensional biological-physical model for the upper ocean: Application to the JGOFS Bermuda Atlantic Time-series Study (BATS) site. *Deep-Sea Research II* 43: 591–624.

Dugdale, R.C., and J.J. Goering. 1967. Uptake of new and regenerated forms of nitrogen in primary productivity. *Limnology and Oceanography* 12: 196–207.

Edwards, A.M., and A. Yool. 2000. The role of higher predation in plankton population models. *Journal of Plankton Research* 22: 1085–1112.

Eppley R.W., and B.J. Peterson. 1979. Particulate organic matter flux and planktonic new production in the deep ocean. *Nature* 282:677–680.

Evans, G.T. 1977. Functional response and stability. *American Naturalist* 111: 799–802.

———. 1988. A framework for discussing seasonal succession and coexistence of phytoplankton species. *Limnology and Oceanography* 33: 1027–1036.

Evans, G.T., and J.S. Parslow. 1985. A model of annual plankton cycles. *Biological Oceanography* 3: 327–347.

Fasham, M.J.R., H.W. Ducklow, and S.M. McKelvie. 1990. A nitrogen-based model of plankton dynamics in the oceanic mixed layer. *Journal of Marine Research* 48: 591–639.

Fasham, M.J.R., J.L. Sarmiento, R.D. Slater, H.W. Ducklow, and R. Williams. 1993. Ecosystem behavior at Bermuda Station "S" and Ocean Weather Station "India": A general circulation model and observational analysis. *Global Biogeochemical Cycles* 7: 379–415.

Friedrichs, M.A.M., and E.E. Hofmann. 2001. Physical control of biological processes in the central equatorial Pacific Ocean. *Deep-Sea Research I* 48: 1023–1069.

Frost, B.W. 1975. A threshold feeding behavior in *Calanus pacificus*. *Limnology and Oceanography* 20: 263–266.

Gin, K.Y.H. 1996. Microbial size spectra from diverse marine ecosystems. Ph.D. dissertation, MIT/WHOI Joint Program in Ocean Science and Engineering.

Gin, K.Y.H., J. Guo, and H.-F. Cheong. 1998. A size-based ecosystem model for pelagic waters. *Ecological Modelling* 112: 53–72.

Hairston, N.G., F.E. Smith, and L.B. Slobodkin. 1960. Community structure, population control, and competition. *American Naturalist* 94: 421–425.

Havlicek, T.D., and S.R. Carpenter. 2001. Pelagic species size distributions in lakes: Are they discontinuous? *Limnology and Oceanography* 46: 1021–1033.

Hofmann, E.E., and C.M. Lascara. 1998. Overview of interdisciplinary modeling for marine ecosystems. Pp. 507–540 in K.H. Brink and A.R. Robinson, editors. *The Sea*. New York: John Wiley and Sons.

Holling, C.S. 1959. The components of predation as revealed by a study of small-mammal predation of the European pine sawfly. *Canadian Entomologist* 91: 293–320.

Honjo, S. 1996. Fluxes of particles to the interiors of the open oceans. Pp. 91–154 in V. Ittekot, P. Schafer, S. Honjo, and P.J. Depetris, editors. *Particle Flux in the Ocean*. New York: John Wiley and Sons.

Hudson, R.J.M., and F.M.M. Morel. 1990. Iron transport in marine phytoplankton: Kinetics of cellular and medium coordination complexes. *Limnology and Oceanography* 36: 1002–1020.

Huisman, J., and F.J. Weissing. 1999. Biodiversity of plankton by species oscillations and chaos. *Nature* 402: 407–410.

———. 2000. Coexistence and resource competition: Reply. *Nature* 407: 694.

Hunter, M.D., and P.W. Price. 1992. Playing Chutes and Ladders: Heterogeneity and the relative roles of bottom-up and top-down forces in natural communities. *Ecology* 73: 724–732.

Hutchinson, G.E. 1961. The paradox of the plankton. *American Naturalist* 95: 137–145.

Kerr, S.R., and L.M. Dickie. 2001. *The Biomass Spectrum.* New York: Columbia University Press.

Kiefer, D.A., and J. Berwald. 1992. A random encounter model for the microbial planktonic community. *Limnology and Oceanography* 37: 457–467.

Landry, M.R., R.T. Barber, R. Bidigare, F. Chai, K.H. Coale, H.G. Dam, M.R. Lewis, S.T. Lindley, J.J. McCarthy, M.R. Roman, D.K. Stoecker, P.G. Verity, and J.R. White. 1997. Iron and grazing constraints on primary production in the central equatorial Pacific: An EqPac synthesis. *Limnology and Oceanography* 42: 405–418.

Landry, M.R., J. Constantinou, M. Latasa, S.L. Brown, R.R. Bidigare, and M.E. Ondrusek. 2000. Biological response to iron fertilization in the eastern equatorial Pacific (IronEx II). III. Dynamics of phytoplankton growth and microzooplankton grazing. *Marine Ecology Progress Series* 201: 57–72.

Landry, M.R., M.E. Ondrusek, S.J. Tanner, S.L. Brown, J. Constantinou, R.R. Bidigare, K.H. Coale, and S. Fitzwater. 2000. Biological response to iron fertilization in the eastern equatorial Pacific (IronEx II). I. Microplankton community abundances and biomass. *Marine Ecology Progress Series* 201: 27–42.

Levin, S.A. 1970. Community equilibria and stability, and an extension of the competitive exclusion principle. *American Naturalist* 104: 413–423.

Loukos, H., B. Frost, D.E. Harrison, and J.W. Murray. 1997. An ecosystem model with iron limitation of primary production in the equatorial Pacific at 140°W. *Deep-Sea Research II* 44: 2221–2249.

Lundberg, P., E. Ranta, V. Kaitala, and N. Jonzén. 2000. Coexistence and resource competition. *Nature* 407: 694.

MacArthur, R.H. 1970. Species packing and competitive equilibrium for many species. *Theoretical Population Biology* 1: 1–11.

———. 1972. *Geographical Ecology.* New York: Harper and Row.

Martin, J.H. G.A. Knauer, D.M. Karl, and W.W. Broenkow. 1987. VERTEX: Carbon cycling in the northeast Pacific. *Deep-Sea Research.* 34: 267–285.

May, R.M. 1975. Some notes on estimating the competition matrix, α. *Ecology* 56: 737–741.

McGillicuddy, D.J., Jr., J.J. McCarthy, and A.R. Robinson. 1995. Coupled physical and biological modeling of the spring bloom in the North Atlantic. I. Model formulation and one dimensional bloom processes. *Deep-Sea Research I* 42: 1313–1357.

Moloney, C.L., and J.G. Field. 1989. General allometric equations for rates of nutrient uptake, ingestion, and respiration in plankton organisms. *Limnology and Oceanography* 34: 1290–1299.

———. 1991. The size-based dynamics of plankton food webs. I. A simulation model of carbon and nitrogen flows. *Journal of Plankton Research* 13: 1003–1038.

Moloney, C.L., J.G. Field, and M.I. Lucas. 1991. The size-based dynamics of plankton food webs. II. Simulations of three contrasting southern Benguela food webs. *Journal of Plankton Research* 13: 1039–1092.

Morel, F.M.M., R.J.M. Hudson, and N.M. Price. 1991. Limitation of productivity by trace metals in the sea. *Limnology and Oceanography* 36: 1742–1755.

Morel, F.M.M., J.G. Reuter, and N.M. Price. 1991. Iron nutrition of phytoplankton and its possible importance in the ecology of ocean regions with high nutrient and low biomass. *Oceanography* 4: 56–61.

Murray, J.D. 1989. *Mathematical Biology.* Berlin: Springer-Verlag.

Pitchford, J.W., and J. Brindley. 1999. Iron limitation, grazing pressure, and oceanic high nutrient-low chlorophyll (HNLC) regions. *Journal of Plankton Research* 21: 525–547.

Platt, T., and K.L. Denman. 1977. Organization in the pelagic ecosystem. *Helgolander Wissenschaftliche Meeresuntersuchungen* 30: 575–581.

Price, N.M., B.A. Ahner, and F.M.M. Morel. 1994. The equatorial Pacific Ocean: Grazer-controlled phytoplankton populations in an iron-limited ecosystem. *Limnology and Oceanography* 39: 520–534.

Raimbault, P., M. Rodier, and I. Taupier-Letage. 1988. Size fraction of phytoplankton in the Ligurian Sea and the Algerian Basin (Mediterranean Sea): Size distribution versus total concentration. *Marine Microbial Food Webs* 3: 1–7.

Sarmiento, J.L., R.D. Slater, M.J.R. Fasham, H.W. Ducklow, J.R. Toggweiler, and G.T. Evans. 1993. A seasonal three-dimensional ecosystem model of nitrogen cycling in the North Atlantic photic zone. *Global Biogeochemical Cycles* 7: 417–450.

Sarmiento, J.L., and J.R. Toggweiler. 1984. A new model for the role of the oceans in determining atmospheric $p\text{CO}_2$. *Nature* 308: 621–624.

Sheldon, R.W., A. Prakash, and W.H. Sutcliff. 1972. The size distribution of particles in the ocean. *Limnology and Oceanography* 17: 329–339.

Siegenthaler, U., and J.L. Sarmiento. 1993. Atmospheric carbon dioxide and the ocean. *Nature* 365: 119–125.

Silvert, W., and T. Platt. 1978. Energy flux in the pelagic ecosystem: A time-dependent equation. *Limnology and Oceanography* 23: 813–816.

Sprules, W.G. 1988. Effects of trophic interactions on the shape of pelagic size spectra. *Internationale Vereinigung für Theoretische und Angewandte Limnologie* 23: 234–240.

Steele, J.H. 1998. Incorporating the microbial loop in a simple plankton model. *Proceedings of the Royal Society of London. Series B.* 265: 1771–1777.

Steele, J.H., and E.W. Henderson. 1981. A simple plankton model. *American Naturalist* 117: 676–691.

———. 1992. The role of predation in plankton models. *Journal of Plankton Research* 14: 157–172.

Takahashi, K., N. Fujitani, M. Yanada, and Y. Maita. 1999. Long-term biogenic particle fluxes in the Bering Sea and the central subarctic Pacific Ocean, 1990–1995. *Deep-Sea Research I* 47: 1723–1759.

Truscott, J.E., and J. Brindley. 1994. Ocean plankton models as excitable media. *Bulletin of Mathematical Biology* 56: 981–998.

15

The Mass Balances of Nutrients in Ecosystem Theory and Experiments: Implications for the Coexistence of Species

John Pastor

Summary

The requirement of mass balance of all transfers of matter and energy into, within, and out of an ecosystem is a major constraint on the behavior of our models and the real world. In ecosystem models, only certain analytical or numerical solutions are allowed that are consistent with the constraint of mass balance. To the extent that an ecosystem model accurately captures the dynamics of the real world within the constraints of mass balance, then these solutions and only these solutions will be observed. For example, if species both respond to and affect the flow of nutrients in different ways, then only certain combinations of species are allowable within a given input-output budget of an ecosystem. Experiments can then be designed to look for examples of these solutions in the real world. I will examine a series of analytical ecosystem models, discuss how mass balance is treated in these models and how it constrains their solutions (including coexistence of species), and review experiments designed to examine these solutions in nature.

Introduction

That ecosystems and the species within them maintain a mass balance of material flow is the hallmark contribution of ecosystem ecology to the larger field of ecology. By mass balance, I mean the quantitative accounting of all pathways of carbon, water, nutrients, or energy flows through an ecosystem. The description and analysis of mass balance in ecosystems depends on the first law of thermodynamics, the conservation of matter (or energy).

Virtually all experiments in ecosystem ecology involve an accounting or manipulation of mass balances of material flows. We prevent or temporarily stop nutrient uptake by plants by clearcutting or trenching plots and then examine the responses of nutrient exports (Likens et al. 1970; Marks and Bormann 1972; Richardson and Lund 1975; Vitousek and Reiners 1975; Vitousek et al. 1979). We stop litter input by enclosing litter in mesh bags and then examine the resulting budgets of carbon and nutrients (e.g., Gosz et al. 1973 and numerous other litterbag experiments). We fertilize ecosystems and then examine the pathways of nutrient flow (Baker et al. 1974; Binkley 1986). We introduce stable isotopes and then follow their fate (see Melin et al. 1983 for a terrestrial example and Hershey et al. 1993 for an aquatic example; other examples reviewed in Peterson and Fry 1987).

In the analyses of these experiments, the accounting procedure often employed explicitly assumes a mass balance by requiring that inputs to any ecosystem or compartment therein must equal outputs plus changes in storage. An early and classic example of such a procedure was the postulation of a missing input of 14 kg/ha of nitrogen per year at Hubbard Brook (Bormann et al. 1977), leading to a search for possible nitrogen fixation pathways. Similarly, so-called mixing models used in isotope experiments (Peterson et al. 1985; Peterson and Fry 1987; Mizutani and Wada 1988; Hershey et al. 1993) are proportional allocations of isotopes among different pools within a mass-balance constraint (see especially O'Leary [1981] for a demonstration of how such mixing model approaches are derived from the mass balance of reactions in organisms).

Besides being an important accounting tool in the analysis of experimental and observational data, mass-balance requirements set powerful constraints on the possible solutions of both analytical and numerical ecosystem models. These constraints specify that only certain analytical or numerical solutions of the model's equations will be allowed. Experimental or observational protocols can then be designed to look for examples or the lack of examples of these solutions in nature. If they are observed, then we gain confidence that the assumptions and processes captured by the model are operative in the real world. If they are not, then we suspect the assumptions, structure, or equations of the model.

In this chapter, I will examine the concept of mass balance in both theoretical and experimental ecosystem ecology. In particular, I will show how the mass balance of nutrients constrains the coexistence of species. If species both respond to and affect the flow of nutrients in different ways, then the number of combinations of species within a given input-output budget of an ecosystem is theoretically constrained. The mass balance of nutrients thus sets constraints on species coexistence and diversity, but the species pool itself may alter nutrient fluxes.

I will examine a series of models that treat resource or nutrient mass balance in increasing detail then I will examine or suggest experiments that test the predictions of these models by manipulating mass balance. These models are differential equation models rather than the more commonly used simulation models of ecosystem ecology. The treatment of mass balance is more apparent in differential equation models than in simulation models because of the former's simplicity. Moreover, recent software advances (MatLab, Mathematica) enable us to find solutions to systems with three or more coupled differential equations, whereas previous generations of theoretical ecologists were limited to solving systems of only two differential equations, such as the classic Lotka-Volterra equations (Kingsland 1985).

I will begin with Tilman's R^* model (Tilman 1982), proceed by adding litter return and decay to this model, then proceed further through the model of Pastor and Cohen (1997), which adds consumers and decomposers, and conclude with the model of Loreau (1998), which partitions the mass balance of nutrient flow through soils into different pools. This sequence represents an elaboration of mass balance in the original R^* model. In the process, I will show both theoretically and experimentally that coexistence of species (and thus species diversity) as well as ecosystem stability requires that mass balance be treated more elaborately than in Tilman's simple R^* model. For each model, I will draw on experimental tests done at the Cedar Creek Long-Term Ecological Research (LTER) site, as well as elsewhere.

First, I will discuss in general how a conservation law mathematically constrains the solutions of a model.

How Conservation of a System Property Constrains the Solutions of Models

Although conservation of momentum is well known to everyone who has taken high school physics, the underlying philosophical and mathematical consequences of this law are deeper than usually appreciated (Jammer 2000). A property of a system, such as momentum, is said to be conserved when it remains the same after a transformation within the system. The property is said to be invariant with respect to the transformation.

The conservation of momentum requires that a system of two particles that collide inelastically and coalesce after the collision (e.g., two wads of chewing gum) have the same total momentum before and after the collision:

$$m_a v_a + m_b v_b = m_{a+b} v_{a+b} \qquad (15.1)$$

Now, imagine that the velocities of the two particles before the collision can be adjusted such that the velocity after collision, v_{a+b}, equals zero. We can then solve Equation 15.1 for the ratio of m_a to m_b as follows:

$$\frac{m_a}{m_b} = \frac{-v_b}{v_a} \qquad (15.2)$$

Setting m_b as one unit of mass and noticing that the sign of v_b is opposite the sign of v_a, we see that the mass of particle a (in units of the mass of particle b) must be equal to the ratios of the absolute values of their velocities. This single, unique solution for m_a is possible only because of the imposition of the conservation of total momentum dictated by the equality in Equation 15.1. Thus, one can predict the mass of particle a simply by measuring the velocities of the two particles before the collision; this model (Eq.15.2) can then be experimentally verified by weighing the masses of the particles before collision.

This example illustrates my general approach in the remainder of this paper. I will examine how mass is conserved in various ecosystem models by requiring that inputs to all compartments equal losses plus changes in storage. I will then examine how this requirement constrains the model's solutions and then discuss experiments that manipulate mass balance to test the model's predictions.

The time-dependent general solutions of these models are often very complicated and difficult to interpret. Therefore, I will instead examine the equilibrium solutions, which are simpler. At equilibrium, the change in storage of any compartment and of the whole ecosystem equals zero. Equilibrium solutions are found by solving the differential equations simultaneously for all compartments when the differential equations are all set to zero. There are usually several such sets of solutions. In this paper, I am interested only in the set in which all compartments have positive, nonzero equilibrium sizes.

Even though most real ecosystems are not at equilibrium, it is important to find and understand the equilibrium solutions because once they are found, the conditions for nonequilibrium behaviors can be specified. Furthermore, the equilibrium solutions are attractors towards which the entire model system is moving, and so the overall trajectories of real-world ecosystems may bear some resemblance to the approach of the model ecosystem towards equilibrium. Finally, many if not all of the qualitative relationships amongst parameters in the equilibrium solutions also hold during the transient nonequilibrium behaviors of the model. However, they are easier to see and interpret in the equilibrium solutions than in the more general but more complicated time-dependent solutions. These relationships can then be tested in real ecosystems by experiments, which manipulate parameter values and hence nutrient fluxes.

In addition to having nontrivial equilibrium solutions (i.e., solutions in which the masses of all compartments are greater than zero), the model must also be stable to perturbations around equilibrium for it to have any credibility and any possibility of being validated against experimental data. The model is stable around equilibrium if, after one of the compartments is partly harvested or augmented (through, for example, adding fertilizer), its size and those of the other compartments does not grow exponentially or decline to extinction but instead returns to the equilibrium values before the perturbation. Stability in this sense in ecosystem models has been extensively discussed by May (1973), Webster et al. (1975), DeAngelis (1992), and Case (2000). A model is said to be stable in this sense if all the eigenvalues of the Jacobian matrix of partial derivatives of all components with respect to each other (sometimes called the "community matrix" in the ecological literature; see May 1973) have negative real parts when the Jacobian is evaluated near the equilibrium values of all components. The eigenvalues are the rates by which the perturbations grow or decay: if they all have negative real parts, the perturbation decays and the system is stable near equilibrium. If some are complex with negative real parts, then the perturbation decays with oscillations. If at least one eigenvalue is positive, then the perturbation grows without bounds and the system is unstable. It is possible that the sizes of the compartments will neither grow nor decay after a perturbation but simply move to a new set of equilibrium values. This is known as neutral stability and occurs when the real parts of the eigenvalues are zero.

The differential equations of a model, and therefore also its equilibrium and time-dependent solutions as well as its response to perturbations near equilibrium, are all expressed in terms of the input and outputs rates of the components. The conservation of matter (mass balance) in an ecosystem model therefore constrains both its equilibrium and nonequilibrium time-dependent solutions. We gain some understanding of how ecosystem processes emerge from the interactions of the components of the ecosystem, including the species that they contain, when we then find corresponding examples of these solutions in the real world.

A note on notation: To clarify the sequential development of these models, I have employed consistent notation throughout. In some cases, this may differ from the notation used by the original authors, but it is not difficult to determine correspondence between the notation here and the original papers by inspection.

Model 1: The $R*$ Hypothesis

Tilman (1982) has proposed a theory of competition for resources that has gained some attention. This theory assumes that different plants compete for a single limiting resource in the soil and that competition works by the winner drawing down the resource pool to a level lower than can be tolerated by the competitor(s).

The theory requires three equations, one each for the two plants and one for the soil resource. Figure 15.1 depicts the flow of nutrients in the model. The equations are

$$\begin{cases} \dfrac{dR}{dt} = Q - eR - u_1 P_1 R - u_2 P_2 R \\[2mm] \dfrac{dP_1}{dt} = P_1(u_1 R - d_1) = u_1 P_1 R - d_1 P_1 \\[2mm] \dfrac{dP_2}{dt} = P_2(u_2 R - d_2) = u_2 P_2 R - d_2 P_2 \end{cases} \qquad (15.3)$$

where P_1 and P_2 are nutrient contents of species 1 and 2; u_1 and u_2 are uptake rates per unit plant species per unit resource per time; d_1 and d_2 are per capita death rates per time; Q is input to R, assumed to be constant, via either exogenous input to the ecosystem or weathering or decay of soil organic matter; and e is the instantaneous leaching rate from R. All nutrients exported from R become either the input to a plant species or are lost from the entire system. Note that when a plant dies, the nutrients are not returned to R via litter decay but are exported from the ecosystem

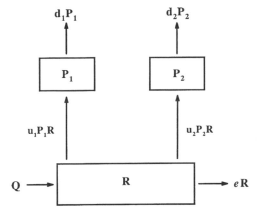

Figure 15.1. Model 1: The $R*$ model of Tilman (1982). R is the inorganic nutrient pool available to plants, and P is the nutrient pool in plant species 1 or 2. Q is the exogenous nutrient input rate, u is the per capita uptake rate for each species, d is the per capita death rate of each species, and e is the fractional export rate (e.g., leaching) from R.

entirely. We will take up this point later in Model 2. Note also that Tilman (1982) uses a Monod function for the uptake term rather than a Lotka-Volterra product term as in Equation 15.3. The Lotka-Volterra formulation of uptake used here greatly simplifies the solutions but does not alter the conclusions drawn here compared with those of Tilman (1982).

Setting the differential equations equal to zero and solving simultaneously yields the equilibrium solutions. These are:

1. $P_1^* = 0, P_2^* = 0, R^* = Q/e$

2. $P_1^* = Q/d_1 - e/u_1, P_2^* = 0, R^* = d_1/u_1$

3. $P_1^* = 0, P_2^* = Q/d_2 - e/u_2, R^* = d_2/u_2$

$$(15.4)$$

When both species are absent, the equilibrium soil resource pool is determined by the ratio of inputs (Q) to outputs (e). There is no solution where two species can coexist.

The surviving species determines the size of the resource pool by the ratio of its death rate, d (which represents export from ecosystem, *not* transfer to litter) to its uptake rate, u. Tilman (1982) argues intuitively that the species that has the lowest $R*$ (i.e., the lowest equilibrium loss rate in relation to uptake) wins by decreasing R to less than the $R*$ for the competitor (see Armstrong and McGehee (1980) or the Appendix [http://www.ecostudies.org/cary9/appendicies.html] for formal proofs). Species decrease $R*$ by either decreasing their death rate (and hence retaining nutrients) or increasing their uptake rate (and hence increasing one of the outputs from R at the expense of the alternative uptake by the competitor). Note that mass balance requires that d and u must be correlated to maintain the same $R*$—as d increases, so must u in the same proportion to maintain the same ratio at equilibrium, and vice versa.

This model implies the following testable predictions: (1) The species with the lowest R^* should drive all competitors to extinction; (2) Increasing nutrient input to the ecosystem should increase the biomass of the competitive winner.

Experimental Tests

The R^* for a limiting nutrient associated with a monoculture of each species can be measured by determining the size of the available nutrient pool in the soil when each species is grown in monoculture. In pairwise competition experiments, the species with a given R^* for the limiting nutrient in monoculture should be outcompeted by any species with a lower R^* in monoculture, and it should therefore decline to extinction. At this point, one would have a monoculture of the surviving species, which would correspond to either equilibrium solution 2 or 3 in Equation 15.4.

Wedin and Tilman (1993) performed an experiment explicitly to test these predictions. In this experiment, the outcome of competition between pairs of four grass species was determined along gradients of nitrogen input to R, nitrogen having previously been shown to be the limiting nutrient to productivity in these systems (Tilman 1984). The R^* for each species was determined by sampling the soil under monocultures and extracting plant available ammonium and nitrate with 1N KCl. The fours species and their R^* values were *Schizachyrium scoparium* (0.096), *Poa pratensis* (0.183), *Agropyron repens* (0.313), and *Agrostis scabra* (0.653). The nitrogen-input gradient was established by creating gardens with three different amounts of the same topsoil mixed into the underlying sand subsoil. Further augmentations of nitrogen input were achieved by adding 6.55 g m^{-2} yr^{-1} of N as NH_4NO_3 to the soil with the greatest amount of topsoil. The outcome of competition was determined by planting pairs of species at different seeding densities and by testing the ability of a competitor to invade an established monoculture of another. Therefore, the experimental tests manipulated mass balance on the pools of NH_4-N and NO_3-N by either pairwise comparisons of species with different uptake rates (u_i) or by creating gradients of nitrogen input rates (Q).

As predicted by the model, *Schizachyrium* caused *Agropyron* and *Poa* to go extinct, but only when both populations were established from seed (a pairwise competition between *Schizachyrium* and *Agrostis* was not established). However, *Agropyron* could coexist with *Poa* even though the R^* of *Agropyron* was almost double that of *Poa*. Furthermore, augmenting nitrogen inputs with fertilizer often increased the biomass of the competitor with higher R^* and actually decreased the biomass of *Schizachyrium*, the species with the lowest R^*.

These experiments only partially corroborated the predictions of Model 1 and in some cases failed completely to corroborate it. Furthermore, species obviously coexist in the real world: while one might argue that this is because equilibrium is not reached, the rarity of monocultures in any grassland (to say nothing of other ecosystems) would argue against that.

Tilman (1988) suggested that species can coexist if one expands the model to include multiple nutrients and set conditions such that each species has the lowest

$R*$ for only one resource. Thus, if the resource ratios of species differ and if there are multiple limiting resources, then species can coexist. Tilman and Pacala (1993) developed this line of reasoning further but admitted that this modification requires as many unique resources as coexisting species, as previously demonstrated by Armstrong and McGehee (1980). Clearly, there are more species than material resources (some 20 or 30 essential elements plus light and water) in even a relatively depauperate temperate ecosystem, to say nothing of a species rich tropical ecosystem. One could argue that resources that are spatially segregated beneath different rooting zones and could be viewed as "different" resources. However, one runs the risk of pushing this argument to a tautology, since no two plants or their roots could ever occupy exactly the same space, and so one could in theory generate as many resources as one needs simply requiring that different individuals occupy different spaces. The theory then becomes unfalsifiable: if the theory does not predict coexistence when in fact it is observed, simply "generate" another "resource" by spatial or temporal segregation. This would effectively generate an infinite number of resources, which is a trivial solution to the problem. We will later, however, explicitly take up the topic of spatial segregation of nutrients in different pools coupled by diffusion in Model 4 (below).

Abrams (1995) points out that the solutions to the $R*$ model or a multiple $R*$ model are likely to be unstable to perturbations near equilibrium. I examined the eigenvalues of the Jacobian matrix of partial derivatives evaluated at the equilibrium solutions 2 and 3 (above); these eigenvalues were surprisingly complicated. While this does not necessarily preclude stability for certain combinations of parameter values, it does greatly restrict the conditions required for stability. In fact, Tilman (1988) had some difficulty producing stable numerical simulations of the $R*$ model using real data for parameter values.

This raises the interesting question of whether unstable models can properly be tested in experiments. If small experimental perturbations (stochastic perturbation from weather events, for example) result in not obtaining the predicted, unstable equilibria, then is the real world "unstable" and is the model validated? Or would we falsely reject the model's (unstable) equilibrium predictions if we do not see the predictions in the experiment itself because we failed to recognize the real-world perturbation? The interpretation of experimental tests of model stabilities (or experimental tests of a model whose instabilities are not well understood) is fraught with ambiguities (DeAngelis et al. 1989). This is not to say that a model is correct simply because it is stable—a stable model must conform to experimental reality by some stringent criteria, statistical or otherwise. Rather, it is to suggest that experimental tests of unstable models are ambiguous. Stability seems to be a criterion we should require a priori of the model before we go ahead and design experimental tests of it. The relationship between model stability and the design of critical experimental tests obviously needs further exploration.

So if Model 1 is only partly corroborated by experiments (Wedin and Tilman 1993), what is wrong with it? Perhaps the problem with this model is that nutrients are exported from the ecosystem when the plant dies, rather than being returned to R in litter. We next turn to a modification of the model, which includes litter return.

Model 2: The $R*$ Hypothesis with Litter Return

Describing and analyzing litter return and decay is one of the distinguishing features of ecosystem ecology. In fact, ecosystem ecology is distinguished from population ecology because mass balance of material requires that one continue to account for biomass even after death. Furthermore, Wedin and Tilman (1990) and Wedin and Pastor (1993) showed that Q in Model 1 is not constant (at least not at Cedar Creek) but changes within four years because of species differences in litter chemistry. Thus, even in the experiments of Wedin and Tilman (1993) discussed above, litter return and its decay must be taken into account.

Rather than have nutrients in plant biomass leave the system at some death rate, Model 1 can be modified to return these nutrients to R at some decay rate k_i (Figure 15.2). The model then becomes:

$$\left.\begin{aligned}
\frac{dR}{dt} &= Q + k_1 d_1 P_1 + k_2 d_2 P_2 - u_1 P_1 R - u_2 P_2 R - eR \\
\frac{dP_1}{dt} &= u_1 P_1 R - k_1 d_1 P_1 \\
\frac{dP_2}{dt} &= u_2 P_2 R - k_2 d_2 P_2
\end{aligned}\right\} \tag{15.5}$$

(For simplicity we neglect to create an explicit litter compartment; the products $k_i d_i$ subsume both litter return and decay. This has no effect on the conclusions we draw from the equilibrium solutions).

At equilibrium, the following solutions hold:

$$\left.\begin{aligned}
&1. \quad P_1^* = 0, P_2^* = 0, R^* = Q/e \\
&2. \quad P_1^* = \left(ed_1 - Qu_1\right)/\left(d_1 u_1 (k_1 - 1)\right), P_2^* = 0, R^* = d_1/u_1 \\
&3. \quad P_1^* = 0, P_2^* = \left(ed_2 - Qu_2\right)/\left(d_2 u_2 (k_2 - 1)\right), R^* = d_2/u_2
\end{aligned}\right\} \tag{15.6}$$

Altering the mass balance of the model to include litter feedback to R still does not allow coexistence of the two species. This is surprising: incorporating litter return and its decay, two of the most cherished and well-studied pathways of nutrient flows in ecosystem ecology, into a model is not sufficient to maintain species coexistence. $R*$ remains the same as for Model 1, but the equilibrium plant species biomasses now include terms for the decay of their litter.

The nutrient feedback through litter does impart some degree of stability near equilibrium for some parameter values. Again, I examined the eigenvalues of the Jacobian matrix of partial derivatives evaluated at equilibrium solution 2 where

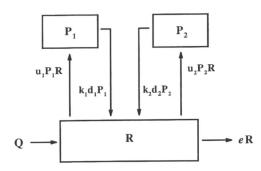

species 1 forms a monoculture. Two of the eigenvalues are purely imaginary, with real parts equal to 0. These impart a neutral stability to the system when the third eigenvalue is not positive (Strogatz 1994). The third eigenvalue is

$$\lambda_3 = \left(u_2 d_1 - u_1 k_2 d_2\right)\big/u_1 \tag{15.7}$$

λ_3 is not positive when

$$u_2 d_1 - u_1 k_2 d_2 \le 0 \tag{15.8}$$

or

$$\frac{u_2 d_1}{u_1 d_2} \le k_2 \tag{15.9}$$

Equation 15.9 sets the conditions under which the monoculture of species 1 will oscillate about equilibrium after a perturbation, such as a small introduction of species 2. Since the other two eigenvalues are purely imaginary, the biomass of species 1 will oscillate around its equilibrium value after a perturbation, neither returning to it nor diverging to infinity. Thus, an introduction of species 2 would not persist as long as this inequality holds.

This inequality is determined partly by the decay rate of species 2. The faster the decay rate, the more likely this inequality is to be maintained. In other words, it is unlikely that species 2 could invade a monoculture of species 1 if it had a high decay rate. However, if species 2 has litter that decays slowly (small k_2), it could reverse this inequality, thereby imparting instability to solution 2 because λ_3 would then be positive. Species 2 could then invade the system and drive species 1 to extinction. The system will then switch to solution 3, a monoculture of species 2, because solution 2 is then unstable.

In this model, the slow decay rate of a plant litter decreases the input to R and may decrease R^* to the point where the competing species with higher R^* may not be able to survive. Similarly, higher uptake rates increase output from R and also therefore decrease R^*. Decay rates as well as uptake rates could be important life-history strategies determining invasion success, a prediction of this model whose experimental test will be discussed shortly.

Thus, even though litter return does not allow for coexistence of species, it does impart at least a weak stability to the system with some restrictions given by relative values of the parameters in Equation 15.9. Similar conclusions have been reached for purely linear models by Webster et al. (1975). Hence, although adding litter feedbacks to Model 1 does not allow for coexistence of species, such feedbacks do provide for some model stability.

Experimental Tests

Since Model 2 does not imply predictions that are qualitatively different from Model 1, the same conclusions regarding the experiments of Wedin and Tilman (1993) hold for Model 2 as for Model 1. In fact, these experiments are more a test of Model 2 than of Model 1 because Wedin and Tilman (1993) did not remove species litter from their plots as technically required by Model 1. Furthermore, Wedin and Tilman (1991) and Wedin and Pastor (1993) showed that the different litters affected the mass balance of R through decay and mineralization within the length of time of the experiment. Therefore, even within the time frame of the experiment, Q was altered through litter return and decay of the constituent species.

It is interesting to note that *Schizachyrium* had the slowest litter decay of any of the species in these experiments (Pastor et al. 1987; Wedin and Tilman 1991; Wedin and Pastor 1993), and so its ability to sometimes outcompete the other species (Wedin and Tilman 1993) is consistent with the inequality of Equation 15.9. However, without numerical values for both uptake and decay rates for all competing species, it is not possible to say for certain whether this is the case.

Why does Model 2 not allow for coexistence of species? After all, the model accounts for at least 90% of the mass balance of materials flowing through ecosystems, the amount being transferred to higher trophic levels usually being no more, and in many cases much less than, 10% of that in the plants. Is species coexistence independent of the mass balance of nutrients flowing through an ecosystem? For this to be the case, then plant species could neither respond to nor affect the fluxes of nutrients through ecosystems, which is clearly false. Alternatively, perhaps the mass balance required for species coexistence is not fully captured by these two models. Two additional modifications to Model 2 might include (1) adding pathways of nutrient flow through other trophic levels (consumers and decomposers); and (2) partitioning some of the pools in Models 1 and 2 more finely. These possibilities are explored in the next two models.

Model 3: Adding Consumers and Decomposers

Some theoretical analyses of the effect of consumers on ecosystems either ignore the role of inorganic resources altogether (Hairston et al. 1960, Oksanen et al. 1981), ignore the return of nutrients to the resource pool via decay of dead biomass (Holt et al. 1994), or assume that plants are homogeneous (Loreau 1995, with further examples in DeAngelis 1992). Pastor and Cohen (1997) examined the case where some plant biomass is consumed before being returned to decomposers, which then transfer them to the inorganic resource pool R (Figure 15.3). This model also assumes an export rate from each compartment and a consumer preference for one plant over another. The product of litter return (d_i in Model 2) and its decay (k_i in Model 2) for each plant species are subsumed into one return-decay parameter (k_i in Model 3). The six differential equations for this model are written by subtracting the sum of the outputs from each compartment from the sum of the inputs:

$$
\left.
\begin{aligned}
\frac{dR}{dt} &= Q + k_D D - u_1 P_1 R - u_2 P_2 R - eR \\[4pt]
\frac{dP_1}{dt} &= u_1 P_1 R - k_1 P_1 - c_1 P_1 C - eP_1 \\[4pt]
\frac{dP_2}{dt} &= u_2 P_2 R - k_2 P_2 - c_2 P_2 C - eP_2 \\[4pt]
\frac{dC}{dt} &= c_1 P_1 C + c_2 P_2 C - k_C C - eC \\[4pt]
\frac{dD}{dt} &= k_1 P_1 + k_2 P_2 + k_C C - k_D D - eD
\end{aligned}
\right\}
\tag{15.10}
$$

Note the maintenance of mass balance: whatever is lost from one compartment is added to another or lost from the system.

The equilibrium solutions are almost all tremendously complicated. However, for the case of coexistence of two plants with the consumer and the decomposer, the equilibrium consumer compartment is

$$
C^* = \frac{u_1(k_2 + e) - u_2(k_1 + e)}{u_2 c_1 - u_1 c_2}
\tag{15.11}
$$

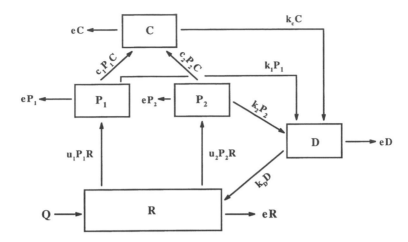

Figure 15.3. Model 3: Model 2 with a consumer (Pastor and Cohen 1997). R is the inorganic nutrient pool available to plants, P is nutrient pool in plant species 1 or 2, C is the nutrient pool in the consumer, and D is the nutrient pool in decomposers. Q is exogenous nutrient input rate, u is the per capita uptake rate per unit R per time for each species, k is the combined per capita litter return rate and decay rate of litter from plant species 1 or 2 and the consumer or nutrient release rate from decomposers, c is the consumption rate of P by C, and e is fractional export rate from all pools.

For C^* to be positive, both the numerator and denominator must be greater than zero, or

$$u_1(k_2 + e) > u_2(k_1 + e) \qquad \text{and} \qquad u_2 c_1 > u_1 c_2$$

or

$$\frac{k_1 + e}{k_2 + e} < \frac{u_1}{u_2} < \frac{c_1}{c_2}$$

(15.12)

The ratios in Equation 15.12 are the preference ratios for the consumer over the two plants (c_1/c_2), the relative differences in uptake rates (u_1/u_2) and the relative differences in litter return and decay rates (k_1/k_2). If we arbitrarily designate producer 1 as preferred by the consumer and the species with faster nutrient uptake and decay, then $k_1 < u_1 < c_1$ (identical inequalities hold for the solution where producer 2 is preferred). This means that not only should the consumer prefer one plant over the other, the producer must also be more discriminatory than the differences in the plants in uptake rates, and the plants must differ more in their uptake rates than in their decay rates. We therefore see increasing discrimination with regard to nutrient flow paths as one moves up the food chain.

We can also ask, what are the regions in parameter space that allow coexistence of the two plants? To do this, one must multiply through the inequalities by u_2 to yield

$$\frac{k_1 + e}{k_2 + e} u_2 < u_1 < \frac{c_1}{c_2} u_2$$

(15.13)

We now have two equations for u_1 in terms of u_2, k_1, k_2, c_1, and c_2. The ratios in Equation 15.13 are now slopes of changes in u_1 with respect to u_2. The region of coexistence of the two plant species lies between the lines defined by the functions of u_2 in Equation 15.13. This region of coexistence can be increased by either: (a) increasing the preference ratio of the consumer for plant species 1 over 2 (i.e., c_1/c_2); (b) decreasing the differences in litter decay rates; or (c) increasing nutrient export (i.e., making the system leaky) because

$$\lim_{e \to \infty} \frac{k_1 + e}{k_2 + e} = 1$$

(15.14)

and so increasing export would decrease this ratio when $k_1 > k_2$. Therefore, a leaky model system promotes coexistence of the two plant species in the presence of a consumer.

Interesting results are obtained when the system is closed ($e = 0$). Then, the equilibrium solution for the consumer coexisting with two plants is

$$C^* = \frac{u_2 k_1 - u_1 k_2}{u_1 c_2 - u_2 c_1}$$

(15.15)

for which the following conditions must hold

$$\frac{k_1}{k_2} > \frac{u_1}{u_2} > \frac{c_1}{c_2}$$

(15.16)

Note that the inequalities are exactly the opposite of that for an open system. Here, the consumer must be *less* discriminatory between the two plants than the plants are different in uptake, and the plants' uptake rates must be *less* different than their decay rates. In other words, if coexistence of the two plants is to be maintained, the consumer must forage more *randomly* between the two plants rather than be more *discriminatory* as in the open system. Therefore, whether the entire ecosystem is open or closed determines the foraging strategy of the consumer—animal behavior must operate in the context of (indeed, is constrained by) biogeochemical cycles.

Moreover, coexistence of the two plants depends on the presence of the consumer—there were no equilibrium solutions where the plants coexisted with each other without the consumer. In particular, there were no equilibrium solutions where the two plant species coexisted with each other in the presence of the decomposer without a consumer. Why is this so? The reason is because the decomposer does not discriminate between the two plant species, as does the consumer. The decomposer is simply a passive recipient of litter from the two plant species and then passes the nutrient back into the resource pool at a rate proportional to its size. In fact, the decomposer could have been omitted from this model altogether without any loss of generality by subsuming each rate of litter return and passage through the decomposer compartment into one term for each plant species. This only highlights the importance of the discriminatory consumer in promoting plant species coexistence in this model.

Pastor and Cohen (1997) show numerically that, for certain forest types containing a deciduous species with high uptake rate, high litter quality, and high consumer preference coexisting with a conifer with the opposite characteristics, the rate of flow of a nutrient through the system is always lower when a consumer is present compared with a system in its absence. Moreover, the system is stable (negative real parts of all eigenvalues) in this region. When the nutrient return by the consumer becomes large enough to increase rates of nutrient flow over that of a system without a consumer, then the system becomes unstable (at least one eigenvalue having positive real part). Therefore, the addition of a consumer to Model 2 is sufficient to allow *both* coexistence of the two plant species *and* stability but at the cost of a decrease in nutrient cycling rate.

Experimental Tests

If Model 3 predicts that consumers promote plant species' coexistence at the expense of a decrease in nutrient cycling rate, then the experimental test of Model 3 is simple: exclude the consumer and measure changes in coexistence and nutrient cycling rate. Pastor et al. (1993) showed that excluding moose from the boreal forests of Isle Royale caused nitrogen mineralization rate to become double that of adjacent browsed plots after 40 years. Moose browsing caused the biomass and density of unbrowsed spruce and lightly browsed balsam fir to increase at the expense of the more heavily browsed deciduous species. The increased dominance of conifers increased the flow of nitrogen along the slowly decomposing litter pathway ($k_2 P_2$). When moose were excluded, rapidly decomposable litter from the deciduous species ($k_1 P_1$) increased the nitrogen mineralization rate. Moose browsing did not cause the deciduous species to go extinct, and so coexistence of the deciduous and conifer functional groups was maintained. However, inside the exclosures, the deciduous species are being succeeded by the conifers (because of lower shade tolerance), and coexistence is not being maintained. Other studies (reviewed in Hobbs 1996) have also found decreases in the nitrogen cycling rate and maintenance of coexistence of plant species in the presence of consumers as predicted by Model 3.

Similar results have been obtained at Cedar Creek (Ritchie et al. 1998). Here, deer graze heavily on nitrogen-fixing legumes with high litter quality. When grazing reduced the legume cover, the abundance of prairies grasses, such as *Schizachyrium* and *Andropogon*, which have slowly decomposing litter (Pastor et al. 1987), increased. The slow decay of litter from these grasses decreases nutrient input to decomposers and eventually reduces R^*. Furthermore, as Wedin and Tilman (1993) showed, *Schizachyrium* and *Andropogon* increase output from R through rapid uptake, thereby also decreasing R^*. Therefore, increased abundance of *Schizachyrium* and *Andropogon* caused by deer grazing on legumes decreases R^* because of both reduced inputs to R and increased uptake from R.

Excluding deer increased the cover of legumes as well as that of woody plants (also browsed by deer) and forbs over a seven-year period. However, graminoid cover declined and graminoids appeared to be headed for extinction in the absence of grazing. Unfortunately, the experiment did not last long enough to determine whether grasses would not continue to coexist with the other groups or would coexist at a reduced density. Without grazing, the increased abundance of legumes and forbs increased R^* by increasing return of nitrogen to R and by increasing dominance by functional groups with higher R^* in monoculture.

In contrast to these results, consumers sometimes increase nitrogen-cycling rates (Hobbs 1996), such as in the Serengeti (McNaughton et al. 1997). In the Serengeti, nitrogen mineralization was greater in grazed areas than in ungrazed areas. However, more than 90% of aboveground plant material was consumed by a succession of waves of consumer populations (McNaughton 1985). While at first these experimental findings appear to disprove the model, in fact such conditions would be met in Model 3 by no selective foraging; that is, if plant species were consumed in proportion to their abundance, selective grazing would not shift species dominance towards the slower (and less preferred) nitrogen cycler. Under such conditions, the consumer grazing succession is treating the plant community as a homogeneous pool. Loreau (1995) showed that consumers increase the nutrient-cycling rate in a model with a single homogeneous plant pool as the return of nutrient returns to decomposers is increasingly dominated by the $k_c C$ pathway.

I know of no experimental tests of the different predictions of the open vs. closed system versions of the model (compare Equations 15.12 and 15.13 with Equations 15.15 and 15.16). Perhaps such experiments could be designed with herbivorous zooplankton and two algal species in flow-through chemostats at equilibrium (which would be a test of Equations 15.12 and 15.13) vs. a closed tank system containing the same species (which would be a test of Equations 15.15 and 15.16).

Model 4: Subdividing R into Several Pools

It has long been known from Lotka-Volterra models that spatial heterogeneity of habitat patches can stabilize competitive exclusion and predator-prey dynamics (Gurney and Nisbet 1978). However, such models do not explicitly consider the mass balance of nutrients; indeed "habitat" is a rather abstract concept in these models.

Huston and DeAngelis (1994) proposed a model whereby nutrient resources in the soil are partially segregated beneath different plant species. In this model, species were able to coexist stably. However, their model did not explicitly include litter return and decay and so suffers from an incomplete description of mass balance.

Loreau (1998) analyzed a model that was developed from that of Huston and DeAngelis (1994) but explicitly included litter return and decay (Figure 15.4). The nutrient pool in the soil is subdivided into a regional pool (R_R) which receives inputs from the surrounding environment and which replenishes local pools (R_i) from which each species i draws for uptake. The differential equations for nutrient flows are the differences between the sums of inputs and sums of outputs for each compartment in Figure 15.4. One advantage of this model is that the number of species need not be specified beforehand. In fact, as we shall see shortly, the model predicts the conditions under which a species can enter the community from the constraints of mass balance.

The equilibrium solution for the regional pool is

$$R_R^* = \frac{Q + d\mu S\sigma\overline{R_1^*}}{1 + d\mu S\sigma}$$

(15.17)

where d is the transport rate of inorganic nutrients between the regional and local pools, S is the total number of species, σ is the average rooting volume of all species, $\overline{R_1^*}$ is the average equilibrium nutrient concentration in the local pools, and μ = $1/(k + e)$, where k is the average decay rate of detritus D into the regional pool, and e is the fractional export rate from the regional pool to the outside environment. Clearly, increasing Q and $\overline{R_1^*}$ increases the size of R_R^*, but the effects of changes in species richness, transport rate of nutrient, root zone occupancy, and μ are more complicated as they appear in both numerator and denominator. However, Loreau (1998) demonstrated a monotonic decline in equilibrium species richness as the transport rate increases. This makes sense because as d increases, there is a greater mixing of all local pools into the regional pool, and the model then reduces to Model 2 in the limit as $d \rightarrow \infty$. Concurrently, productivity increases with increased transport rate because the nutrient becomes available to all species at a greater rate as faster transport replenishes all local pools by passage through the

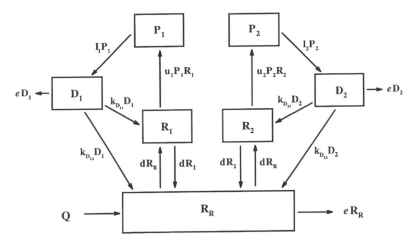

Figure 15.4. Model 4: Model 2 with subdivision of R into regional and local pools (Loreau 1998). R_R is the regional inorganic nutrient pool, R is the local inorganic nutrient pools available to different species, P is the nutrient pool in plant species 1 or 2, *and D* is the different detrital pools of species 1 or 2. Q is exogenous nutrient input rate, u is the per capita uptake rate per unit R per time for each species, l is the per capita litter return rate from plan species 1 or 2, k is the decay rate of detritus from plant species 1 or 2, d is the transport rate of inorganic nutrient in soil, and e is fractional export rate from R_R or D.

regional pool. Thus, as transport between soil nutrient pools increases, productivity also increases but species richness decreases.

From this equilibrium (Equation 15.7), Loreau (1998) was able to show that persistence of a species i requires that it satisfy the following mass balance equation

$$Q + d\mu S\sigma\left(\overline{R}_1^* - R_i^*\right) - R_i^* > 0 \qquad (15.18)$$

The first two terms represent the nutrient input to the ecosystem (Q) plus nutrient input to the regional pool from all local pools other than from species i (e.g., $\overline{R}_1^* - R_i^*$). The third term represents the size of species i local pool, which is replenished by detrital losses from species i. Loreau (1998) further showed that this equation determines the maximum stable species richness imposed by the mass balance of nutrients through the system. This is a remarkable contribution to the current debate on the relationship between richness and ecosystem properties because, in principle, it establishes rigorous and measurable criteria by which we can experimentally determine the maximum species richness for a given rate of nutrient cycling through the ecosystem. Thus, richness and the nutrient-cycling rate can both be predicted a priori from the constraints of mass balance and species life history traits.

Experimental Tests

A logical first test of Model 4 is to determine if different species access different local pools of nutrients. Local pools of nutrients can be segregated both in space (implicit in Loreau's model) but also in time (which would require that the uptake terms in Figure 15.4 be time-dependent rather than constant). By following the mass balance of an ^{15}N tracer injected into different soil depths at different times during the growing season, McKane et al. (1990) showed that *Schizachyrium*, *Poa*, and other species at Cedar Creek do have segregated local nitrogen pools.

A full test of the model requires determination of the transport rates of the inorganic nutrient through the soil, which can also be done using tracers, although the experiments could be technically difficult. The rooting volumes of different species also need to be determined through, for example, careful excavation of roots from the sides of soil pits. While determining the values of parameters of Equation 15.18 may be difficult, it is not impossible. If this can be done, then the current large-scale experimental test of the relationship between richness, productivity, and nutrient retention at Cedar Creek (Tilman et al. 1996, 1997) may prove to be exactly the test required of Loreau's model.

Discussion

Studies on the coexistence of species have been pursued on an industrial scale throughout this century. It appears that at least as many factors have been proposed as there are ecologists. The more commonly proposed factors include disturbance, habitat heterogeneity, differences in dispersal rates, and differences in colonization rates (for an initial discussion, see the papers in Ricklefs and Schluter 1993).

These factors certainly do influence species coexistence and may prove to be the more important determinants of it. But although the conservation of material flows is required by the more fundamental laws of physics, mass-balance constraints and the effects that different species have on material flows have rarely been explicitly and rigorously considered in the debate on species coexistence. As we have seen with Model 2, mass balance via litter return is at least required for stability, if not for coexistence. Moreover, as demonstrated with Models 3 and 4, simply adding a consumer or partly segregating nutrient pools in soil along with adding litter return and decay allows for stable coexistence of species, at least in the model.

Thus, the mass balance of material flows considered at the proper trophic or spatial level of resolution is a sufficient condition for stable coexistence, and I would argue, because of the more fundamental laws of physics, a necessary requirement of any model of coexistence of species in ecosystems. This is not to say that disturbance, dispersal, and the other factors should be ignored and are unimportant. Rather, it may be profitable to consider how these factors affect species coexistence and richness within the framework of how they alter the mass balance of nutrients.

This approach to determining the constraints on species coexistence has important consequences for the design of experiments. First, nutrient flows through ecosystems can be altered in relatively straightforward ways; in fact, such an experiment is the basic experiment in ecosystem ecology. Secondly, by imposing mass balance on the analysis of experimental data, one can determine whether all flows are accounted for (see especially Bormann et al. 1977). Thus, within the framework of mass balance depicted by each of these or other models, one can experimentally test the importance that alternative pathways of nutrient flow may have on species coexistence. Third, because the models describe specific mechanisms (litterfall, decay, transport, leaching, uptake) rather than phenomenological parameters (i.e., competition coefficients in Lotka-Volterra models), manipulative experiments can be explicitly designed to test whether these mechanisms are operative.

Adding consumers or partitioning the nutrient resource pools are two of the many ways that additional complexity can be added to a nutrient-cycling model. As we have seen, both actions result in stable coexistence of species. Other elaborations of the basic $R*$ model may also result in stable coexistence. No one model can a priori be chosen over others: whether a model reflects the real world has to be decided by experimental tests of model predictions or implications. Furthermore, both Model 3 and Model 4, as well as others, may be validated to some degree because they are not necessarily mutually exclusive. In fact, de Mazancourt and Loreau (2000) analyze a model that is a combination of Models 3 and 4. The point here is not that Models 3 and 4 are the only correct models of species coexistence consistent with the mass balance of nutrient flows, but that Models 1 and 2 are too simple to allow for stable coexistence and must be discarded as sufficient descriptions of how species might coexist in an ecosystem.

Vitousek et al. (1998) claim that a model similar to Model 2 is the "standard" model of ecosystem ecology, at least with respect to the more traditional problem of the controls on input-output budgets. Such a model may be sufficient for capturing the main features of input-output balance on a whole ecosystem, but it is not sufficient for addressing the question of species diversity and coexistence in the context of material flows through an ecosystem. As ecosystem ecologists, we need to expand our thinking beyond Model 2 if we are to make a contribution to the species richness–ecosystem property debate. Considerations of Models 3 and 4 suggest that our "standard model" of an ecosystem and experiments designed from it should include at least the effect of consumers and/or segregation of different pools of nutrients in soils on mass balances of material flows in ecosystems.

Acknowledgments. I would like to thank Charlie Canham, Jon Cole, and Bill Lauenroth for organizing this Cary Conference and inviting this paper. I am grateful to Ed Rastetter, who helped me clarify the stability of Model 2 by his penetrating observations at this conference. Thanks also to an anonymous reviewer for a particularly helpful review and to Bill Lauenroth for editing the paper. Finally, I thank Gene Likens for making the Institute of Ecosystem Studies such a pleasant place to discuss science and for stimulating my continued interest in the mass balance of material flows through ecosystems by establishing the Hubbard Brook Experiment with Herb Bormann almost forty years ago.

References

Abrams, P.A. 1995. Monotonic or unimodal diversity-productivity gradients: What does competition theory predict? *Ecology* 76: 2019–2027.

Armstrong, R.A., and R. McGehee. 1980. Competitive exclusion. *American Naturalist* 115: 151–170.

Baker, J.D., G.L. Switzer, and L.E. Nelson. 1974. Biomass production and nitrogen recovery after fertilization of young loblolly pines. *Soil Science Society of America Journal* 38: 958–961.

Binkley, D. 1986. *Forest Nutrition Management.* New York: John Wiley and Sons.

Bormann, F.H., G.E. Likens, and J.M. Melillo. 1977. Nitrogen budget for an aggrading northern hardwood forest ecosystem. *Science* 196: 981–983.

Case, T.J. 2000. *An Illustrated Guide to Theoretical A.D. Ecology.* Oxford: Oxford University Press.

DeAngelis, D.L. 1992. *Dynamics of Nutrient Cycling and Food Webs.* London: Chapman and Hall.

DeAngelis, D.L., P.J. Mulholland, A.V. Palumbo, Steinman, M.A. Huston, and J.W. Elwood. 1989. Nutrient dynamics and food web stability. *Annual Review of Ecology and Systematics* 20: 71–95.

de Mazancourt, C., and M. Loreau. 2000. Effect of herbivory and plant species replacement on primary production. *American Naturalist* 155: 735–754.

Gosz, J.R., G.E. Likens, and F.H. Bormann. 1973. Nutrient release from decomposing leaf and branch litter in the Hubbard Brook Forest, New Hampshire. *Ecological Monographs* 43: 173–191.

Gurney, W.S.C., and R.M. Nisbet. 1978. Predator-prey fluctuations in patchy environments. *Journal of Animal Ecology* 47: 85–102.

Hairston, N.G., F.E. Smith, and L.B. Slobodkin. 1960. Community structure, population control, and competition. *American Naturalist* 94: 421–425.

Hershey, A.E., J. Pastor, B.J. Peterson, and G.W. Kling. 1993. Stable isotopes resolve the drift paradox for *Baetis* mayflies in an arctic river. *Ecology* 74: 2315–2326.

Hobbs, N.T. 1996. Modification of ecosystems by ungulates. *Journal of Wildlife Management* 60: 695–713.

Holt, R.D., J. Grover, and D. Tilman. 1994. Simple rules for interspecific dominance in ecosystems with exploitative and apparent competition. *American Naturalist* 144: 741–771.

Huston, M.A., and D.L. DeAngelis. 1994. Competition and coexistence: The effects of resource transport and supply rates. *American Naturalist* 144: 954–977.

Jammer, M. 2000. *Concepts of Mass in Contemporary Physics and Philosophy.* Princeton: Princeton University Press.

Kingsland, S.E. 1985. *Modeling Nature: Episodes in the History of Population Biology.* Chicago: University of Chicago Press.

Likens, G.E., F.H. Bormann, N.M. Johnson, D.W. Fisher, and R.S. Pierce. 1970. Effects of forest cutting and herbicide treatment on nutrient budgets in the Hubbard Brook Watershed-Ecosystem. *Ecological Monographs* 40: 23–47.

Loreau, M. 1995. Consumers as maximizers of matter and energy flow in ecosystems. *American Naturalist* 145: 22–42.

———. 1998. Biodiversity and ecosystem functioning: A mechanistic model. *Proceedings of the National Academy of Sciences USA* 95: 5632–5636.

Marks, P.L., and F.H. Bormann. 1972. Revegetation following forest cutting: mechanisms for return to steady-state nutrient cycling. *Science* 176: 914–915.

May, R.M. 1973. *Stability and Complexity in Model Ecosystems.* Princeton, NJ: Princeton University Press.

McKane, R.B., D.F. Grigal, and M.P. Russelle. 1990. Spatiotemporal differences in ^{15}N uptake and the organization of an old-field plant community. *Ecology* 71: 1126–1132.

McNaughton, S.J. 1985. Ecology of a grazing ecosystem: The Serengeti. *Ecological Monographs* 53: 291–320.

McNaughton, S.J., F.F. Banyikawa, and M.M. McNaughton. 1997. Promotion of the cycling of diet-enhancing nutrients by African grazers. *Science* 278: 1798–1800.

Melin, J., H. Nonmik, U. Lohm, and J. Flower-Ellis. 1983. Fertilizer nitrogen budget in a Scots pine ecosystem attained by using root-isolated plots and ^{15}N technique. *Soil Science Society of America Journal* 35: 346–349.

Mizutani, H., and E Wada. 1988. Nitrogen and carbon isotope ratios in seabird rookeries and their ecological implications. *Ecology* 69: 340–349.

Oksanen, L. S.D. Fretwell, J. Aruda, and P. Niemala. 1981. Exploitative ecosystems in gradients of primary production. *American Naturalist* 131: 424-444.

O'Leary, M.H. 1981. Carbon isotope fractionation in plants. *Phytochemistry* 20: 553–567.

Pastor, J., and Y. Cohen. 1997. Herbivores, the functional diversity of plants species, and the cycling of nutrients in ecosystems. *Theoretical Population Biology* 51: 165–179.

Pastor, J., B. Dewey, R.J. Naiman, P.F. McInnes, and Y. Cohen. 1993. Moose browsing and soil fertility in the boreal forests of Isle Royale National Park. *Ecology* 74: 467–480.

Pastor, J., M.A. Stillwell, and D. Tilman. 1987. Little bluestem litter dynamics in Minnesota old fields. *Oecologia* 72: 327–330.

Peterson, B.J., and B. Fry. 1987. Stable isotopes in ecosystem studies. *Annual Review of Ecology and Systematics* 18: 293–320.

Peterson, B.J., R.W. Howarth, and R.H. Garritt. 1985. Multiple stable isotopes used to trace the flow of organic matter in estuarine food webs. *Science* 227: 1361–1363.

Richardson, C.J., and J.A. Lund. 1975. Effects of clear-cutting on nutrient losses in aspen forests on three soil types in Michigan. Pp. 673–686 in F.G. Howell, J.B. Gentry, and M.H. Smith, editors. *Mineral Cycling in Southeastern Ecosystems.* Washington, DC: U.S. Energy Research and Development Administration.

Ricklefs, R.E., and D. Schluter, editors. 1993. *Species Diversity in Ecological Communities: Historical and Geographic Perspectives.* Chicago: University of Chicago Press.

Ritchie, M.E., D. Tilman, and J.M.H. Knops. 1998. Herbivore effects on plant and nitrogen dynamics in oak savanna. *Ecology* 79: 165–177.

Strogatz, S.H. 1994. *Nonlinear Dynamics and Chaos.* Reading, MA: Perseus Books.

Tilman, D. 1982. Resource Competition and Community Structure. *Monographs in Population Biology.* Princeton: Princeton University Press.

———. 1984. Plant dominance along an experimental nutrient gradient. *Ecology* 65: 1445–1453.

———. 1988. *Plant Strategies and the Dynamics and Structure of Plant Communities.* Monographs in Population Biology. Princeton: Princeton University Press.

Tilman, D., J. Knops, D. Wedin, P. Reich, M. Ritchie, and E. Siemann. 1997. The influence of functional diversity and composition on ecosystem processes. *Science* 277: 1300–1302.

Tilman, D., and S. Pacala. 1993. The maintenance of species richness in plant communities. Pp. 13–25 in R.E. Ricklefs and D. Schluter, editors. *Species Diversity in Ecological Communities: Historical and Geographic Perspectives.* Chicago: University of Chicago Press.

Tilman, D., D. Wedin, and J. Knops. 1996. Productivity and sustainability influenced by biodiversity in grassland ecosystems. *Nature* 379: 718–720.

Vitousek, P.M., J.R. Gosz, C.C. Grier, J.M. Melillo, W.A. Reiners, and R.L. Todd. 1979. Nitrate losses from disturbed ecosystems. *Science* 204: 469–474.

Vitousek, P.M., L.O. Hedin, P.A. Matson, J.H. Fownes, and J. Neff. 1998. Within-system element cycles, input-output budgets, and nutrient limitations. Pp. 432–451 in M.L. Pace and P.M. Groffman, editors. *Successes, Limitations, and Frontiers in Ecosystem Science.* New York: Springer-Verlag.

Vitousek, P.M., and W.A. Reiners. 1975. Ecosystem succession and nutrient retention: A hypothesis. *BioScience* 25: 376–381.

Webster, J.R., J.B. Waide, and B.C. Patten. 1975. Nutrient recycling and the stability of ecosystems. Pp. 1–27 in F.G. Howell, J.B. Gentry, and M.H. Smith, editors. *Mineral Cycling in Southeastern Ecosystems.* Springfield, VA: U.S. Energy Research and Development Administration, National Technical Information Service.

Wedin, D.A., and J. Pastor. 1993. Nitrogen mineralization dynamics in grass monocultures. *Oecologia* 96: 186–192.

Wedin, D., and D. Tilman. 1990. Species effects on nitrogen cycling: a test with perennial grasses. *Oecologia* 84: 433–441.

―――. 1993. Competition among grasses along a nitrogen gradient: Initial conditions and mechanisms of competition. *Ecological Monographs* 63: 199–229.

Part III

The Role of Models in Environmental Policy and Management

16

The Role of Models in Ecosystem Management

Graham P. Harris, Seth W. Bigelow, Jonathan J. Cole, Hélène Cyr, Lorraine L. Janus, Ann P. Kinzig, James F. Kitchell, Gene E. Likens, Kenneth H. Reckhow, Don Scavia, Doris Soto, Lee M. Talbot, and Pamela H. Templer

Introduction

It is now quite clear that our planet faces a large number of major environmental and natural resource management problems. Biodiversity loss, land-use change including habitat fragmentation, and destruction of ecosystems go on apace as the human population rises (e.g. World Resources Institute 2000; Likens 1991). Habitat fragmentation is leading to the loss of biodiversity and the loss of function and ecosystem services on a large scale. We now realize that forms of western agriculture also lead to environmental damage, including loss of biodiversity, salinization, soil erosion, leakage of nutrients and deleterious effects on water quality in receiving waters (Likens 1992). Landscape failure is rife, leading to phenomena like acidification of lakes and rivers (Likens et al. 1972; Charles 1991; Turner and Rabalais 1991; Vitousek et al. 1997), phosphorus leakage from agriculture (Likens and Bormann 1974; Sharpley et al 1991), increased nitrogen deposition and groundwater pollution (Agren and Bosatta 1988; Aber et al. 1989; Likens 1992; Stoddard 1994; Galloway 1995), eutrophication of estuaries and coastal areas (Howarth 1988; Caraco and Cole 1999) and dryland salinity (Kotb et al. 2000). Environment and natural resource managers are facing enormous, growing, and complex problems that frequently cut across physical, philosophical, institutional, local, and state boundaries.

Impediments to effective management at these complex scales often come from a lack of comprehensive planning (e.g., on a watershed- or landscape-scale) and lack of understanding of the full range of impacts of various actions or inaction (Walters 1986). Environmental conflicts frequently arise from attempts to manage multiple uses of ecosystems and their services. Mobbs and Dovers (1999) have recently summarized these problems as being characterized by

- Failure to consider the possibility of irreversible impacts, and the urgency of setting policy

- Lack of attention to connectivity between problems: cumulative rather than discrete impacts
- Uncertainty and ignorance: highly variable temporal and spatial scales of problems, lack of uncontested research methods, policy instruments and management approaches, sheer novelty of environmental and policy problems
- Systemic causes for environmental problems being thoroughly embedded in patterns of production, consumption, settlement, and governance
- The lack of defined policy, management and property rights, and roles and responsibilities, couples with intense demands for increased community participation in both policy formulation and actual management

In this environment, it is necessary to clearly define key questions, delineate their scale and scope, and agree on desirable outcomes. However, this is rarely easy because of the difficulty in bridging gaps among various interests in tackling these large-scale, complex problems. For these problems, the packaging of essential ecosystem science into "useful" information requires simplicity and clarity and, often, the ability to paint simple conceptual pictures for a wide constituency of the public, managers, and policy makers. This ability can illuminate the debate, better define options, and bring various interests together around common ideas and values. A goal for ecological science is to find ways to compress vast stores of knowledge and information into scientifically sound but understandable packages.

It is clear that management decisions will be made and economic and regulatory policy will be developed whether or not ecological science is involved. Further, most of those actions will be based on forecasts (predictions, projections, expected outcomes) that may or may not have the rigor of a scientific foundation. The question before the ecological community is how to ensure that the best science is made available for developing forecasts and informing policy.

Models in a Management Context

Models have played key roles in informing public debate and informing management decisions for environmental and natural resource management issues. In cases where ecological sciences have played a substantial role in the decision-making process, quantitative and/or qualitative models were often an integral component. For example, population and ecosystem models can provide internally consistent frameworks for integrating anecdotal and informal knowledge, as well as results from disparate basic research efforts, into tools that can explore answers to policy-maker and manager "what if" questions. While ecological forecasts are, by their very nature, risky and imprecise, these tools are frequently how ecosystem science is brought to bear on applied problems (see Chapter 7).

Some examples include

- Population management and stock harvesting (Hilborn and Walters 1992; Botsford et al 1997)
- Managing the effects of nutrient loads on lakes (Vollenweider 1968; Carpenter et al. 1995) and estuaries (Valiela et al. 2000), and of preemption of freshwater on estuarine salinity (Livingston 1991)
- Managing the effects of acid rain by controlling the emissions of oxides of sulphur and nitrogen (Likens 1989b; Likens 1992; Weathers and Lovett 1998; Driscoll et al. 2001)
- Forecasting the effects of human population growth on resource use and painting various future scenarios (Sala et al. 2000)

The Role of Ecosystem Models In Management

Environmental managers and policy makers need formal synthesis and integration of ecosystem science either when there is a need for the resolution of an imminent existing problem or where there is a lack of knowledge about possible future unforseen problems.

Two ways to effectively bring this science to the policy making process are through integrated assessments and adaptive management. Both approaches require close collaboration among institutions, managers, science, and the public, including the definition of goals, values, and ethics and the willingness to move forward and make decisions in an uncertain environment. While these two processes have different roots, their impacts and approaches are similar, and they both rely on the ability to estimate future impacts of potential actions— that is, to forecast.

Combining these two approaches, an efficient strategy is to

(1) Document the status and trends in properties of concern
(2) Describe social, economic, and ecologic causes and consequences of those trends
(3) Forecast the social, economic, and ecologic impacts of various policy options
(4) Provide technical guidance for implementing those options
(5) Take appropriate action, if necessary
(6) Put in place monitoring and research to track results of those actions
(7) Repeat the sequence over time

Step 1 is perhaps the easiest to perform where appropriate monitoring programs are in place. Step 2 relies on the synthesis and application of extant knowledge about how the natural system functions, including how humans interact within those systems. Once policy or management actions have been selected, the applied sciences, engineering, and economic disciplines can provide the guidance called for in Step 4. Taking action (Step 5) is the realm of incentives, regulation, and enforcement. Step 3—forecasting impacts of policy options—is most often the weakest link in these analyses, and it is through improvements in this step that modeling can help bring ecological sciences to bear on environmental policy making.

The Nature of Ecosystem Models

From the examples outlined above, it is clear that there are parsimonious ecosystem models that capture sufficient aspects of reality to act as a guide to management action and policy development. Why this is so is beyond the scope of this chapter, but ecosystem science is progressing to the point where some basic rules are emerging. At certain defined levels of complexity, the emergent properties of ecosystems may indeed be "simpler than we think" (Harris 1999). There is a real need to define uncertainty and predictive power (Chapter 8) and the limits to each model's capability, and then to move forward in an adaptive mode, recognizing the risks and uncertainties. Refusing to act because of lack of complete understanding or fear of making a mistake is a virtual assurance of failure of the management system (Gunderson et al. 1995). It is worth stressing here that in a changing world of adaptive agents, inaction often carries more risk than action—so clear definitions of uncertainty can be a useful guide to management action.

Ecosystem models should

- Be consistent with accepted theory, based on well-documented fundamental science, and integrate across disciplines and scales, as appropriate
- Make predictions about properties that people care about and that are based on inputs that can be measured and managed
- Quantify uncertainty (the inverse of information content) of the predictions based on uncertainty or variability in model inputs, parameters, and structure
- Be questioned and continuously improved through research and through their use in integrated assessments and adaptive management

Building From the Fundamental Science Base

Extant models build from and depend on historical investments in the fundamental sciences. Future models will likewise depend upon continued investments and advances in the disciplines. However, there exists a need for a new kind of science to bridge the gap between fundamental, curiosity-driven research and investments in the application of research. Science directed specifically toward improving models may be one way to bridge the gap. This problem facing ecosystem scientists is not unique. Gibbons et al. (1994) termed this the rise of mode-II science, where mode I was the traditional form of science characterized by incremental advancements within disciplines and publication of papers in the scientific literature. Mode-II science is defined as "science in the context of its application" where the goal is to make a difference or, at least, to have a defined outcome in the form of changed policy or management practice.

They do not suggest mode II would, or should, replace mode I, but that mode II is growing in importance in today's society. When used in the contexts of integrated assessment and adaptive management, as outlined above, mode II science can form a useful bridge between mode I science and policy.

Mode-II science is practiced in an arena where the community has a growing role and a demand for knowledge and empowerment: where nongovernmental organizations (NGOs) are powerful knowledge brokers and where there is widespread lack of trust. Institutional failure is frequently encountered when resource-management problems do not conveniently equate with ecosystem or biophysical boundaries. Watershed management is a classic example of an issue that frequently cuts across philosophical, geographic, and institutional boundaries. Thus, there is a critical, growing need for integrated catchment models and management (e.g., Groffman and Likens 1994).

Making Predictions That Matter

While there are many reasons for building ecosystem models, when building them for use in management, inputs should be properties that managers can control and their predictions (outputs) should be of properties that people care about. For example, for estuarine eutrophication problems, inputs would include properties that influence nutrient loads (land-use patterns, sewage treatment plant discharges, atmospheric sources), in addition to others that are not within managers' control (e.g., temperature, solar radiation, ocean fluxes). The output of these models would be properties like water clarity and fish abundance rather than primary production or nutrient and chlorophyll concentrations. While these latter properties are clearly important and may be necessary to reach the endpoints, they may not be what matters most to people and policy makers.

Quantifying Uncertainty or Information Content of the Forecasts

Ecosystem models are, of necessity, abstraction from reality (Chapter 2). Any ecological prediction or forecast must therefore be couched in terms of risk and uncertainty (Chapters 9, 12, and 19) and must be regarded as provisional. This is precisely why an adaptive management framework is essential. As knowledge improves and forecasts become more certain, management actions can be altered and updated to improve the chances of obtaining the desired result. If the ecosystem management is set in an uncertain socioeconomic framework, then decisions about social and economic policy, as they affect ecosystem management, can also be updated over time (Likens 1989a).

We recognize that ecosystem models cannot be fully validated (Chapter 2), but that problem is probably no worse than other uncertainties associated with large-scale ecosystem management. Ecologists and managers must recognize that ecosystems are open systems which constantly exchange energy and materials across their boundaries and which are set in a global context of constant variability and change (Oreskes et al. 1994). Ecosystems are also set in a socio-economic context of changing human populations and management practices. The whole has been likened to the properties of complex adaptive systems (Harris 1999 and many others), where the emergent properties of the whole are determined by the internal pandemonium of interactions between active and adaptive agents (Holland 1998).

Learning From, Improving, and Applying Models

Models—their structure, inputs, parameters, and time and space domains—should be continually challenged both in terms of their ability to capture reality and their utility in management contexts. In this context, the growing demands from the scientific community and policy institutions, along with technological change, will drive advances in modeling techniques for both the natural and social sciences. One area of rapid change is the introduction of spatially explicit models in the context of geographical information systems (GIS). Large scale GIS-based ecosystem models that incorporate socioeconomic factors—such as large scale watershed models incorporating aspects of economics and the behavior of rural communities—lie at the cutting edge of modeling and management practice. Issues of spatial "emergence" and validation for models that produce novel behaviors are yet to be addressed.

While the relative merits of simple and complex models are continually debated and are discussed elsewhere in this volume (Chapters 4, 5, and 25), most likely, hierarchies of modeling approaches will be used. Simple problems may be addressed by ANOVA and regression models (Carpenter et al. 1995). Simple models may be used where there are simple, regular emergent properties of systems, such as Vollenweider's loading models (Vollenweider 1968; Caraco and Cole 1999), or where small-scale experiments in pots or plots can be used. More complex problems may require mechanistic and simulation models. These are frequently used in situations where more traditional experimental approaches are inappropriate for reasons of spatial scale or duration or the need for more process detail (Walters et al. 1992: Curnutt et al. 2000). Finally, and now more commonly, adaptive statistical and modeling approaches are being used to incorporate temporal changes in model structures or the behavior of the constituent agents. Both informal and formal methods, such as adaptive changes to model structure and agent-modeling techniques, will lead to robust models forecasts. Whatever approach is used within a management context, it is necessary to iterate: take action, monitor performance and do experiments, updating knowledge and forecasts as new information about system behavior becomes available.

Coda

Ecosystem management is, in fact, now a way of life. Some would even argue that we have passed the era of ecologically sustainable development and have entered an era of managed sustainability (Hilborn et al. 1995). In this environment, adaptive management techniques are essential. Ecosystem modeling, with all its limitations, will be an integral part of the management processes. Most important is the need for courage among those willing to recognize that modelling is an essential component of the scientific process (Hilborn and Mangel 1997). There is much to be learned and we do not agree with Horgan (1996) that this is the "end of science"—this is just the beginning of a new phase of a much more complex interaction between science and society that must lead to a new era of successful ecosystem management.

References

Aber, J., K.J. Nadelhoffer, P. Steudler, and J.M. Melillo. 1989. Nitrogen saturation in northern forest ecosystems. *BioScience* 39(6): 378–386.

Ågren, G.I., and E. Bosatta. 1988. Nitrogen saturation of terrestrial ecosystems. *Environmental Pollution* 54: 185–197.

Botsford, L.W., J.C. Castilla, and C.H. Peterson. 1997. The management of fisheries and marine ecosystems. *Science* 277: 509–515.

Caraco, N.F., and J.J. Cole. 1999. Human impact on nitrate export: Analysis using major world rivers. *Ambio* 28: 167–178

Carpenter, S.R., S.W. Chisholm, C.J. Krebs, D.W. Schindler, and R.F. Wright. 1995. Ecosystem experiments. *Science* 269: 324–327

Charles, D.F., editor. 1991. *Acidic Deposition and Aquatic Ecosystems: Regional Case Studies*. New York: Springer-Verlag.

Curnutt J.L., J. Comiskey, M.P. Nott, and L.J. Gross. 2000. Landscape-based spatially explicit species index models for Everglades restoration. *Ecological Applications* 10(6): 1849–1860.

Driscoll, C.T., G. Lawrence, A. Bulger, T. Butler, C. Cronan, C. Eagar, K. Fallon Lambert, G.E. Likens, J. Stoddard, and K.C. Weathers. 2001. Acidic deposition in the northeastern United States: Sources and inputs, ecosystem effects, and management strategies. *BioScience* 51(3): 180–198.

Galloway, J.N. 1995. Acid deposition: Perspectives in time and space. *Water, Air, and Soil Pollution* 85: 15–24.

Gibbons, M., C. Limoges, H. Nowotny, S. Schwartzmann, P. Scott, and M. Trow. 1994. *The New Production of Knowledge.* London: Sage Publications.

Groffman, P.M., and G.E. Likens, editors. 1994. *Integrated Regional Models: Interactions Between Humans and Their Environment.* New York: Chapman and Hall.

Gunderson, L.H., C.S. Holling, and S.S. Light, editors. 1995. *Barriers and Bridges to the Renewal of Ecosystems and Institutions.* New York: Columbia University Press.

Harris, G.P. 1999. This is not the end of limnology (or of science): The world may well be a lot simpler than we think. *Freshwater Biology* 42: 689–706

Hilborn, R., and M. Mangel. 1997. *The Ecological Detective: Confronting Models with Data.* Princeton: Princeton University Press.

Hilborn, R., and C.J. Walters. 1992. *Quantitative Fisheries Stock Assessment: Choice, Dynamics, and Uncertainty.* New York: Chapman and Hall.

Hilborn, R., C.J. Walters, and D. Ludwig. 1995. Sustainable exploitation of renewable resources. *Annual Review of Ecology and Systematics* 26: 45–67.

Holland, J. 1998. *Emergence: From Chaos to Order.* Reading, MA: Addison-Wesley.

Horgan, J. 1996. *The end of science.* London: Abacus, Little, Brown.

Howarth, R.W. 1988. Nutrient limitation of net primary production in marine ecosystems. *Annual Review of Ecology and Systematics* 19: 89–110.

Kotb, T.H.S., T. Wantanabe, Y. Ogino, and K.K Tanji. 2000. Soil salinization in the Nile Delta and related policy issues in Egypt. *Agricultural Water Management* 43(2): 239–261.

Likens, G.E., editor. 1989a. *Long-Term Studies in Ecology: Approaches and Alternatives.* New York: Springer-Verlag.

———. 1989b. Some aspects of air pollution on terrestrial ecosystems and prospects for the future. *Ambio* 18(3): 172–178.

———. 1991. Human-accelerated environmental change. *BioScience* 41(3): 130.

———. 1992. *The Ecosystem Approach: Its Use and Abuse.* Excellence in Ecology, Vol. 3. Oldendorf/Luhe, Germany: Ecology Institute.

Likens, G.E., and F.H. Bormann. 1974. Linkages between terrestrial and aquatic ecosystems. *BioScience* 24(8): 447–456.

Likens, G.E., F.H. Bormann, and N.M. Johnson. 1972. Acid rain. *Environment* 14(2): 33–40.

Livingston, R.J. 1991. Historical relationships between research and resource management in the Apalachicola River estuary. *Ecological Applications* 1(4): 361–382.

Mobbs, C., and S. Dovers. 1999. Social, economic, legal, policy and institutional R and D for natural resource management: Issues and direction for LWRRDC. Land and Water Research and Development Corporation, Occasional Paper 01/99, Canberra ACT.

Oreskes, N., K. Shrader-Frechette, and K. Belitz. 1994. Verification, validation and confirmation of models in the earth sciences. *Science* 263: 641–646.

Sala, O.E., F.S. Chapin III, J.J. Armesto, E. Berlow, J. Bloomfield, R. Dirzo, E. Huber-Sanwald, L.F. Huenneke, R.B. Jackson, A. Kinzig, R. Leemans, D.M. Lodge, H.A. Mooney, M. Oesterheld, N.L. Poff, M.T. Sykes, B.H. Walker, M. Walker, and D.H. Wall. 2000. Global Biodiversity Scenarios for the Year 2100. *Science* 287: 1770–1774.

Sharpley, A.N., S.J. Smith, J.R. Williams, O.J. Jones, and G.A. Coleman. 1991. Water quality impacts associated with sorghum culture in the Southern Plains. *Journal of Environmental Quality* 20(1): 239–244.

Stoddard, J.L. 1994. Long-term changes in watershed retention of nitrogen. Pp. 223–284 in L.A. Baker, editor. *Environmental Chemistry of Lakes and Reservoirs*. Advances in Chemistry Series, Vol. 237. Washington, DC: American Chemical Society.

Turner, R.E., and N.N. Rabalais. 1991. Changes in Mississippi River water quality this century: Implications for coastal food webs. *BioScience* 41(3): 140–147.

Valiela, I., G. Tomasky, J. Hauxwell, ML. Cole, J. Cebrián, and K.D. Kroeger. 2000. Operationalizing sustainability: Management and risk assessment of land-derived nitrogen loads to estuaries. *Ecological Applications* 10(4): 1006–1023.

Vitousek, P.M., J. Aber, R. Howarth, G. Likens, P. Matson, D. Schindler, W. Schlesinger, and D. Tilman. 1997. Human alteration of the global nitrogen cycle: Source and consequences. *Ecological Applications* 7(3): 737–750.

Vollenweider, R.A. 1968. *Scientific Fundamentals of Lake and Stream Eutrophication, with Particular Reference to Phosphorus and Nitrogen as Eutrophication Factors*. Technical Report DAS/DSI/68.27. Paris, France: OECD.

Walters, C.J. 1986. *Adaptive Management of Renewable Resources*. New York: MacMillan.

Walters, C., L. Gunderson, and C. S. Holling, 1992: Experimental policies for water management in the Everglades. *Ecological Applications* 2(2): 189–202.

Weathers, K.C., and G.M. Lovett. 1998. Acid deposition research and ecosystem science: synergistic successes. Pp. 195–219 in M.L. Pace and P.M. Groffman, editors. *Successes, Limitations, and Frontiers in Ecosystem Science*. New York: Springer-Verlag.

World Resources Institute. 2000. *World Resources 2000–2001. People and Ecosystems: The Fraying Web of Life*. New York: Elsevier.

17

The Role of Models in Addressing Critical N Loading to Ecosystems

Bridget A. Emmett and Brian Reynolds

Summary

The impact of N deposition in terrestrial and freshwater ecosystems has become of increasing concern in many parts of Europe. This awareness has provided a policy impetus to synthesize the available data into a coherent framework from which future changes may be forecast and emission reductions agreed. The development of various modeling approaches has been a crucial component in achieving this aim. To date, these approaches have included steady-state mass-balance models, which attempt to quantify the long term sinks for N in an ecosystem below which no deleterious effects are observed, and dose-response models based on empirical data. While there are limitations and uncertainties associated with these approaches, the synthesis of current understanding and effective simplification of knowledge has provided the underpinning science for the N component of the 1999 Gothenberg Protocol. For the future, a need to develop dynamic models to enable the forecasting of the magnitude and direction of change has been identified. Dynamic models have not been used extensively in the past for N critical loading due to a lack of suitable models for the majority of ecosystems and because of the limited availability of regional data at the level of complexity required to parameterize dynamic models. Data availability has now improved, and model development is underway in various modeling groups to improve the capability of models in forecasting future impacts of N deposition.

Introduction

Nitrogen (N) emissions have been increasing in many areas of the world due to the release of ammonia and N oxides associated with the intensification of agriculture and combustion of fossil fuels (Galloway 1995). In Europe, this has led to increased N loading to many natural and seminatural terrestrial and freshwater systems from less than 5 kg N/ha/yr in areas of low pollution to 50–100 kg N/ha/yr in areas of intensive agriculture or industry. Recently, emissions and deposition have leveled off or declined (Tarrason and Schuag 2000). Following concern that increased N deposition may contribute to both the acidification and N enrichment of receiving systems (Nilsson and Grennfelt 1988; Skeffington and Wilson 1988), there has been an array of studies to identify ongoing changes in ecosystem dynamics and to forecast future changes given different scenarios of N loading. Much of the research into effects on ecosystems has been brought together in various recent reviews, and some have concluded that there is evidence for a relationship between N deposition and N status of freshwater and terrestrial environments in some areas (Bobbink et al. 1998; van der Hoek et al. 1998; Langan 1999; Stoddard et al. 1999; Bertills and Näsholm 2000; Driscoll et al. 2001).

In freshwater systems, higher than expected nitrate concentrations have been reported in forested areas with high N deposition although there is considerable scatter in the relationship (e.g., Henriksen and Brakke 1988; Stoddard 1994; Dise and Wright 1995). However, in long-term monitoring studies there is no consistent relationship between trends in nitrate leaching and those of N deposition (Stoddard et al. 1999; Wright et al. 2001). Various factors have been proposed to explain the often weak or negative relationship between N deposition and nitrate leaching. These include climate trends, such as extreme droughts and frost (Mitchell et al. 1996; Aber and Driscoll 1997) and the North Atlantic Oscillation (Monteith et al. 2000), changes in species composition (Lovett and Reuth 1999), net accumulation of biomass (Vitousek and Reiners 1975; Emmett et al. 1993), and site history (Magill et al. 1997). These factors may affect the magnitude of sources and sinks for N in the forest through their influence on the rate of N transformations in the soil and the net uptake of N into the biomass. Changes in soil N transformations may be of particular importance, since strong relationships with nitrate leaching and mineralization and/or nitrification have been reported (e.g. McNulty et al. 1990). Because rates of mineralization and nitrification are generally greater in soils with a high N content or a low C/N ratio (McNulty et al. 1990; Wilson and Emmett et al. 1999), nitrate leaching has also been found to be related to these soil variables (Emmett et al. 1995; Matzner et al. 1997; Gundersen et al. 1998). The influence of N deposition on nitrate leaching from forests becomes clearer once these other internal site factors are taken into account (e.g., Dise et al. 1998). For non-forested systems such as heathlands and acid grasslands, elevated nitrate concentrations have also been reported in some high N-deposition areas. Unfortunately, the mechanistic controls on nitrate leaching from nonforested systems are even less well understood than for forested systems (Curtis et al. 2000).

In terrestrial ecosystems, increased N deposition has been associated with changes in vegetation composition. Changes reported include a decline in species characteristic of nutrient-limited conditions and an increase in nitrophilous species across a range of ecosystem types (review by Bobbink et al. 1998). Evidence is drawn from a range of studies, including repeated sampling of permanent plots (e.g., Falkengren-Grerup 1986), sampling across gradients (e.g., Pitcairn et al. 1998), physiological studies (e.g., Padgc•: and Allen 1999), and evidence from manipulation studies in which N inputs have been either experimentally increased or decreased to systems (e.g., Tilman 1987; Bobbink 1991; Theodose and Bowman 1997; Boxman et al. 1998; Lee and Caporn 1998). Changes in plant nutritional status or vegetation composition can result in secondary changes in animal populations (van Tol et al. 1998; Ball et al. 2000; Haddad et al. 2000), which can further accelerate changes in community structure (Heil and Bobbink 1993).

Within Europe, there has been a policy impetus to synthesize the plethora of experimental and observational data into a coherent framework from which forecasts of future changes in ecosystem structure, function, and dynamics in response to changes in N deposition can be attempted. The approach adopted was based on the critical-load concept, which had already been used as a basis for negotiation of sulphur emission reductions under the Second Sulphur Protocol of the Convention on Long-Range Transboundary Air Pollution (Nilsson and Grennfelt 1988). It was hoped that critical loads would again provide effective simplification and a predictive tool enabling information to be transferred in a useful format to environmental managers and policy makers. The benefits of the critical load approach compared to emission-based, cost-benefit, or transboundary-limit approaches is that it is effects-based and geographically explicit so areas at potential risk from pollutant deposition can be identified and the likely direction and, in some cases, magnitude of changes recognized. The concept is also simple enough to be universally applicable and widely understood by nonspecialists while the relatively modest data requirements are often available at the broad, regional scale. The development of various modeling approaches has been a crucial component in developing the concept for N deposition. However, as forecasts are made for the long term (decades), it is difficult to test their validity within the time framework required to inform policy formulation and protect ecosystems at risk (Chapter 7). It has therefore been emphasized that the uncertainties and limitations of knowledge available should be recognized when evaluating critical-load maps (Skeffington 1999). The purpose of this chapter is to outline the methods used in calculating the N critical loading for ecosystems to date, describe some of their limitations, and demonstrate how model development is crucial if we are to improve our capability to forecast the impacts of N deposition.

The Critical-Load Approach

At an important workshop in Skokloster, Sweden in 1988 (Nilsson and Grennfelt 1988), the scientific basis for the critical-load approach was developed and a working definition for a critical load given as "A quantitative estimate of an

Effect

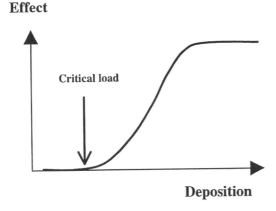

Critical load

Figure 17.1. Theoretical
dose-response curve
showing a threshold, or
"critical load" at which
effects are observed.

Deposition

exposure to one or more pollutants below which significant harmful effects on specified sensitive elements of the environment do not occur according to present knowledge" (pp. 225-268).

Harmful effects may refer to changes in ecosystem structure and function, the development of N saturation, accelerated acidification of soil, or the reduced sustainability of managed systems. The most appropriate concept is selected depending on the ecosystem being considered. Thus, the concept of sustainability may be used for managed coniferous forests while changes in ecosystem structure and function (e.g., biodiversity) may be more appropriate for wetland or heathland systems. N saturation as indicated by the onset of nitrate leaching or acidification of soil may be appropriate to protect linked freshwater systems. The N deposition below which no effect or change is observed (Figure 17.1) is defined as the critical load.

Different methods can be used to calculate the critical load. These include steady-state mass-balance models, empirical data from ecosystem studies, and dynamic modeling. In Europe, mass-balance approaches have generally been used to calculate critical loads for soil and freshwater while empirical data have usually been used to calculate critical loads to protect ecosystem structure and function. The empirical and mass-balance approaches used are recognized as having limitations, crucially in that they cannot predict the timescale over which effects might occur. They are a pragmatic choice because of the lack of suitable dynamic models for the majority of systems and the limited availability of regional data at the level of complexity required to parameterize dynamic models. Once the critical loads are calculated, these can be combined with spatial-deposition data sets to determine the reductions in N deposition required to protect terrestrial and freshwater ecosystems from acidification and eutrophication. Recently, this process was completed in time to provide the basis of negotiations for the 1999 United Nations Economic Commission for Europe

(UNECE) Multi-Pollutant Multi-Effect (Gothenberg) Protocol produced under the Convention on Long-Range Transboundary Air Pollution (see: http://www.unece.org/env/lrtap/welcome.html). Since the negotiation of the Gothenberg Protocol, greater attention has been focused on the development of appropriate dynamic models that will forecast the magnitude and timing of change on the ecosystem component selected (nitrate leaching, ecosystem structure, declining sustainability), while including feedbacks and interactions with other drivers such as the availability of other nutrients, climate, and site history.

Modeling Approaches

The Empirical Approach

A synthesis of empirical data from ecosystem studies is used to generate a dose-response function for eutrophication or N enrichment for each ecosystem type. Based on the available evidence, expert knowledge is then used to identify a range of N-deposition values for change or "damage" (e.g., loss of species, change in species dominance) that may occur in the long term. This range of deposition values is defined as the critical-load range for the specified ecosystem type (Table 17.1). The effect of N varies between different ecosystems, thus indicators of critical-load exceedance (i.e., when deposition is in excess of the critical-load range), also differ between ecosystem types. As there is only limited data available for some ecosystem types, the degree of confidence for each critical-load range is indicated for each ecosystem type. This approach has generally been used when biodiversity or ecosystem structure is to be protected, although it can also be used to protect against N saturation.

A range of critical-load values is produced for each ecosystem type because local conditions may influence the sensitivity of an ecosystem to deposited N. For example, phosphorus limitation may depress the sensitivity of a calcareous grassland system to N deposition because a phosphorus deficit decreases the competitive advantage of the grass species *Brachypodium pinnatum,* which can assume dominance in calcareous grasslands under high N loading (Wilson et al. 1995). If phosphorus limitation were known to occur in a particular region, a critical-load value at the higher end of the range would be selected, indicating a lower sensitivity to N deposition. Values at the high end of the critical-load range would also be selected under heavy grazing pressure or other management practices, such as burning, which remove N or affect the rate of N cycling (Hornung et al. 1995).

While there has been a great deal of effort to review the available data and determine critical-load values at UNECE workshops using this approach, there are some potential problems as identified by Skeffington (1999). The studies available may be biased toward the more sensitive subcommunities within an ecosystem type. This may lead to an underestimation of critical-load values. Alternatively, it is inevitable in relatively short-term studies that there is the

Table 17.1. Summary of empirical critical loads for nitrogen deposition to (semi-) natural freshwater and terrestrial ecosystems.

Ecosystem Type	Critical Load for Nitrogen Deposition (kg N/ha/yr)	Indication of Critical Load Exceedance
Trees and forest ecosystems		
Coniferous trees (acidic) (low nitrification rate)	10–15 ##	Nutrient imbalance in trees
Coniferous trees (acidic) (moderate/high nitrification rate)	20–30 #	Nutrient imbalance in trees
Deciduous trees	15–20 #	Nutrient imbalance in trees; increased shoot/root ratio
Acidic coniferous forests	7–20 ##	Changes in ground flora and mycorrhizas; increased N leaching
Acidic deciduous forests	10–20 #	Changes in ground flora and mycorrhizae
Calcareous forests	15–20 (#)	Changes in ground flora
Acidic forests	7–15 (#)	Changes in ground flora and increase in N leaching
Forests in humid climates	5–10 (#)	Decline in lichens and increase in free-living algae
Heathlands		
Lowland dry heathlands	15–20 ##	Transition from heather to grass; functional change (litter production; flowering; N accumulation)
Lowland wet heathlands	17–22 #	Transition from heather to grass
Species-rich heaths/acid grassland	10–15 #	Decline in sensitive species
Upland Calluna heaths	10–20 (#)	Decline in heather dominance, mosses, and lichens; N accumulation
Arctic and alpine heaths	5–15 (#)	Decline in lichens, mosses, and evergreen dwarf shrubs

Table 17.1. *Continued*

Ecosystem Type	Critical Load for Nitrogen Deposition (kg N/ha/yr)	Indication of Critical Load Exceedance
Species-rich grasslands		
Calcareous grasslands	15–35 (#)	Increased mineralization, N accumulation and leaching; increase in tall grass, change in diversity
Neutral-acid grasslands	20–30 #	Increase in tall grass, change in diversity
Montane-subalpine grass-lands	10–15 (#)	Increase in tall grami-noids, change in diversity
Wetlands		
Mesotrophic fens	20–30 #	Increase in tall grami-noids, decline in diversity
Ombrotrophic bogs	5–10 #	Decline in typical mosses, increase tall graminoids, N accumulation
Shallow soft-water bodies	5–10 ##	Decline in isoetid species

Source: Werner and Spranger 1996.
Note: Values are rated as: ## reliable, # quite reliable, and (#) expert judgement.

potential for missing longer-term impacts, thus leading to an overestimation of critical-load values. For some ecosystems, particularly Mediterranean and alpine systems, there is a paucity of data making critical load values for these systems less reliable (Table 17.1). Finally, the high dose rates of N applied and the step function characteristic of manipulation experiments may lead to unreliable conclusions.

The Steady-State Mass-Balance Model

The steady-state mass-balance model attempts to define the threshold (i.e. the critical load for N deposition, CL [N]) by setting "acceptable" limits to the major long-term sinks for N in an ecosystem. It is assumed that N lost to these long-term sinks can be replaced by N deposition without undesirable effects, while N in excess of these sinks will cause enrichment to some component of the ecosystem, resulting in deleterious effects to ecosystem functioning. This is analogous to the N saturation concept first described by Ågren (1983), which has since been defined in various ways including where atmospheric N inputs exceed the biotic requirements of the ecosystem (Aber et al. 1989).

Sinks included in the mass-balance model are long term accumulation of N in the soil, removal of N in harvested plant biomass and losses through denitrification and leaching as shown in Equation 17.1 (Werner and Spranger 1996), where the terms are expressed in units of kg N ha^{-1} yr^{-1}:

$$CL\ (N) = N_{up} + N_i + N_{de} + N_{le}$$ (17.1)

where

N_{up} = N removed in plant biomass from normal harvesting operations

N_i = long-term N immobilization rate in the soil

N_{de} = denitrification rate at the critical load

N_{le} = acceptable N leaching

The plant-uptake term (N_{up}) should only include the net removal of N from the site in biomass that is sustainable (i.e., does not exceed the supply of other nutrients). N immobilization values (N_i) should be sustainable in the long term without causing deleterious effects to ecosystem functioning. Values for denitrification and leaching are generally limited to values observed in low deposition areas or at the critical load to ensure terrestrial systems are not protected at the expense of the atmosphere or freshwater systems. Losses due to volatilization, erosion, and fire may also be included if they are important and will have the effect of increasing the critical load. If N fixation is important, then this can be included, but it should be a negative term because it is a source, not a sink, for N.

In addition to eutrophication, N may also contribute to the acidification of ecosystems. When the critical load for acidification is calculated, the N component is set so that the combined total acid loading from sulphur and N deposition does not exceed sources of alkalinity within the system (e.g., mineral weathering, non-seasalt base cation deposition) minus an acceptable loss of acid neutralizing capacity (ANC) from the system. In calculating the critical load, the mass balance allows for long-term sinks for N ($N_i + N_{up} + N_{de}$) and the acidifying effect of net base cation removal in harvest products (Eq. 17.2) (Werner and Spranger 1996):

$$CL(S + N) = BC_w - BC_{up} + N_i + N_{de} + N_{up} - ANC_{le(crit)}$$ (17.2)

where

BC_w = base cation release from weathering

BC_{up} = net removal of base cations from the system

$ANC_{le(crit)}$ = acceptable or critical ANC leaching

The critical ANC leaching term can be determined in relation to any one of a number of critical chemical criteria that relate soil chemical conditions to plant response. The most commonly used are critical Ca/Al ratio, critical pH, and critical Al concentration. The choice of criteria depends on factors such as data availability for a particular ecosystem and its suitability. However, Hall et al. (2001a) point out that widely differing critical loads can be calculated for the same system depending on the choice of critical chemical criteria, pointing to a need for rigorous justification of choice of criteria when the models are applied across large regions such as Europe.

Using this approach, critical loads for acidity have been calculated for ex-perimental sites (e.g., Reynolds et al. 1998), regions (e.g., Sverdrup et al. 1992), and Europe as a whole (e.g., Posch et al. 1999). It has been emphasised that values should be calculated for the long term to include natural cycles such as stand rotations. These will differ from critical-load values calculated for short-term periods, which do not include an aggrading or harvesting period (Pardo and Driscoll 1996). In addition, when critical loads are used to identify areas at risk by overlaying the values with deposition data, a significant problem emerges: deposition estimates for the different N species are highly dependent on a series of deposition models. These models are themselves under develop-ment, and current uncertainties in their output may significantly alter the area identified as being at risk from excess N deposition. At present, there are too few independent data sources to allow for a definitive estimate of this uncer-tainty (RGAR 1997).

Both of these mass-balance approaches assume that the N cycle and acidity balance can be simplified into a few key processes, which are crucial for deter-mining the damage response, and thus can be used to generate predictive equa-tions. As Skeffington (1999) pointed out, however, there are many questions which remain, including the ecological reality of thresholds, the importance of the most sensitive component, the controversy surrounding the critical chemical criteria (such as the calcium to aluminium ratio), major uncertainties concern-ing key parameters (such as weathering rates and long-term N immobilization rates in the soil), and the lack of biotic interactions in the mass-balance models.

Dynamic Modeling

The development of dynamic models may overcome some of the concerns ex-pressed by Skeffington (1999) and others relating to abiotic and biotic interac-tions. In addition, policy needs have changed and questions concerning the time scale and magnitude of change have been raised. The cost of emission reduc-tions increases significantly if they are required quickly (Skeffington 1999), and there is a need to assess the likely time scale of recovery to enable an evaluation of the success of current abatement strategies. In addition, the ecological bene-fits of emission reductions that do not reduce deposition below the critical load are also of interest.

Ågren and Bosatta (1988) were one of the first to specifically attempt the modeling of N saturation of the plant and soil components of a terrestrial eco-system. With respect to the plant sink, they emphasized the importance of the supply of other limiting nutrients in determining the effective magnitude of the plant sink, but they suggested that this could be quantified. For the soil system, they identified a C/N ratio threshold of soil organic matter below which N saturation may occur. This has subsequently been confirmed in studies by Matzner and Grosholz (1997) for *Picea abies* Karst. (Norway spruce) systems in Germany and for a broader range of conifer forests across Europe (e.g., Dise et al. 1998; Gundersen et al. 1998). The time taken to reach the C/N ratio threshold is dependent on site history, due to the influence of litter quality on the quality of soil organic matter and inputs and removal of N from the system. Ågren and Bosatta (1988) concluded that the soil would set the limits for the possibilities of absorbing N deposition but predicted that the inherent variability in soils would cause serious complications when attempting to predict this capacity. This has indeed been the major problem for many modelers attempting to forecast N-immobilization capacity.

Since the early attempt by Ågren and Bosatta (1988), a range of models has been used in an attempt to forecast the impacts of N in ecosystems. Geochemical models have been used to assess the potential contribution of N deposition to the acidification of ecosystems (e.g., SMART [Posch et al. 1993], MAGIC [Cosby et al. 1985], and SAFE [Warfvinge et al. 1993]). In these models, N processes are very simple and generally limited to net fluxes of N (Posch and de Vries 1999). In models specifically developed to represent the N cycle such as NIICE (van Dam and van Breemen et al. 1995), SOILN (Johnsson et al. 1987), MERLIN (Cosby et al. 1997), and PnET-CN (Aber and Driscoll 1997), internal cycling of N is included with a variable degree of complexity (Figure 17.2). However, irrespective of their complexity, immobilization of N in most models is dependent on the C:N ratio of the active soil organic pool, topsoil layer, or microbial biomass. Parameterizing this function a priori remains a problem due to the uncertainty surrounding the definition of the active layer and how this function may be affected by other variables such as pH and climate.

In Europe, only three models have been used for testing the impact of different N-deposition scenarios in regional applications: SMART, SAFE, and MAGIC. These models and others are now being developed to improve their ability to forecast the impacts of N deposition on soils and water quality. To succeed, it is likely that they will need to include a coupling between the N and carbon cycles, a mechanistic component for both N immobilization/remobilization and weathering, and require data that exist or are easily obtainable at the regional scale. Now that the limitations of current models have become clearer, there are several models under development that include at least some of these requirements. However all model development would benefit from a greater understanding of the underlying controls of the N-immobilization capacity of soils for a range of ecosystem types.

Linking Soil and Vegetation Responses in Models

Most of the dynamic models described above are limited to forecasting the impact of N deposition on soil N processes and fluxes of N. The association

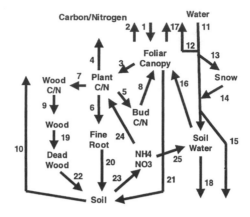

1. Gross Photosynthesis
2. Foliar Respiration
3. Transfer to Mobile C
4. Growth and Maint. Resp.
5. Allocation to Buds
6. Allocation to Fine Roots
7. Allocation to Wood
8. Foliar Production
9. Wood Production
10. Soil Respiration

11. Precipitation
12. Interception
13. Snow-Rain Partition
14. Snow-melt
15. Fast Flow
16. Water Uptake
17. Transpiration
18. Drainage

19. Wood Litter
20. Root Litter
21. Foliar Litter
22. Wood Decay
23. Mineralization
24. N Uptake
25. To Soil Solution

Figure 17.2. A schematic diagram of the interactions in the PnET-CN model (Aber and Driscoll 1997).

between these processes and changes in the competitive balance between species and vegetation composition is not well understood for most systems. One exception is for Dutch heathlands, where the causal links between N deposition, increase in N availability, and change in species competitive balance is well understood and has been successfully modeled for two separate heathland types (Berendse 1988; Heil and Bobbink 1993). One of these models, CALLUNA (Heil and Bobbink 1993), includes an important biotic interaction with the activity of the heather beetle. The competitive ability of the dominant shrub, *Calluna vulgaris,* is reduced in high-deposition areas due to increased herbivory pressure, enabling the expansion of the invasive grass species, *Deschampsia flexuosa.* This sequence of events has resulted in the conversion of heathland to grass-dominated systems in many areas of the Netherlands. The development of the dynamic CALLUNA model may have practical benefits for land managers because it can be used to plan the frequency of turf cutting to remove a proportion of the N store to maintain the species balance in favor of the desired shrub species (Bobbink et al. 1992).

However, there is a need for a more generic approach to link soil and vegetation responses, which will enable the future possible effects of N deposition on species performance and vegetation composition to be examined. Response functions to site factors can be developed for individual species and groups of species, but doing so is very costly and time-consuming and is limited to indicator or the most sensitive species for most habitats (see Empirical Approach section). An alternative approach is to use Ellenberg indicator values as a means of assessing the likely competitive advantage between different species under low or high N availability as projected by dynamic soil models (e.g., Latour and Reiling 1993; Latour et al. 1994). Ellenberg indicator values were created by the German biologist Heinz Ellenberg, who estimated the response of about 3,000 central European vascular plant species to light, nutrient availability, moisture, salinity, and acidity (Ellenberg 1988). Ellenberg considered the indicator value for nutrient availability to be primarily an indicator of a plant's response to N. Consistently high indicator values for species present in an ecosystem indicate the importance of an environmental factor (light, N, moisture, salinity, or acidity) in defining the vegetation composition. Values lie on a scale of 1 to 9 and may require modification for different geographical areas because some species have different ecological requirements across their range (van der Maarel 1993). Attempts have been made to translate these Ellenberg values into measurable soil variables either by relating mean Ellenberg values to empirically measured soil variables across a large number of sites (e.g., Ertsen et al. 1998) or by using experimental data for soils and the response of representative species of different families (e.g., Hansson 1995, using data from Tilman 1987).

However, there is some concern about the ecological interpretation of Ellenberg's N value. A variable response between mean Ellenberg values and soil N variables has been reported (e.g., Diekmann and Falkengren-Grerup 1998; Ertsen et al. 1998, Liu and Bråkenhielm 1998). Diekmann and Falkengren-Grerup (1998) suggest that a functional N index for species based on relationships, derived from N mineralization studies, provides an improved basis for species' responses to N availability. In addition, while N influences plant performance, other factors such as moisture, acidity, and structural or successional changes will also be critical factors in determining vegetation composition. Thus multistress models are required and are under development including SMART-MOVE (Figure 17.3), which links together a dynamic soil model that predicts changes in moisture, nutrient availability and soil acidity and the likely species response to these changing soil factors (Latour et al. 1994); various forest-growth models with linked carbon, nutrient and water cycles (e.g., PnET-CN [Aber and Driscoll 1997] and CENW [Kirschbaum 1999]), and successional models, such as SUMO, which model the growth of five functional vegetation types and their response to N availability, light, and management (Wamelink et al. 2000). These can be coupled together in model chains although the uncer-

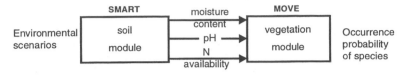

Figure 17.3. A schematic presentation of the model SMART-MOVE (Latour et al. 1994).

tainty of model output becomes increasingly difficult to determine (Schouwenberg et al. 2000).

Uncertainties and Benefits of the Critical-Load Approach

Some scientists remain concerned that our understanding of critical N loading of ecosystems, and thus our capacity to forecast change under future deposition scenarios, is highly uncertain (Bull and Hall 1998). Barkman (1997) has reviewed the reasons why the uncertainties are often neglected. Pressures include the desire to avoid blurring priorities, the need for simplification and ease of communication, and the concern about political manipulation of uncertainty. Tradition and difficulties in assessing uncertainties may also play a role. Significant benefits however are missed by this neglect of uncertainty. These include safety margins in policy development and the management of uncertainties to prevent surprises (Chapter 7). It could also be argued that there is an ethical responsibility to create better awareness of uncertainties.

Uncertainties in critical-load predictions are now being assessed (e.g., Hall et al. 2001b; Suutari et al. 2001) but notwithstanding these and other caveats, much has been achieved in the willingness of some scientists to use their knowledge to address policy needs. The synthesis of current understanding and effective simplification of knowledge in empirical and mass-balance models has provided the underpinning science for the 1999 Gothenberg Protocol to abate acidification, eutrophication, and ground-level ozone. The protocol calls for a 41% cut in NO_x and a 17% cut in ammonia emissions, a 63% cut in S, and a 41% cut in VOC by 2010 based on 1990 levels. Assuming successful implementation across Europe, the estimated areas at risk from acidification will be reduced from 93 to 15 million hectares and from 165 to 108 million hectares for eutrophication (UNECE 1999). There have also been valuable benefits for science in that there has been a synthesis of available data in European data centers, and the development of numerous cross-border and multidisciplinary collaborations, and there has been a focus on our lack of understanding of some key processes, in particular soil N immobilization and weathering. Critical needs have been identified, such as the requirement for mechanistic models that can be used at the regional scale with a priori capabilities to forecast the fate of N deposition and this development work is being supported in some areas. An improved understanding of the relation between soil status and vegetation response is also required, as are improved deposition models.

References

Aber, J.D., and C.T. Driscoll. 1997. Effects of land use, climate variation, and N deposition on N cycling and C storage in northern hardwood forests. *Global Biogeochemical Cycles* 11: 639–648.

Aber, J.D., K.J. Nadelhoffer, P. Steudler, and J.M. Melillo. 1989. Nitrogen saturation in northern forest ecosystems. *BioScience* 39: 378–386.

Ågren, G.I. 1983. Model analysis of some consequences of acid precipitation on forest growth. Pp. 233–244 in *Ecological Effects of Acid Deposition.* Report PM 1636. Stockholm: Swedish National Environment Protection Board.

Ågren, G.I., and E. Bosatta. 1988. Nitrogen saturation of terrestrial ecosystems. *Environmental Pollution* 54: 185–197.

Ball, J.P., K. Danell, and P. Sunesson. 2000. Response of herbivore community to increased food quality and quantity: An experiment with nitrogen fertilizer in a boreal forest. *Journal of Applied Ecology* 37: 247–25.

Barkman, A. 1997. *Applying the Critical Loads Concept: Constraints Induced by Data Uncertainty.* Reports in Ecology and Environmental Engineering, no. 1. Lund University.

Berendse, F. 1988. De nutriëntbalans van droge zandgrondvegetaties in verband met de eutrofiëring via de lucht. Deel 1. Een simulatiemodel als hulpmiddel bij het beheer van vochtige heidevelden. Wageningen: Centrum voor Agrobiologisch Onderzoek.

Bertills, U., and T. Näsholm. 2000. *Effects of Nitrogen Deposition on Forest Ecosystems.* Report No. 5067. Stockholm: Swedish Environmental Protection Agency.

Bobbink, R. 1991. Effects of nutrient enrichment in a Dutch chalk grassland. *Journal of Applied Ecology* 28: 28–41.

Bobbink, R., D. Boxman, E. Fremstad, G. Heil, A. Houdijk, and J. Roelofs. 1992. Critical loads for nitrogen eutrophication of terrestrial and wetland ecosystems based upon changes in vegetation and fauna. Pp. 111–159 in P. Grennfelt and E. Thörnelöf, editors. *Critical Loads for Nitrogen.* Report of the Lökeborg Workshop. Nord (Miljörapport) 41. Copenhagen: Nordic Council of Ministers.

Bobbink, R., M. Hornung, and J.G.M. Roelofs. 1998. The effects of air-borne nitrogen pollutants on species diversity in natural and semi-natural European vegetation. *Journal of Ecology* 86: 717–738.

Boxman, A.W., P.J.M. van der Ven, and J.G.M. Roelofs. 1998. Ecosystem recovery after a decrease in nitrogen input to a Scots pine stand at Ysselsteyn, the Netherlands. *Forest Ecology and Management* 101: 155–163.

Bull, K.R., and J.R. Hall. 1998. Setting international targets for controlling emissions of pollutants: Now and in the future. Pp. 581–590 in K.W. Van der Hoek, J.W. Erisman, S. Smeulders, J.R. Wisniewski, and J. Wisniewski, editors. *Proceedings of the First International Nitrogen Conference.* Noordwijkerhout, The Netherlands. 23–27 March 1998. Oxford: Elsevier.

Cosby, B.J., R.C. Ferrier, A. Jenkins, B.A. Emmett, R.F. Wright, and A. Tietema. 1997. Modelling the ecosystem effects of nitrogen deposition at the catchment scale: Model of Ecosystem Retention and Loss of Inorganic Nitrogen (MERLIN). *Hydrology and Earth System Sciences* 1: 137–158.

Cosby, B.J., J.N. Hornberger, J.N. Galloway, and R.F. Wright. 1985. Modelling the effects of acid deposition: Assessment of a lumped parameter model of soil water and streamwater chemistry. *Water Resources Research* 22: 1283–1291.

Curtis, C., T. Allott, J. Hall, R. Harriman, R. Helliwell, M. Hughes, M. Kernan, B. Reynolds, and J. Ullyett. 2000. Critical loads of sulphur and nitrogen for freshwaters in Great Britain and assessment of deposition reduction requirements with the First-order Acidity Balance (FAB) model. *Hydrology and Earth System Sciences* 4: 125–140.

Diekmann, M., and U. Falkengren-Grerup, U. 1998. A new species index for forest vascular plants: Development of functional indices based on mineralization rates of various forms of soil nitrogen. *Journal of Ecology* 86: 269–283.

Dise, N.B., E. Matzner, and M. Forsius. 1998. Evaluation of organic horizon C:N ratio as an indicator of nitrate leaching in conifer forests across Europe. *Environmental Pollution* 102: 453–456.

Dise, N.B., and R.F. Wright. 1995. Nitrogen leaching from European forests in relation to nitrogen deposition. *Forest Ecology and Management* 71: 153–162.

Driscoll, C.T., G.B. Lawrence, A.J. Bulger, T.J. Butler, C.S. Cronan, C. Eagar, K.F. Lambert, G.E. Likens, J.L. Stoddard, and K.C. Weathers. 2001. Acidic deposition in the northeastern United States: Sources and inputs, ecosystem effects, and management strategies. *BioScience* 51 (3): 180–198.

Ellenberg, H. 1988. *Vegetation Ecology of Central Europe.* 4th Edition. Cambridge: Cambridge University Press.

Emmett, B.A., B. Reynolds, P.A. Stevens, D.A. Norris, S. Hughes, J. Gorres, and I. Lubrecht, I. 1993. Nitrate leaching from afforested Welsh catchments: Interactions between stand age and nitrogen deposition. *Ambio* 22: 386–394.

Emmett, B.A., P.A. Stevens, and B. Reynolds. 1995. Factors influencing nitrogen saturation in Sitka spruce stands in Wales UK. *Water, Air and Soil Pollution* 85: 1629–1634.

Ertsen, A.C.D., J.R.M. Alkemade, and M.J. Wassen. 1998. Calibrating Ellenberg indicator values for moisture, acidity, nutrient availability and salinity in the Netherlands. *Plant Ecology* 135: 113-124.

Falkengren-Grerup, U. 1986. Soil acidification and vegetation changes in deciduous forest in southern Sweden. *Oecologia.* 70: 339–347.

Galloway, J.N. 1995. Acid deposition: Perspectives in time and space. *Water, Air and Soil Pollution* 85: 15–24.

Gundersen, P., I. Callesen, and W. de Vries. 1998. Nitrate leaching in forest ecosystems is related to forest floor C/N ratios. *Environmental Pollution* 102: 403–407.

Haddad, N.M., J. Haarstad, and D. Tilman. 2000. The effects of long-term nitrogen loading on grassland insect communities. *Oecologia* 124: 73–84.

Hall, J., B. Reynolds, J. Aherne, and M. Hornung, M. 2001a. The importance of selecting appropriate criteria for calculating acidity critical loads using the Simple Mass Balance equation. *Water, Air and Soil Pollution: Focus* 1:29–41.

Hall, J., B. Reynolds, S. Langan, M. Hornung, F. Kennedy, and J. Aherne. 2001b. Investigating the uncertainties in the Simple Mass Balance Equation for acidity critical loads for terrestrial ecosystems in the United Kingdom. *Water, Air and Soil Pollution: Focus* 1: 43–56.

Hansson, J. 1995. *Modelling Effects of Soil Solution BC/Al-Ratio and Available N on Ground Vegetation Composition.* Reports in Ecology and Environmental Engineering. Lund University Report no. 3.

Heil, G.W., and R. Bobbink. 1993. CALLUNA: A simulation model for evaluation of impacts of atmospheric nitrogen deposition on dry heathlands. *Ecological Modelling* 66: 61–182.

Henriksen, A., and D.F. Brakke. 1988. Increasing contribution of nitrogen to the acidity of surface waters in Norway. *Water, Air and Soil Pollution* 42: 183–201.

Hornung, M., M.A. Sutton, and R.B. Wilson. 1995. *Mapping and Modelling of Critical Loads for Nitrogen: A Workshop Report.* Grange-over-Sands, Cumbria, UK 24–26 October 1994. Edinburgh, UK: Institute of Terrestrial Ecology.

Johnsson, H., L. Bergström, P.E. Jansson, and K. Paustrian. 1987. Simulation of nitrogen dynamics and losses in a layered agricultural soil. *Agriculture, Ecosystems and Environment* 18: 333–356.

Kirschbaum, M.U.F. 1999. CenW: A forest growth model with linked carbon, energy, nutrient and water cycles. *Ecological Modelling* 188: 17–59.

Langan, S., editor. 1999. *The Impact of Nitrogen Deposition on Natural and Semi-Natural Ecosystems.* Vol. 3. of Environmental Pollution. Dordrecht: Kluwer.

Latour, J.B., and R. Reiling. 1993. A multiple stress model for vegetation "Move": A tool for scenario studies and standard-setting. *The Science of the Total Environment* (Supplement) 1513–1526.

Latour, J.B., R. Reiling, and W. Slooff. 1994. Ecological standards for eutrophication and desiccation: Perspectives for a risk assessment. *Water, Air and Soil Pollution* 78: 265–277.

Lee, J.A., and S.J.M. Caporn. 1998. Ecological effects of atmospheric reactive nitrogen on semi-natural terrestrial ecosystems. *New Phytologist* 139: 127–134.

Liu, Q., and S. Bråkenhielm. 1998. Ecological indication of forest plants: Application on environmental monitoring and assessment in Sweden. Department of Environmental Assessment. SLU, Uppsala.

Lovett, G.M., and Rueth, H. 1999. Soil nitrogen transformation in beech and maple stands along a nitrogen deposition gradient. *Ecological Applications* 9: 1330–1344.

Magill, A.H., J.D. Aber, J.J. Hendricks, R.D. Bowden, J.M. Melillo, and P.A. Steudler. 1997. Biogeochemical response of forest ecosystems to simulated chronic nitrogen deposition. *Ecological Applications* 7: 402–415.

Matzner, E., and C. Grosholz. 1997. Beziehung zwischen NO_3-Austragen, C/N-Verhältnissen de Auflage und N-eintrogen in Fichtenwald (*Picea abies* Karst). *Ökosystemen Mitteleuropas. Forstw. Cbl.* 116: 39–44.

McNulty, S.G., J.D. Aber, T.M. McLellan, and S.M. Katt. 1990. Nitrogen cycling in high-elevation forests of the northeastern U.S. in relation to N deposition. *Ambio* 19: 38–40.

Mitchell, M.J., C.T. Driscoll, J.S. Kahl, G.E. Likens, P.S. Murdock, and L.H. Pardo. 1996. Climatic control of nitrate loss from forested watersheds in the northeast United States. *Environment, Science and Technology* 30: 2609–2612.

Monteith, D.T., C.D. Evans, and B. Reynolds. 2000. Are temporal variations in the nitrate content of UK upland freshwaters linked to the North Atlantic Oscillation? *Hydrological Processes* 14: 1745–1749.

Nilsson, J., and P. Grennfelt, editors. 1988. *Critical loads for Sulphur and Nitrogen.* Report of the Skokloster workshop. Miljörapport 15: 225-268. Copenhagen: Nordic Council of Ministers.

Padgett, P.E., and E.B. Allen. 1999. Differential response to nitrogen fertilization in native shrubs and exotic annuals common to Mediterranean coastal sage scrub of California. *Plant Ecology* 144: 93–101.

Pardo, L.H., and C.T. Driscoll. 1996. Critical loads for nitrogen deposition: Case studies at two northern hardwood forests. *Water, Air and Soil Pollution* 89: 105–128.

Pitcairn, C.E.R., I.D. Leith, L.J. Sheppard, M.A. Sutton, D. Fowler, R.C. Munro, S. Tang, and D. Wilson. 1998. The relationship between nitrogen deposition, species composition and foliar nitrogen concentrations in woodland flora in the vicinity of livestock farms. *Environmental Pollution* 102: 41–48.

Posch, M., P.A.M. de Smet, J.P. Hettelingh, and R.J. Downing. 1999. *Calculation and Mapping of Critical Threshold in Europe: Status Report 1999.* Rijksinstituut voor Volksgezondheid en Milieu (RIVM) Report No. 259101009. Coordination Center for Effects, National Institute of Public Health and the Environment, Bilthoven, Netherlands.

Posch, M., and W. de Vries. 1999. Derivation of critical loads by steady-state and dynamic soil models. Pp. 213–234 in S. Langan, editor. *The Impact of Nitrogen on Natural and Semi-Natural Ecosystems.* Dordrecht: Kluwer.

Posch, M., G.J. Reinds, and W. de Vries. 1993. SMART—*A Simulation Model for Acidification's Regional Trends: Model Description and User Manual.* Mimeograph Series of the National Board of Waters and the Environment 477. Helsinki, Finland.

Review Group on Acid Rain (RGAR) 1997. *Acid Deposition in the United Kingdom 1992–1994.* Fourth Report of the Review Group on Acid Rain, AEA Technology, Harwell.

Reynolds, B., E.J. Wilson, and B.A. Emmett. 1998. Evaluating critical loads of nutrient nitrogen and acidity for terrestrial systems using ecosystem-scale experiments (NITREX*). Forest Ecology and Management* 101: 81–94.

Schouwenberg, E.P.A.G., H. Houweling, M.J.W. Jansen, J. Kros, and J.P. Mol-Dijkstra. 2000. *Uncertainty Propagation in Model Chains: A Case Study in Nature Conservancy.* Alterra rapport 001. Alterra, Wageningen: Green World Research.

Skeffington, R.A. 1999. The use of critical loads in environmental policy making: A critical appraisal. *Environmental Science and Technology,* (June 1): 245–252A.

Skeffington, R.A., and E.J. Wilson. 1988. Excess nitrogen deposition: Issues for consideration. *Environmental Pollution* 54: 159–184.

Stoddard, J.L. 1994. Long term changes in watershed retention of nitrogen: Its causes and aquatic consequences. Pp. 223–284 in L.A. Baker, editor. *Environmental Chemistry of Lakes and Reservoirs.* ACS Advances in Chemistry Series, no. 237. Washington, DC: American Chemical Society.

Stoddard, J.L., D.S. Jeffries, A. Lükewille, T.A. Clair, P.J. Dillon, C.T. Driscoll, M. Forsius, D.T. Monteith, P.S. Murdock, S. Patrick, A. Rebsdorf, B.L. Skjelkvåle, M.P. Stainton, T. Traaen, H. van Dam, K.E. Webster, J. Wieting, and A. Wilander. 1999. Regional trends in aquatic recovery in North America and Europe. *Nature* 401: 575–578.

Suutari, R., W. Schöpp, and M. Posch. 2001. Uncertainty analysis of ecosystem protection in the framework of integrated assessment modelling. Pp. 55–62 in *Modelling and Mapping of Critical Thresholds in Europe,* CCE Status Report 2001. Bilthoven, The Netherlands: Rijksinstituut voor Volksgezondheid en Milieu (RIVM).

Sverdrup, H., P. Warvinge, T. Frogner, A.O. Haoya, M. Johansson, and B. Andersen. 1992. Critical loads for forest soils in the Nordic Countries. *Ambio* 21: 348–355.

Tarrason, L., and J. Schuag. 2000. *Transboundary Acidification and Eutrophication in Europe.* EMEP Summary Report 2000. EMEP/MSC-W 1/2000. Oslo, Norway: Norwegian Meteorological Institute.

Theodose, T.A., and W.D. Bowman. 1997. Nutrient availability, plant abundance, and species diversity in two alpine tundra communities. *Ecology* 78: 1861–1872.

Tilman, D. 1987. Secondary succession and the pattern of plant dominance along experimental nitrogen gradients. *Ecological Monographs* 57: 189–214.

United Nations Economic Commission for Europe (UNECE). 1999. *The 1999 Gothenberg Protocol to Abate Acidification, Eutrophication and Ground Level Ozone.* New York and Geneva: United Nations.

van Dam, D., and N. van Breemen. 1995. NIICE: A model for cycling of nitrogen and carbon isotopes in coniferous forest ecosystems. *Ecological Modelling* 79: 255–275.

van der Hoek, K.W., J.W. Erisman, S. Smeulders, J.R. Wisniewski, and J. Wisniewski, editors. 1998. *Proceedings of the First International Nitrogen Conference.* Noordwijkerhout, The Netherlands. 23-27 March 1998. Oxford: Elsevier.

van der Maarel, E. 1993. Relations between sociological-ecological species groups and Ellenberg indicator values. *Phytocoenologia* 23: 343–362.

van Tol, G., H.F. van Dobben, P. Schmidt, and J.M. Klap. 1998. Biodiversity of Dutch forest ecosystems as affected by receding groundwater levels and atmospheric deposition. *Biodiversity and Conservation* 7: 221–228.

Vitousek, P.M., and W.A. Reiners. 1975. Ecosystem succession and nutrient retention: A hypothesis. *Bioscience* 25: 376–381.

Wamelink, G.W.W., J. Mol-Dijkstra, H.F. van Dobben, J. Kros, and F. Berendse. 2000. *Eerste fase van de ontwikkeling can Successie Model SUMO 1.0. Verbering van de vagetatiemodellering in de Natuurplanner.* Alterrarapport 045. Alterra, Wageningen: Green World Research.

Warfvinge, P., U. Falkengrengrerup, H. Sverdrup, and B. Andersen. 1993. Modeling long-term cation supply in acidified forest stands. *Environmental Pollution* 80: 209–221.

Werner, B., and T. Spranger. 1996. *Mapping Critical Levels/Loads.* Berlin: Federal Environmental Agency.

Wilson, E.J., and B.A. Emmett. 1999. Factors influencing nitrogen saturation in forest ecosystems: Advances in our understanding since the mid 1980s. Pp. 123–152 in S. Langan, editor. *The Impact of Nitrogen Deposition on Natural and Semi-Natural Ecosystems.* Dordrecht: Kluwer.

Wilson, E.J., T.C.E. Wells, and T.H. Sparks. 1995 Are calcareous grasslands in the UK under threat from nitrogen deposition: An experimental determination of a critical load. *Journal of Ecology* 83: 823–832.

Wright, R.F., C. Alewell, J.M. Cullen, C.D. Evans, A. Marchetto, F. Moldan, A. Prechtel, and M. Rogora. 2001. Trends in nitrogen deposition and leaching in acid sensitive streams in Europe. *Hydrology and Earth Systems Sciences* 5: 299–310.

18

The Role of Models in Addressing Coastal Eutrophication

Anne E. Giblin and Joseph J. Vallino

Summary

Eutrophication models have become increasingly important tools for scientists and managers attempting to determine the best and most cost-effective way to assess and control the flow of nutrients into estuaries and coastal waters. However, use of models has been controversial, in part because there is not yet agreement in the scientific community over what criteria should be used to judge models used for management decisions. In this chapter we discuss ways in which eutrophication models could be better evaluated for use in a management context. Rather than being judged on their ability to match current conditions, we advocate robust tests of how these models perform in response to perturbations in nutrient inputs or other stressors.

We note that there have only been a few attempts to systematically compare different eutrophication models with common data sets, something that is now being widely employed for models being developed for use in other management contexts. Mesocosms, and other large-scale experiments, could also be used more effectively to test model responses to changes in conditions. Finally, tracer studies are now becoming a powerful tool for model development and testing.

Introduction

Coastal areas throughout the world receive large amounts of nutrients from sewage, fertilizer, animal wastes, and atmospheric deposition (NRC 2000). These nutrients stimulate primary production that can lead to the excessive production of organic material, or eutrophication (Nixon 1995). Excessive nutrient enrichment has had detrimental effects on many coastal ecosystems, including depletion of oxygen in bottom waters, blooms of nuisance or toxic algae, decreased water clarity, losses of finfish and shellfish, and the loss of submerged aquatic vegetation. The problem is widespread: for example, symptoms of eutrophication are present in nearly one third of the estuaries in the United States (Bricker et al. 1999).

There has been a concerted effort in many areas to prevent the eutrophication of clean coastal and estuarine systems and to reverse the eutrophication of highly impacted ones through the management of nutrient inputs. As part of this effort, models have become increasingly important tools for scientists and managers attempting to determine the best and most cost-effective way to assess and control the flow of nutrients into estuaries and coastal waters (e.g., Boesch et al. 2001). In this chapter, we will briefly discuss some of the successes and problems with two classes of models that have seen widespread use for the management of eutrophication in the coastal zone. The first are models that are used to calculate nutrient inputs from the watershed. The second are numerical "water quality models" used to calculate nutrient, carbon, and oxygen dynamics in the water column and sediments. The goal of this paper is to explore how such models could be developed, tested, and used in a management context more effectively.

Nutrient-Input Models

A variety of approaches have been used to assess nutrient loading from watersheds, and several papers have reviewed the major models available (Deliman et al. 1999; NRC 2000; Valiela et al. 2002). The field has advanced rapidly and many models now use a hybrid approach to calculating nutrient inputs. The models, however, tend to fall into three categories: (1) complex simulation models, which have been developed to predict both hydrologic and water quality processes in the watershed under dynamic or steady-state conditions (e.g., HSPF; Bicknell et al. 1993); (2) "spreadsheet models," which use land-use data to generate export coefficients or mean concentrations to calculate loads (Valiela et al. 1997); or (3) statistical models, which use empirical data to develop relationships between characteristics in the watershed and nutrient export (e.g., SPARROW; Preston et al. 1998; Caraco and Cole 1999). There are many examples of models of each type being successfully applied in a variety of watershed types, but in nearly all cases there needs to be some adjustment to the parameters when the models are applied to a different location.

Nutrient-input models are judged based upon how well they can predict observed nutrient concentrations in streams or groundwater entering coastal systems. If total nutrient loading were the only objective, this would be sufficient, but if mitigation efforts are to be undertaken, it is probably equally, if not more important, to know the specific source of nutrient loading. For example, some models developed for the northeastern United States may have greatly underestimated the importance of nitrogen from precipitation and erroneously assumed

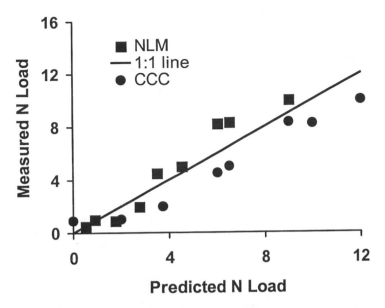

Figure 18.1. A comparison of the predicted versus the measured N loading to 9 sub-estuaries using 2 different models of N inputs from watersheds (Valiela et al. 2002).

a very high nitrogen export from undisturbed forests (NRC 2000). Another important consideration is to determine whether the model accurately reflects how nutrient inputs to the receiving waters change with changes within the watershed.

Applying models that were developed using different structures and assumptions to the same watershed and comparing how they partition nitrogen sources is one way to examine the robustness of the models' conclusions. Valiela et al. (2002) compared a number of different models used to estimate land-derived nitrogen inputs to a series of nine small sub-estuaries in Waquoit Bay having very different population densities. The predicted nitrogen loads were compared to nitrogen loads that had been calculated from nitrogen concentrations measured in groundwater and recharge estimates. Some of the models were developed for use in these watersheds, which consist of unconsolidated sands and glacial till, while others were developed in other watersheds.

Process, spreadsheet, and statistical models were included in the comparison. In general many of the models were judged to be quite "responsive": that is, they predicted a change in loading that correlated well with the observed loading change. In contrast, there was a very large variability in the predictions of the percentage of nitrogen coming from different sources. For example, both the Waquoit Bay nitrogen-loading model (NLM; Valiela et al. 1997) and Cape Cod Commission model (CCC; Eichner and Cambareri 1992) are quite accurate

in their prediction of nitrogen loading (Figure 18.1) and show excellent responsiveness, but they differ in how they predict the N is derived. The NLM predicts that 47% of the N inputs are from wastewater, 32% come from the atmosphere, and 21% is attributable to fertilizer. In contrast, the CCC predicts that
83% comes from wastewater while only 1% comes from the atmosphere. Overall, the range of predictions for wastewater-derived N varied from 13 to 83%
(Valiela et al. 2002). From a management point of view, this large range in the
estimate of the percentage of N coming from a controllable source creates grave
problems when attempting to develop a strategy of nitrogen mitigation.

Linked Hydrodynamic-Water Quality Models

Because of the dynamic nature of estuarine and coastal ecosystems, there is
great interest in simulation models that include transport and stratification. It is
now common, and nearly expected, that models of the coastal ecosystems will
include a sophisticated treatment of the hydrodynamics of the system. One of
the most common types of eutrophication models, which has been widely applied by managers, is the coupled hydrodynamic–water quality model. Water
quality models calculate parameters such as dissolved oxygen, chlorophyll *a*,
and particulate carbon on hourly or daily time scales using nutrients, light, and
temperature as drivers.

Water quality models differ in the complexity of the hydrodynamics
schemes employed as well as the kinetic expressions used to describe water
quality variables (reviewed in NRC 2000). Many are derived from a common
model or contain common components. For example, the CE-QUAL-ICM
model being used in Chesapeake Bay (Cero and Cole 1993; Cero 2000) and the
ECOM/*EM model applied to Massachusetts Bay and the New York Apex
(HydroQual 1991, 1995) both contain a common sediment diagenesis submodel
(DiToro and FitzPatrick 1993) and many key features of water column dynamics. These models, as do most models designed to examine eutrophication problems (e.g., Humborg et al. 2000), share common features in that they use light,
temperature, and nutrient concentrations and a kinetic framework to calculate
phytoplankton production. Phytoplankton may be broken down into several
functional groups. In the Massachusetts Bay model, phytoplankton were broken
down into a winter group (diatoms) and a summer group that differed in their
silica requirements. The Chesapeake Bay model has three groups: diatoms,
cyanobacteria, and green algae. Production is lost through respiration (modeled
as a function of temperature) and predation. Most eutrophication models do not
model higher trophic levels but treat predation as a loss term that is a function
of temperature and that may also be a function of phytoplankton type. Lost production enters one of several possible dissolved or particulate detrital pools that
differ in their decomposition rates.

Current eutrophication models have several limitations (NRC 2000). In general, the models are much more successful in predicting the seasonally averaged conditions in the estuary than in predicting the highs and lows that are often of greatest concern to management. The inability to model toxic and nuisance algae is another serious limitation. All of the nutrient-based models have been criticized for their very limited ability to deal with higher trophic levels. In spite of these limitations the models have seen widespread use, and eutrophication models have been successful in helping guide management decisions in a number of locations.

Case Studies: Massachusetts Bay and Chesapeake Bay

A linked hydrodynamic–water quality model was used to predict water quality in the Boston Harbor–Massachusetts Bay region given a variety of locations for the municipal wastewater outfall (HydroQual 1995). The modeling effort was supported by an extensive monitoring effort. Simulations were run with both the existing harbor surface wastewater discharge and with a proposed location offshore into deeper (30–35m) water in Massachusetts Bay. The simulations showed lower chlorophyll conditions in both the harbor and the bay for most of the year after relocation. This was because the model predicted that with the harbor discharge most of the nutrients were already reaching the bay. With relocation to the bay, nutrients from the outfall remain below the photic zone for much of the year, leading to somewhat lower overall production than before. Mass-balance calculations, stable isotopic results, and simple box models of N processes in the harbor (Giblin et al. 1997; Kelly 1999) also indicated that there was a significant export of N from the harbor into the bay with the harbor outfall and provided corroboration of the model predictions of current conditions.

Other scenarios were conducted in which the model was used to determine the importance of atmospheric deposition and secondary treatment on chlorophyll and oxygen levels. Finally, the model was used to calculate the percentage of the nitrogen coming from anthropogenic sources compared to oceanic sources. The model calculated that more than 90% of the nutrients entering the Massachusetts Bay system came from oceanic sources, indicating the small role that anthropogenic nutrients may currently play in controlling productivity in this system as a whole.

Overall, the model results were useful in both management and public forums. The strong effort to corroborate some portions of the model predictions by independent means was very helpful in gaining support for the use of the model. Eventually the outfall was moved.

The original impetus to develop a linked hydrodynamic–water quality model of Chesapeake Bay was to better understand contemporary and historical trends in bottom-water anoxia (Cero 2000). After initially developing a two-dimensional vertically averaged model, the decision to proceed with a full three-dimensional model was made, and the model was completed in 1984

(HydroQual 1987). The initial model did not include sufficient detail in sediment processes, so the model was coupled with a sediment diagenesis model (DiToro and FitzPatrick 1993) and a watershed model. Subsequently an atmospheric model was added. The model has been used extensively for scenario testing (Cero 2000), especially for examining how the reduction of nutrient loads will affect bottom-water anoxia. However, use of this model for managing nutrient inputs to Chesapeake Bay has been controversial, in part because there is not yet agreement in the scientific community over what criteria should be applied for judging models used in management decisions (Blankenship 2000).

Differences between Water Quality Models and Other Food Chain Models

The treatment of zooplankton is the one area where the current generation of coastal and estuarine eutrophication models differs significantly from pelagic food chain models being developed by the oceanic research community. While the impetus for many of the oceanic models was research, models are now also being developed with management issues in mind. The oceanic models have always placed a much greater emphasis on including higher trophic levels as specific components in the model, either in relatively simple formulations that use a single zooplankton compartment (e.g., Guillaud et al. 2000) or by simulating complex pelagic food webs that include microzooplankton and bacteria (Tett and Wilson 2000). Originally, pelagic food chain models, unlike the coastal eutrophication models, did not have a benthic component, but more recent models of large systems, including that of the North Sea, are attempting to incorporate pelagic and benthic processes as well as multilevel food-chain dynamics (e.g., Baretta et al. 1995).

The Future of Models Used to Address Coastal Eutrophication

Current water quality models, which generally focus on algal biomass and carbon and oxygen dynamics, do not address all the issues that are of concern. The loss of specific habitats, such as submerged aquatic vegetation, is particularly important in many shallow systems. Habitat models for these ecosystems are currently under development, and managers have begun to use them to test restoration goals (Wetzel 1996). These models are also being coupled to water quality models.

Higher trophic levels offer a great challenge. Attempts to use models to analyze the affects of nutrients on higher trophic levels have focused largely on single species or single functional groups and have had limited successes. Eutrophication models coupled to even fairly simple food web models have yet to become fully functional. Recently, progress has been made coupling bioenergetic models of higher trophic levels with spatially explicit models of physical and biological parameters such as oxygen, salinity, temperature, and prey density (Dermers 2000).

Testing and Using Eutrophication Models for Management: Issues and Approaches

As models grow in size and complexity, there are more terms that can be "tuned" to fit the available data. When the number of model parameters exceeds the number of data input parameters it is possible to develop a model that can fit the observed data with more than one parameter set (Aber 1997; Chapter 11). Consequently, model predictions will vary depending upon the parameter set chosen to fit the data (Beck 1987). Oreskes et al. (1994; Chapter 2) have suggested that models cannot be validated or verified but only confirmed by demonstrated agreement between observation and prediction and that such confirmation is only partial. One solution to this problem is to be sure that managers recognize the uncertainty in the models and use models as tools for exploring questions rather than for generating answers (Deegan et al. 2001; Chapters 7 and 16). While this would reduce the misuse of models, it still leaves open the question as to what sorts of criteria should be used to judge the performance of a model when the goal is investigating possible future management actions (Chapter 7).

We suggest that an important criterion for management models is that the model response to the stressor of interest, such as a change in nutrient loading, be confirmed in some manner outside of the calibration data set. Intercomparisons of models across a range of systems that receive different level of nutrient inputs is one way to test such models. Experimental systems, such as mesocosms, offer another. A third method, which is only now being widely applied, is the use of stable nitrogen isotopes as tracers of processes to provide independent confirmation of nitrogen flows within models.

Intercomparisons

Multisite intercomparisons of complex models are expensive and difficult to perform but can prove to be exceedingly valuable in demonstrating the state of the science (Chapter 12). Managers will have more confidence using models in areas where there is substantial agreement among models. Areas where there are substantial differences highlight research priorities. Both the terrestrial-ecosystem-modeling community (e.g., Kicklighter et al. 1999) and the global-climate-modeling community have benefited from intercomparisons. Policy makers in the global-change arena now commonly assess the impact of policy options using climate predictions generated from several different general circulation models.

Comparisons of nutrient-input models, such as that carried out by Valiela et al. (2002) should become more widespread. Intercomparisons of water quality models is more problematic because the biological models are now coupled to complex hydrodynamic models, and the hydrodynamic models themselves differ between systems. However, a first step would be to test only the water quality portion of the model using the available site-specific hydrodynamic model.

There are now sufficient systems where there are both good hydrodynamic models and a wealth of monitoring data where an intersite, intermodel comparison could be undertaken. Ideally the comparison would include models with a variety of structures and complexity (Chapter 12).

Experimental Systems

Experimental systems offer another way to test the biological portion of eutrophication models in isolation from the hydrodynamic component. A modified version of the eutrophication model used in the Chesapeake, Long Island Sound, and Massachusetts Bays programs, was used to model more than two years of data from a series of mesocosm experiments run at the University of Rhode Island Marine Ecosystem Research Laboratory (MERL). The mesocosm experiments were done to examine the effects of increased nutrient inputs to coastal waters with loadings calculated to mimic from 1 to 32 times the current area loading to Narragansett Bay (Nixon et al. 1986). The model reproduced water-column oxygen quite well and captured the overall seasonal cycles of chlorophyll *a* in all treatments. However, the model tended to overpredict the average chlorophyll *a* levels while missing the extreme high and low values (NRC 2000). This result was similar to that observed when the model was applied to natural settings and suggested that the biological component of the model, rather than the hydrodynamic component, was responsible for the model response curve being smoothed relative to nature. The sediment portion of the model was able to match the observed distribution in sediment oxygen demand quite well, but the modeled fluxes differed from the data for silica and nitrate fluxes (DiToro 2001). As a consequence, the model did not reproduce some aspects of the water-column nutrient dynamics. This type of comparison of model to data over a range of nutrient values helped demonstrate the responsiveness of the model to changes in nutrient inputs. Direct comparisons of model output to a range of conditions can give managers a much better feel for the strengths and limitations of the model.

Experimental systems can also be used to distinguish limitations in model structure from problems with parameter values. Vallino (2000) compared a simple food web model (Figure 18.2) to data from four experimental pelagic mesocosms: control (bag A); a large single addition of dissolved organic carbon (DOC) to reach a concentration of 300 μM (bag B); daily additions of dissolved inorganic nutrients to reach target concentrations of nitrogen (5 μM), phosphorus (0.5 μM), and silica (7 μM), (bag C); and a one time addition of DOC with daily nutrient additions (bag D). The experiment was run for 21 days and a host of variables were measured daily.

The model was designed to focus on the importance of organic matter production and consumption and to specifically examine the potential role of DOC in altering the competition between bacteria and phytoplankton for nutrients. Unlike most eutrophication models, bacteria and zooplankton (heterotrophs) were specifically modeled. The model contained 29 parameters governing the growth kinetics of the organisms and decomposition of the organic matter pool,

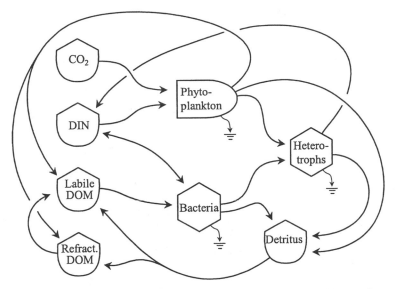

Figure 18.2. The structure of the foodweb model developed by Vallino (2000) and tested in several mesocosms.

and 10 parameters were used to specify the initial conditions of the state variables. Three of the parameters were fixed and the other 36 were set using data assimilation to determine the optimum values. The parameters were fit to the data from the nutrient + DOC bag (bag D) and then applied to all other treatments.

The optimized parameter set was able to do an excellent job simulating conditions in bag D as evidenced by the correspondence between modeled and measured concentrations of chlorophyll a, DOC, and DIN (Figure 18.3). Unlike most eutrophication models, the model captured the full range of primary production (not shown) and chlorophyll a values although the experiment was run for only 21 days. When the same calibration was applied to the DOM-only bag (bag B) the correspondence between the model and the data was quite good (Figure 18.3). The model's fit to the control bag (A) may or may not be considered adequate depending upon the application. The model completely fails to capture the chlorophyll a, DIN and DOC dynamics observed in the nutrient-alone bag (C) (Figure 18.3).

To examine the effect of using a specific treatment on parameterizing the model, the model was reparameterized using data from the nutrient-only treatment (bag C). Again the model was able to do an excellent job mimicking the behavior of the bag from which it was calibrated (Figure 18.4). The fit to the control bag (A) is poor, but the model did capture the low chlorophyll a values

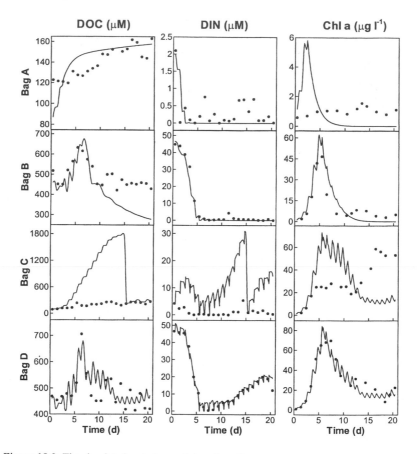

Figure 18.3. The simulated and observed data from four mesocosms, Bag A (control), Bag B (DOM added), Bag C (inorganic nutrients added daily), and Bag D (DOM added and daily additions of inorganic nutrients). All parameters were fit only using the data from Bag D. Lines indicate model output while symbols show data.

of this treatment relative to the nutrient addition bags (Figure 18.4). Where the model completely failed is in simulation of either treatment when DOC was added (bags B and D; Figure 18.4).

Why does the model fail to capture the dynamics of all of the bags using a single parameter set? When parameterized individually, the model could reproduce specific treatments with good accuracy and precision. This suggests that the problem is not parameter uncertainty but rather that the structure of the model is not appropriate for modeling across a range of treatments that include both high DOC and low DOC values (Vallino 2000). Therefore, we conclude

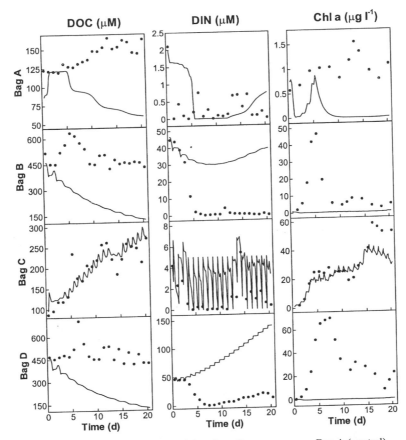

Figure 18.4. The simulated and observed data from four mesocosms, Bag A (control), Bag B (DOM added), Bag C (inorganic nutrients added daily) and Bag D (DOM added and daily additions of inorganic nutrients. All parameters were fit only using the data from Bag C. Lines indicate model output while symbols show data.

that this model is inadequate as a prognostic tool. This type of insight would not be possible without comparing the model to different treatments.

Experimental systems can provide the range of conditions needed to vigorously evaluate the responsiveness of eutrophication models. They will not replace field data or cross-system comparisons, but their utility should not be underestimated.

Tracers

Stable isotopic tracers are now being used in a variety of ways to develop and evaluate models. In some cases, there is a strong natural isotopic signal that can be exploited. For example, different models used in calculating nutrient loading to the Waquoit Bay sub-estuaries calculate very different source strengths even when the total N-loading predictions are similar (Valiela et al. 2002). Studies have shown that nitrogen from on-site septic systems is considerably heavier than that derived from atmospheric or fertilizer sources. The $\delta^{15}N$ signal of dissolved N in the groundwater entering the sub-estuaries allows for an independent estimate of the wastewater source strength (McClelland and Valiela 1998) that then can be used to further evaluate the predictions of the different models.

Although there are places when natural isotopic signals can be exploited as a natural tracer with success, there are many places where they cannot. However, the lower cost of isotopes has now made it possible to carry out tracer additions to natural systems. An extensive comparison of the N processing of streams and estuaries using tracer studies revealed a great deal about N processing in these systems (Peterson et al. 2001) and led to the development of better models of nutrient dynamics in streams. A nitrogen isotope tracer addition to forests has helped improve our ability to model N cycling in forests (Currie et al. 1999).

Model Parameterization

Rigorous model–data comparisons are the key to creating confidence in the utility of the model in the management community, yet in many areas there are disagreements over the form that this comparison should take (Chapter 12). One frequent recommendation is that models should include realistic, empirically based parameterizations that would tie the model to field and laboratory observations (Aber 1997; NRC 2000). The counterargument is that the model is a simplification of the system (Chapter 2). Compartments and flows may represent aggregations that have no exact equivalent in the real world and as such, cannot be individually calibrated from laboratory data (Vallino 2000; Chapter 13). Indeed, Wallach and Genard (1998) have shown that a better fit to observations can be obtained when parameter values are allowed to exceed their expected ranges in highly aggregated models. Therefore calibrating the model using data-assimilation techniques helps remove the parameter uncertainty that cannot be removed through laboratory experiments or observations. Additional information about the sensitivity of the parameters in the model and codependence of parameters can be investigated through these techniques (discussed in detail in Vallino 2000).

An analysis of the parameter values calculated using data assimilation techniques under different experimental treatments can also lead to insights about the model and help identify areas where the model structure is not adequate, or where compartment aggregation leads to an inappropriate representation of the

Table 18.1. A comparison of the values of some parameters determined using data from either the DOM + DIN addition experiment (Bag D) or the DIN only addition (Bag C) for the model shown in Figure 18.2 (Vallino 2000). See text for details.

Parameter	Bag D	Bag C
Maximum specific DOC uptake rate by bacteria	49.9 d^{-1}	39.2 d^{-1}
Half-saturation constant of DOC consumption by bacteria	48.8 μM	24.6 μM
Mortality of bacteria	48.4 d^{-1}	49.3 d^{-1}
Decomposition of detritus	49.6 d^{-1}	50.0 d^{-1}
Maximum growth efficiency of heterotrophs	0.151	0.866
Half-saturation constant of H feeding on A and B	200 μM	0.31 μM

system. For example in the food web model (Figure 18.2) the best fit for the model to the DOC + DIN treatment resulted in growth kinetics for bacteria that would be considered extreme and unrealistically high decomposition rates for detritus (Table 18.1).

When the model was recalibrated to the DIN-alone bag, growth parameters were still high and the values showed little change from the initial calibration (Table 18.1). This analysis suggests that the bacteria and detritus compartments are actually composed of multiple compartments and that when aggregated as single compartments, a high turnover is necessary to mimic the system. Had the model successfully been able to mimic all nutrient and DOC treatments with a single calibration, we suggest that it should not have be rejected as a potential tool for exploring management options simply because of the parameter values, which by virtue of their aggregation may in fact not be comparable to laboratory derived values.

In contrast, there were large changes in the parameters between the model fits to bag C versus bag D governing zooplankton growth kinetics (Table 18.1). In bag C, which did not receive DOM additions, zooplankton growth parameters favor a more efficient growth behavior. This is perhaps understandable, since the microbial loop, which can support higher trophic levels, is less important in bag C. When the parameters from the fit to bag C are used to describe the growth in bag D, the zooplankton are able to readily graze down the phytoplankton when given the additional support provided by the microbial loop, which in turn is supported by the increase in DOM availability. In essence, the model underestimates the importance of the microbial loop when fit to bag C and overemphasizes it in bag D. This indicates that the model structure governing the microbial loop is not well articulated in the current model, hence the lack of model robustness between treatments.

Conclusions

Modelers and managers are well aware that models are simplifications of the systems they are designed to mimic. However, it may not always be obvious where the multiple processes are hidden within the model structure. A calibrated model may appear to represent a single explicit process firmly based upon empirical data, and yet when performing in a calibrated model, this process may actually represent a quite complex set of processes that are lumped together and no longer completely represent the observations. In this case, the fit between model output and observations may be satisfactory only under a limited range of conditions.

Successful use of models in a management context does not require that they perfectly mimic nature (Chapter 2), yet models are often evaluated by how well they match current observations over seasonal or annual timescales. In fact, the ability of a model to match current conditions may be a poor predictor of how the model will respond to changes in conditions outside the range of current observations. If the goal is to understand the response of the system to stressors, a different criterion for the testing and evaluation of eutrophication models may be warranted.

We need to find robust ways to test how models perform in response to perturbations and to evaluate how realistic the response appears. To date there has been little attempt to systematically compare different models with common data sets. Such an exercise, especially if carried out using data from a range of coastal ecosystems, would improve our understanding of the strengths and weaknesses of current eutrophication models. The coastal-eutrophication-modeling community could also make much better use of mesocosms, and other large-scale experiments to test model response to changes in conditions. Tracer studies offer another way of testing models and are only now being done on a widespread basis. A tighter connection between the experimentalists and the modeling community in designing and carrying out more of these types of experiments could facilitate the development of both better models and an improved understanding of how systems respond to perturbation.

Acknowledgments. Financial support to the authors was provided by NSF grant OCE-9726921.

References

Aber, J. 1997. Why don't we believe models? *Bulletin of the Ecological Society of America* 78: 232–233.

Baretta, J.W., W. Ebenhoh, and P. Ruardij. 1995. The European regional seas ecosystem model, a complex marine ecosystem model. *Netherlands Journal of Sea Research* 33: 233–246.

Beck, M.B. 1987. Water quality modeling: A review of the analysis of uncertainty. *Water Resources Research* 23: 1393–1442.

Bicknell, R.R., J. Imhoff, J.L. Kittle, Jr., and R.C. Johanson. 1993. *Hydrological Simulation Program-Fortran: User's Manual for Release 10.* Environmental Protection Agency, Office of Research and Development: Athens, GA.

Blankenship, K. 2000. Scientists working to improve Bay water model. *Bay Journal* 10: n(2).

Boesch, D.F., R.B. Brinsfield, and R.E. Magnien. 2001. Chesapeake Bay eutrophication: Scientific understanding, ecosystem restoration, and challenges for agriculture. *Journal of Environmental Quality* 30: 303–320.

Bricker, S.B., C.G. Clement, D.E. Pirhalla, S.P. Orlando, and D.G.G. Farrow. 1999. *National Estuarine Eutrophication Assessment: Effects of Nutrient Enrichment in the Nation's Estuaries.* Special projects office and the National Centers for Coastal Ocean Science, National Oceans Service, NOAA: Silver Springs, MD.

Caraco, N.F., and J.J. Cole. 1999. Human impact on nitrate export: An analysis using major world rivers. *Ambio* 28: 167–170.

Cero, C.F., and T. Cole. 1993. Three dimensional eutrophication model of Chesapeake Bay. *Journal of Environmental Engineering* 119: 1006–1025.

Cero, C.F. 2000. The Chesapeake Bay eutrophication model. Pp. 363–404 in J.E. Hobbie, editor. *Estuarine Science: A Synthetic Approach to Research and Practice.* Washington DC: Island Press.

Currie, W.S., K.J. Nadelhoffer, and J.D. Aber. 1999. Soil detrital processes controlling the movement of ^{15}N tracers to forest vegetation. *Ecological Applications* 9: 87–102.

Deegan, L.A., J. Kremer, T. Webler, and J. Brawley. 2001. The use of models in integrated resource management in the coastal zone. Pp. 295–305 in von Bodungen, B., and R.K. Turner, editors. *Science and Integrated Coastal Management.* Berlin: Dahlem University Press.

Deliman P.N., R.H. Glick, and C.E. Ruiz. 1999. *A Review of Watershed Water Quality Models.* Vicksburg, MS: US Army Corps of Engineers, Waterways Experiment Station.

Dermers, E., S.P Brandt, K.L. Barry, and J.M. Jech. 2000. Spatially explicit models of growth rate potential: Linking estuarine fish production to the biological and physical environment. Pp. 363-404 in J.E. Hobbie, editor. *Estuarine Science: A Synthetic Approach to Research and Practice.* Washington DC: Island Press.

DiToro, D.M. 2001. Sediment Flux Modeling. *Wiley Interscience.* New York: John Wiley and Sons.

DiToro, D.M., and J. FitzPatrick. 1993. *Chesapeake Bay Sediment Flux Model*. Prepared for the U.S. Environmental Protection Agency and U.S. Army Engineer District, Baltimore. Mahwah, NJ: HydroQual Inc.

Eichner, E.M., and T.C. Cambareri. 1992. *Nitrogen loading*. Cape Cod Communication Technical Bulletin 91-001. Barnstable, Massachusetts, USA.

Giblin, A.E., C.S. Hopkinson, and J. Tucker. 1997. Benthic metabolism and nutrient cycling in Boston Harbor, Massachusetts. *Estuaries* 20: 346–364.

Guillaud, J-F., F. Andrieux, A. Menesguen. 2000. Biogeochemical modeling in the Bay of Seine (France): An improvement by introducing phosphorus in nutrient cycles. *Journal of Marine Systems* 25: 369–386.

Humborg, C., K. Fennel, M. Pastuszak, W. Fennel. 2000. A box model approach for a long-term assessment of estuarine eutrophication, Szczecin Lagoon, southern Baltic. *Journal of Marine Systems* 25: 387–403.

HydroQual Inc. 1987. *A Steady-State Coupled Hydrodynamic/water Quality Model of the Eutrophication and Anoxia Processes in Chesapeake Bay*. Prepared for the U.S. EPA Chesapeake Bay Program: Mahwah, NJ.

———. 1991. *Water Quality Modeling Analysis of Hypoxia in Long Island Sound*. Prepared for the management committee on Long Island Sound Estuary Study and New England Interstate Water Pollution Control Commission: Mahwah, NJ.

———. 1995. *A Water Quality Model for Massachusetts and Cape Cod Bays: Calibration of the Bays Eutrophication Model (BEM)*. Mahwah, NJ.

Kelly, J. 1999. Nitrogen flow and the interaction of Boston Harbor with Massachusetts Bay. *Estuaries* 20: 365–380.

Kicklighter, D.W., M. Bruno, S. Dönges, G. Esser, M. Heimann, J. Helfrich, F. Ift, F. Joos, J. Kaduk, G.H. Kohlmaier, A.D. McGuire, J.M. Melillo, R. Meyer, B. Moore III, A. Nadler, I.C. Prentice, W. Sauf, A. Schloss, S. Sitch, U. Wittenberg, and G. Würth. 1999. A first-order analysis of the potential role of CO_2 fertilization to affect the global carbon budget: A comparison study of four terrestrial biosphere models. *Tellus* 51B: 343–366.

McClelland, J.W., and I. Valiela. 1998. Linking nitrogen in estuarine producers to land-derived sources. *Limnology and Oceanography* 43: 577–585.

National Research Council (NRC). 2000. *Clean Coastal Waters: Understanding and Reducing the Effects of Nutrient Pollution*. Ocean Studies Board and Water Science and Technology Board, Commission on Geosciences, Environment, and Resources, NRC. Washington, DC: National Academy Press.

Nixon, S.W. 1995. Coastal marine eutrophication: A definition, social causes, and future concerns. *Ophelia* 41: 199–219.

Nixon, S.W., C. Oviatt, J. Frithsen, and B. Sullivan. 1986. Nutrients and the productivity of estuarine and coastal marine ecosystems. *Journal of the Limnology Society of South Africa*. 12 (1/2): 43–71.

Oreskes, N., K. Shrader-Frechette, and K. Belitz. 1994. Verification, validation, and confirmation of numerical models in the earth science. *Science* 263: 641–646.

Peterson, B.J., W.M. Wollheim, P.J. Mulholland, J.R. Webster, J.L. Meyer, J. L. Tank, E. Marti, W.B. Bowden, H.M. Valett, A.E. Hershey, W.H. McDowell, W.K. Dodds, S.K. Hamilton, S. Gregory, and D.D. Morral. 2001. Control on nitrogen export from watersheds by headwater streams. *Science* 292: 86–90.

Preston, S.D., R.A. Smith, G.E. Schwartz, R.B. Alexander, and J.W. Brakebill. 1998. *Spatially Referenced Regression Modeling of Nutrient Loading in the Chesapeake Bay Watershed.* Proceedings of the First Federal Interagency Hydrologic Modeling Conference: Las Vegas, NV.

Tett, P., and H. Wilson. 2000. From biogeochemical to ecological models of marine microplankton. *Journal of Marine Systems* 25: 431–446.

Turner, R.K. 2000. Integrating natural and socio-economic science in coastal management. *Journal of Marine Systems* 25: 447–460.

Valiela, I., J.L. Bowen, and K.D. Kroeger. 2002. Assessment of models for estimation of land-derived nitrogen loads to shallow estuaries. *Applied Geochemistry* 17(7): 935–953.

Valiela, I., G. Collins, J. Kremer, K. Lajtha, M. Geist, B. Seely, J. Brawley, and C.H. Sham. 1997. Nitrogen loading from coastal watersheds to receiving estuaries: New method and application. *Ecological Applications* 7: 358–380.

Vallino J.J. 2000. Improving marine ecosystem models: Use of data assimilation and mesocosm experiments. *Journal of Marine Research* 58: 117–164.

Wallach, D., and M. Genard. 1998. Effect of uncertainty in input and parameter values on model prediction error. *Ecological Modeling* 105: 337–345.

Wetzel, R.L. 1996. Ecosystem processes and modeling of lower Chesapeake submerged aquatic vegetation and littoral zones. Briefing paper to the Chesapeake Bay Program on Living Resources subcommittee.

19

Quantitative Models in Ecological Toxicology: Application in Ecological Risk Assessment

James T. Oris and A. John Bailer

Summary

In this chapter, we present an overview of the Ecological Risk Assessment (ERA) process and review the use of quantitative models in three areas. The first involves the development and use of bioaccumulation models in exposure assessment. These models range from simple, single-organism models to full-scale ecosystem models incorporating multiple trophic levels. The second area involves the determination of toxicity endpoints that are used in the hazard assessment component of ERA. Short-term lethal (i.e., acute) effects of toxicants have most often been assessed using point estimates derived from probit or logistic regression analyses. Long-term, sublethal (i.e., chronic) effects of toxicants historically have been evaluated using procedures to derive no-observable-effect concentrations (NOECs). The use of NOECs in ERA has come under serious criticism, and many investigators have promoted the use of regression models to calculate the concentration associated with a specific impact relative to control organisms. The third area involves the characterization of risk associated with a particular toxicant. Risk characterization in its simplest form is accomplished deterministically by evaluating the quotient of the predicted or measured exposure concentration divided by a reference toxicity value (also a concentration) determined from the hazard assessment component. If this quotient is greater than or equal to 1, significant risk is presumed. Due to the drawbacks of this approach, however, probabilistic methods of evaluating risk have been developed. These methods involve evaluating joint probability distributions between exposure and effect levels for a variety of exposure scenarios and for a range of species. Use of advanced-level bioaccumulation models, the use of likelihood-based endpoint estimates in effects assessment, and the use of probabilistic models in risk characterization all provide more realistic evaluations of ecological risk. However, these methods are computation- and data-

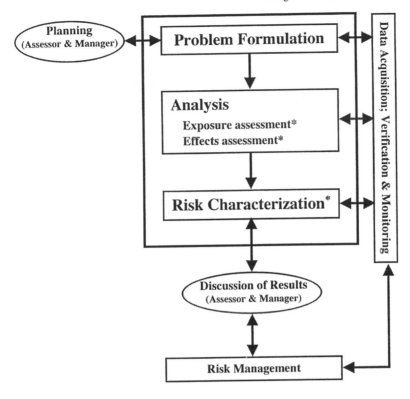

Figure 19.1. U.S. Environmental Protection Agency's Framework for Ecological Risk Assessment (redrawn from EPA 1992). Components of the process marked with an asterisk (*) involve quantitative modeling and are the focus of this chapter.

intense, and thus are more expensive and time-consuming compared to traditional methods. Regardless of the methods chosen, qualitative judgments must still be made in reference to biologically significant effects and acceptable levels of risk in populations and communities.

A Framework for Ecological Risk Assessment

In response to the development and use of an increasing number of synthetic chemicals and the resulting degradation of environmental quality in the United States, the National Environmental Policy Act (NEPA) was passed in 1969. The U.S. Environmental Protection Agency (EPA) was formed with the passage of the Environmental Quality Improvement Act (EQIA) of 1970, and a series of legislative acts were developed to address the control and regulation of environmental hazards. In order to fulfill the requirements of these laws and regulations, quantitative models for exposure and effects of toxicants were developed and used to determine the risks associated with exposure to environmental stressors. In 1983, the National Research Council (NRC) published a paradigm (NRC 1983) that was used in the development of EPA's human risk assessment guidelines. More recently, as a result of many years of discussion on the process of ecological risk assessment, the U.S. EPA's Framework for Ecological Risk Assessment was developed (Figure 19.1) (EPA 1992, 1996). The framework is a multicomponent process for systematically estimating risks associated with a particular stressor in the environment. The outcome of an ecological risk assessment is a qualitative or quantitative appraisal of the actual or potential effects of environmental stressors on plants and animals other than humans and domesticated species. In some cases, humans may be considered to be part of the ecosystem, but they are not the focus of an ecological risk assessment. There are three main steps to a basic assessment (Figure 19.1): (1) problem formulation, (2) analysis, and (3) risk characterization. Quantitative models are used extensively in the analysis and Risk Characterization steps of Ecological Risk Assessment (ERA). These three steps lead to the process of risk management, in which rules or policy regarding the stressor is made.

Problem Formulation

The first component of ecological risk assessment is problem formulation. This is a systematic planning component that establishes the goals, breadth, and focus of the assessment. There is two-way communication between scientists and policy makers during this component, thus linking the assessment to the regulatory and policy needs of the decision makers. This component leads to the development of a conceptual model of the fates and effects of a stressor. Based on this model and on the regulatory and policy needs of the assessment, two types of endpoints are determined in the problem formulation component. Assessment endpoints are typically qualitative in nature and describe the desired outcome of regulatory or policy implementation. Measurement endpoints are then quantitatively assessed to determine the achievement level of the assessment endpoint. The problem formulation component thus directs the focus of the study such that the final outcome will result in relevant information for the policy maker.

Analysis

The second component of ecological risk assessment is the quantitative analysis of the potential exposure and effects of a stressor or stressors in an ecosystem. The general approach during the analysis component is separated into two areas: exposure assessment and effects assessment. Exposure assessment involves the prediction and/or measurement of stressor levels at an ecological target as defined in the problem formulation step. Targets of interest are typically organisms but may include target tissues or other ecological components such as air, soil, sediments, or water. Tools used in the exposure assessment component include a large number of predictive, quantitative fate, and distribution models that have been developed over the past thirty years. Exposure assessment results in the prediction of the magnitudes of exposure as well as spatial and temporal distribution of the stressor. Effects assessment determines the relationship between the level of stressor exposure and level of adverse effect on the target ecological component. Statistical models ranging from simple to complex have been used to characterize levels of effect. Because of the assumptions and judgments necessary in the statistical determination of effects endpoints, the effects characterization component has been one of the most challenging areas of ecological risk assessment and will provide one of the focal areas of this chapter.

Risk Characterization

The characterization of risks associated with a particular stressor given a set of assessment and measurement endpoints is the final scientific component of ecological risk assessment. It is this component that evaluates the likelihood or probability of adverse effects based on the results of the exposure and effects assessment. Risk estimation is an integration of the stressor-response and exposure profiles, and it should include an analysis of the uncertainties associated with both stressor-response measurements and exposure predictions. Risk estimation in its simplest form is accomplished deterministically by evaluating the quotient of the predicted or measured exposure concentration divided by a reference toxicity value (such as an LC_{50}, the concentration associated with 50% lethality) determined from the effects assessment component (Barnthouse et al. 1982). If this quotient is greater than or equal to 1, significant risk is presumed. However, this calculation does not result in a true estimate of risk. In addition, since it is calculated by using single, deterministic endpoint values, information about the toxicity-response distribution and error distribution cannot be incorporated into the quotient. Because of these limitations, probabilistic methods of evaluating risk have been developed. These methods are often described as "probabilistic risk assessment" and involve simultaneously evaluating probability distributions between exposure and effect levels for a variety of exposure scenarios and for a range of species (Newman 1998). Risk description involves detailing a summary of ecological impacts and an interpretation of the ecological significance of those impacts. In order to be most effective, this summary should be described within the context of the assessment endpoints as defined in the problem formulation component (ECOFRAM 1999).

Risk Management

The information provided during all components, summarized in the risk characterization component, is provided to policy makers for their use in risk management. Risk management utilizes the results from risk assessments, regulatory and economic principles, and political considerations in a decision-making process that sets policy.

Quantitative Models in Ecological Risk Assessment

Quantitative models are used extensively in both the analysis and the risk characterization components of ERA. The following sections highlight areas that have received the most development and are areas of current discussion and modification or controversy.

Exposure Assessment Models

The EPA, primarily through the Center for Exposure Assessment Modeling (CEAM), has developed a series of models that are available for use in the assessment of the fate, dynamics, and speciation of chemicals in the environment (http://www.epa.gov/ceampubl/) (Table 19.1).

Table 19.1. Quantitative models used in ecological risk assessment.

Model	Description
US EPA	Models used in the Exposure Assessment Component of Ecological Risk Assessment.
CORMIX	The Cornell Mixing Zone Expert System models aqueous toxic or conventional pollutant discharges into diverse water bodies. A related set of models, PLUMES, is intended for use with plumes discharged to marine and some freshwater bodies.
3DFEMWATER and 3DLEWASTE	Three-Dimensional Finite Element Model of Water Flow Through Saturated-Unsaturated Media (3DFEMWATER) and Three-Dimensional Lagrangian-Eulerian Finite Element Model of Waste Transport Through Saturated-Unsaturated Media (3DLEWASTE) are related simulation models that provide estimates of flow and transport in three-dimensional, variably saturated porous media.
EXAMS	The EXposure Analysis Modeling System is a steady-state simulation model that estimates exposure, fate, and persistence following release of an organic chemical into an aquatic ecosystem. It has been used extensively to screen and identify synthetic organic chemicals likely to adversely impact aquatic systems (Burns 1997).
FGETS	The Food and Gill Exchange of Toxic Substances is a dynamic simulation model that predicts whole-body concentration and time to lethality of nonionic, nonmetabolized, organic chemicals in fish that are accumulated from either water only or water and food together.

Table 19.1. *Continued*

Model	Description
GCSOLAR	This model provides estimates of direct photolysis rates and half-lives of pollutants in the aquatic. Estimates of photolysis rates can be used as input variables for other models (e.g., EXAMS).
GENEEC	The GENeric Estimated Environmental Concentration (Parker et al. 1995) is a meta-model of PRZM-EXAMS and is recommended for use (ECOFRAM 1999) as a first-tier screening tool to estimate environmental concentrations for a pesticide in an edge-of-field water body.
HSCTM2D	The Hydrodynamic, Sediment, and Contaminant Transport Model (HSCTM2D) is a simulation model of the hydrodynamics of sediment and contaminant transport in rivers or estuaries.
HSPF	The Hydrological Simulation Program is a simulation model of watershed hydrology and water quality for both conventional and toxic organic pollutants and allows the integrated simulation of land and soil contaminant runoff processes.
MINTEQA2	MINTEQA2 is a chemical speciation model used to calculate the equilibrium composition of dilute aqueous solutions (e.g., metals speciation) in the laboratory or in natural aqueous systems.
MULTIMED	The Multimedia Exposure Assessment Model simulates the movement of contaminants released from a waste disposal facility.
PRZM	The Pesticide Root Zone Model is a package of simulation models that predicts pesticide transport and transformation in the crop root and unsaturated zone and is a standard model in the FIFRA mandated risk assessment of pesticides (ECOFRAM 1999).
PATRIOT	The Pesticide Assessment Tool for Rating Investigations of Transport allows analyses of ground water vulnerability due to leaching of pesticides on a regional, state, or local level.
QUAL2E	The Enhanced Stream Water Quality Model is a simulation

Table 19.1. *Continued*

Model	Description
	model used as a water-quality planning tool to study waste load impacts of conventional pollutants in branching streams and well-mixed lakes.
SWMM	The Storm Water Management Model is a simulation model of urban runoff events, including rainfall, snowmelt, surface and subsurface runoff, flow routing through drainage network, storage, and treatment.
SMPTOX3	The Simplified Method Program—Variable Complexity Stream Toxics Model calculates water column and streambed toxic substance concentrations resulting from point-source discharges into streams and rivers and is used for performing waste load allocations for toxic chemicals.
WASP	The Water Quality Analysis Simulation Program models contaminant fate and transport in surface waters, simulating biochemical oxygen demand, dissolved oxygen dynamics, nutrient dynamics and eutrophication, bacterial contamination, and organic-chemical and heavy-metal contamination.

With the exception of the FGETS model, these provide simulations and estimates of environmental concentrations of chemicals but do not provide much information regarding the bioavailability and uptake dynamics of chemicals into organisms. Since it is typically the amount of chemical that is taken up by an organism and the interaction with specific biological target tissues that causes toxicity (e.g., liver, kidney, nervous tissue), it is often important to model the uptake and distribution of chemicals within plants and animals. In addition, many toxicants are stored and concentrated within organisms, and the trophic transfer of chemicals can be an important mechanism of exposure. Thus, a large body of bioaccumulation models has been developed to predict uptake, distribution, and elimination of chemicals in organisms.

The assessment and modeling of bioaccumulation is an important step in ERA. A variety of approaches have been used to conduct these assessments, but the underlying assumption of bioaccumulation models is that the amount of a chemical actually accumulated by the organism and the ensuing interaction with an internal target site is the critical estimate of environmental exposure. These assessments are often made by determination of a so-called bioaccumulation factor (BAF). A BAF is defined as the steady-state concentration of a chemical in an organism, accounting for all mechanisms of uptake and release, relative to the concentration in the external environment (Newman 1998). Specific categories of BAFs include the bioconcentration factor (BCF—the concentration in the organism relative to concentration only from uptake from water), and the biomagnification factor (BMF—the concentration in the organism relative to concentration from uptake from food). Accumulation from sediment sources has been termed the biota-sediment accumulation factor (BSAF) (Mackay and Fraser 2000).

The simplest and earliest BAF models were empirical and not directly linked to any biological mechanisms. Neeley et al. (1974), Veith et al. (1979), Mackay (1982), Meylan et al. (1999), and others, developed regression models of measured BAF values compared to known physicochemical properties of bioaccumulating substances. The most common property used in these models has been a measure of lipid solubility: the octanol-water partition coefficient (K_{ow}). Log(K_{ow}) values range from less than 0 to greater than 10 for organic chemicals. The earliest models included a single, linear term. Recently, empirical models have been refined to include multiple parameters based on chemical class (e.g., ionic vs. neutral organics) and log(K_{ow}) range (Meylan et al. 1999). U.S., Canadian, and European guidelines recognize the importance of this measurement and require reporting of measured or estimated BAF values for all new chemical products (ECOFRAM 1999). The critical roles that these "simple models" have played in chemical safety assessments strongly support the view of Pace (in Chapter 4 of this volume).

Single-organism, mechanistic models represent a next level of sophistication in BAF estimation (Mackay and Fraser 2000). Mechanistic models describe organisms, or portions thereof, as discrete mass-balance compartments. Uptake and release mechanisms are defined based on the characteristic of the organism, the chemical, and the environment. Rate constants are determined for each mechanism to estimate fluxes of chemical into and out of each compartment. Bioaccumulation can thus be estimated under equilibrium or nonequilibrium conditions. As an example for an aquatic organism, a basic, one-compartment model treats an individual organism as a discrete and homogeneous mass-balance compartment with one diffusion-based uptake and one diffusion-based elimination mechanism from water:

$$\frac{dC_a}{dt} = K_u \cdot C_w - K_e \cdot C_a \qquad (19.1)$$

where:

K_u is the rate constant for uptake (ml(water) $\cdot$ g^{-1}(organism) $\cdot$ hr^{-1})

K_e is the rate constant for elimination (hr^{-1})

C_w is the concentration of compound in water (g[toxin] $\cdot$ ml^{-1})

C_a is the concentration of compound in animal (g[toxin] $\cdot$ g^{-1}(organism))

$\frac{dC_a}{dt}$ describes the net flux of compound into (or out of) the animal at any time t

$K_u \cdot C_w$ describes the uptake flux

and

$K_e \cdot C_a$ describes the elimination flux

If the environmental concentration can be assumed to be constant, this model can be represented by

$$C_a = K_u / K_e (1 - e^{-K_e t})$$ (19.2)

Since at equilibrium, $\frac{dC_a}{dt} = 0$, $K_u/K_e = C_a/C_w$. Thus, based on the definition of a BAF as C_a/C_w, K_u/K_e provides an estimate of BAF. An advantage of this type of model is that one can estimate a BAF experimentally without the need to reach chemical equilibrium conditions between the organism and the environment. Practically, these types of models are developed using data collected from short-term laboratory experiments and utilizing nonlinear regression techniques to obtain estimates of rate constants. Once estimated, BAF values can be compared statistically to other BAF values using a variety of established methods (Bailer et al. 2000b).

More complicated mechanistic models can be developed to account for multiple uptake and elimination mechanisms and for cases when C_w is not constant (Gibaldi and Perrier 1982). MacKay and Fraser (2000) define two potential uptake and six potential elimination mechanisms for a typical organism. Barber et al. (1991) developed such a model for fish, the Food and Gill Exchange of Toxic Substances (FGETS), that estimates uptake and accumulation of nonmetabolized organic chemicals from food and water (Table 19.1).

The most sophisticated mechanistic modeling approach for single organisms is based on physiological characteristics. These models have been generally referred to as physiologically based toxicokinetic (PBTK) models (Gibaldi and Perrier 1982; Newman 1995). PBTK models represent an organism as a series of volumetric compartments connected to one another by the bulk flow of a fluid (e.g., blood for an animal model). Compartment types, connections, and perfusion rates are determined by physiological relevance. Bulk flow of fluid between compartments (and thus transfer of chemical from compartment to compartment) can be measured under a variety of environmental and physiological conditions. Beyond the physiological reality imposed by such model structure, PBTK models have the advantage that they can be parameterized to account for wide-ranging environmental conditions and can be allometrically scaled to different size classes or types of organisms (Bailer and Dankovic 1997). Compartment volumes are determined in part by the partitioning characteristics of the chemical and bulk fluid exchange in the compartment. These are represented as volumes of distribution (Vd) of the chemical, expressed relative to an equivalent volume of the bulk fluid, and provide a measure of accumulation of chemical in a particular compartment. In this model formulation, measurement of Vd is equivalent to the BAF (Mackay and Fraser 2000).

Mackay has promoted and makes an elegant argument for the use of a fugacity-based approach (as opposed to diffusion-based rate constants) to mechanistic BAF modeling (Mackay and Paterson 1982, Diamond et al. 1994, Hung and Mackay 1997, Mackay and Fraser 2000, Sharpe and Mackay 2000, Cousins and Mackay 2001). Fugacity represents the partial pressure of a chemical and can be used to estimate chemical equilibrium concentrations between two immiscible phases (e.g., water and animal). One advantage of the fugacity approach is that the calculations inherently account for the actual bioavailable fraction of chemical in the environment. This can also be accomplished with diffusion-based models, but it is not done very often. If bioavailable fractions are assessed, however, the two approaches are equivalent.

The most complicated mechanistic models estimate accumulation at a larger scale, incorporating multiple trophic levels. These so-called food web models have been used successfully to predict polychlorinated biphenyl (PCB) dynamics in the Great Lakes and have been used as environmental management tools to examine the effects of introduced species and other environmental alterations on the dynamics of chemical pollutants in aquatic ecosystems (e.g., Morrison et al. 1998; Watras et al. 1998; Morrison et al. 1999; Iannuzzi and Ludwig 2000; Sharpe and Mackay 2000). The earliest food web models deterministically assigned organisms to trophic levels, but recent refinements include the use of $\delta^{15}N$ to assess the trophic status of different populations of organisms (Cabana and Rasmussen 1994; Vander Zanden and Rasmussen 1996; Branstrator et al. 2000). Recently, probabilistic methods have been introduced into food web models, accounting for variability and uncertainty in each portion of the models (Iannuzzi and Ludwig 2000). The importance of considering variability and uncertainty in environmental measurements has been discussed by Håkanson (Chapter 8). Additional refinements have also included the coupling of resource assessment models (e.g., nutrient loading and dynamics) with food web contaminant models (Koelmans et al. 2001).

Mackay has recommended utilizing a range of bioaccumulation models in the ERA process, proceeding in a tier-wise fashion from most simple to most complex (Mackay and Fraser 2000). It has been suggested that this tiered assessment scheme would provide a cost-effective and scientifically sound approach to the assessment of bioaccumulation.

Effects Assessment Models

Statistical models are used to define dose-response or concentration-response relationships. Driven by the requirements of the ERA procedure, these models are used to determine doses or concentrations of chemicals that are considered to cause minimally adverse impacts on the environment or organisms under assessment. Lethality endpoints are typically assessed using binary regression models such as logistic or probit regression. In ERA, sublethal endpoints traditionally have been analyzed using a so-called hypothesis-testing approach. Using this approach, statistical endpoints are employed as environmental quality criteria and include no-observed-effect concentrations (NOECs) and lowest-observed-effect concentrations (LOECs). The NOEC (LOEC) is the highest (lowest) concentration with responses that does not (does) statistically differ from responses in the control group (Weber et al. 1989). To obtain these endpoints, ANOVA models are fit to toxicity test data and then an appropriate multiple-range test (e.g., Dunnett's test) is used to compare the responses of organisms from all test concentrations back to the control response. The use of the ANOVA approach in environmental assessments (Chapter

12) and the use of the NOEC/LOEC as endpoints (Crump 1984; Chapman et al. 1996; Crane and Newman 2000; Yanagawa and Kikuchi 2001) have received much criticism. Most arguments against the use of these endpoints include their design dependency and the issue of statistical power of the toxicity tests. In addition, the NOEC is often misinterpreted as an absolute no-effect level, where exposures less than this concentration are viewed as safe. Experiments are typically designed to control type I error rates. The consequences of committing a type II error when using the NOEC as an endpoint are severe (i.e., saying that something is not toxic when it, in fact, is toxic).

Several alternatives to the NOEC/LOEC approach have been debated. These include bioequivalency testing (Shukla et al. 2000) and a variety of regression-based methods (Bruce and Versteeg 1992; Bailer and Oris 1993, 1997; Wang and Smith 2000). All of these methods require that the experimenter specify a biologically significant level of impact as an endpoint (e.g., 20% inhibition of growth). Some ecological risk assessors consider this designation to be a flaw in the methods. However, this decision is also implicitly made in the design of toxicity tests when the NOEC/LOEC approach is used in that the standard experimental protocols are assumed to have sufficient power to detect a particular effect size. For example, the NOEC was observed to be close to a 25% inhibition response in bioassays used to evaluate reproductive toxicity in aquatic systems (Oris and Bailer 1993). Therefore, even though the NOEC/LOEC approach may have intuitive appeal to ecological risk assessors, it most likely possesses statistical properties that are undesirable.

Among the regression-based methods is a nonparametric, interpolation procedure promoted by the EPA, the inhibition concentration associated with p level of effect estimate (ICp method; Norberg-Kling 1993). In this method, sublethal toxicity tests are conducted using the same design as for the NOEC/LOEC approach. Data are analyzed by linear interpolation between the two concentrations that span the specified p level of effect. The assumptions of this method are (1) that the response of organisms to the toxicant is monotonically inhibited over the range of concentrations tested, and (2) that the response of organisms to the toxicant is linear between concentration levels. If the assumption of monotonicity is violated, successive treatment responses are averaged with the control response until a monotonic pattern is achieved. Thus, this method cannot accommodate the frequently observed pattern of an enhanced response of organisms to very low concentrations of toxicant. This response has been referred to as a hormetic response (Stebbing 1982). The assumption of linearity is not tested. The properties of the ICp approach have been studied (Bailer et al. 2000a), demonstrating that the ICp tends to produce confidence limits that are too narrow and cannot estimate toxicity values for hormetic hazards.

A more recent parametric alternative for potency estimation has been proposed by Bailer and Oris (1993, 1997). This approach uses a model of adverse effect as a function of concentration of toxicant specified relative to the control response. Parameters are fit using Generalized Linear Models (GLiMs), taking advantage of using a single-model formulation to fit models on different response scales (e.g., dichotomous, counts, continuous). Toxicity endpoints are determined by inversion of the model at a specified level of effect. This approach has been referred to as the relative inhibition associated with p level of effect (RIp method). The RIp method is essentially an inverse regression problem in which the model employed is linked directly to the measurement levels of the response of interest. An appealing aspect of this approach is the flexibility that is available in the linkage of the model to the response. Using an appropriate link function (e.g., probit, logit, log, identity) and error distribution (e.g., binomial, gamma, Poisson, normal), this approach can be used to model responses ranging from mortality to counts of young produced to growth responses. In addition, the models specified using the RIp approach can easily accommodate hormetic responses in toxicity tests (Bailer and Oris 1998). Recent comparisons of the different endpoint-estimation approaches (Bailer et al. 1999; Bailer et al. 2000a; Hughes et al. 2001) demonstrated the advantages of regression-based techniques and, in particular, the RIp method. The main disadvantage of the regression approach is that it is more computationally intense compared to other methods.

Risk Characterization Models

Once levels of exposure and levels of effect are estimated, the risk associated with a particular chemical or stressor can be characterized. Risk characterization is accomplished in numerous ways, ranging from qualitative, best professional judgments to fully implemented field studies (Murray and Claassen 1999). Quantitative models used for risk characterization can either be deterministic (methods that utilize a quotient approach) or probabilistic (methods that utilize probability distributions of species sensitivities and of all the factors involved in the determination of exposure and effect).

The quotient method (Barnthouse et al. 1982) has been applied to risk characterization in many assessments, but it is deemed appropriate only for screening level assessments (Suter 1993, ECOFRAM 1999). In this method, environmental exposure levels either are directly measured or are estimated using fate models to determine the Environmental Concentration (EC). For example, a recent scientific forum recommended the use of the GENEEC model (Table 19.1) for screening level exposure assessments of pesticide compounds (ECOFRAM 1999). Toxicity endpoints are determined either by conducting a set of toxicity tests or by estimating of toxicity based on structural similarity to chemicals of known toxicity. An application factor (i.e., an uncertainty factor) is applied to the toxicity value depending on the discriminatory level of the toxicity test (e.g., 1000× for estimated toxicity, 100× for acute toxicity test results, 10× for sublethal toxicity test results, 1× for field-exposure toxicity tests) to determine a Critical Concentration (CC). The CC is thus considered a maximum allowable concentration that can be ob-

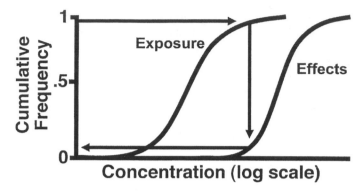

Figure 19.2. Comparison of probability distributions for predicted or measured environmental exposure levels and for predicted or measured concentration effect distribution. Arrows indicate that the top 5% of exposures cause an adverse effect in the most sensitive 5% of the population (if single species) or 5% of tested species (if species sensitivity distribution).

served without adverse environmental impact. Using the CC value for the most sensitive species, risk characterization is achieved by calculating the quotient of EC/CC. If this quotient is equal to or greater than 1.0 (i.e., EC ≥ CC), the risk assessor concludes that there is a high probability of observing an adverse environmental impact. This approach has appeal in its simplicity, but it cannot account for error associated with measurements of exposure or toxicity. While it incorporates uncertainty, albeit crudely, using application factors, it does not explicitly account for variability. In addition, it is easy to misinterpret the quotient as an estimate of risk, when in fact it simply indicates how close the environmental exposure level is compared to a toxicity value. While most risk assessment scientists recognize the shortcomings of the quotient approach, there is not strong agreement on methods, and there are no standardized methods used for higher level risk characterizations (Hamer 2000). Because of cost and time considerations, a large number of risk assessments continue to use the quotient method as a primary means of risk characterization (e.g., Hall and Anderson 1999; Suter et al. 1999; Wenning et al. 2000).

Probabilistic methods incorporate elements of variability and uncertainty in measurements of exposure and response. The simplest methods use single probability density functions to describe the pattern of either exposure or effect. Hill et al. (2000) describe probabilistic methods of exposure analysis, examining the probability density function of contaminant concentrations and compare this exposure distribution to a point estimate for an effects benchmark or critical concentration. The outcome of this risk characterization is statements such as "it is expected that 25% of exposures will exceed the effects benchmark." Suter (1993) discusses the use of probability density functions to extrapolate from acute to chronic effects

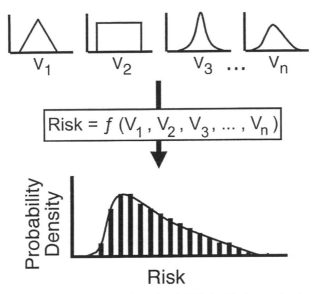

Figure 19.3. Diagrammatic representation of probabilistic risk characterization. Known error distributions of *n* variables ($V_1 \ldots V_n$) are used to calculate a risk distribution from Monte Carlo simulations.

or from surrogate to target species. In addition, concentration-response relationships can be used to set benchmark effect levels or critical concentrations along with their associated 95% confidence levels. Probability density functions can also be constructed for species sensitivity distributions across multiple species (Knoben et al. 1998; Newman et al. 1999). These can be compared to point estimates of exposure values to characterize risk with statements such as "this exposure level will adversely affect 15% of species tested." Another approach is to compare probability distributions for both exposure and effect (Figure 19.2). This can be done with single-species concentration-response relationships or with multiple-species sensitivity distributions (Solomon et al. 2000). The outcomes of these analyses are risk-characterization statements such as "it is expected that the top 5% of exposures will adversely affect the most sensitive 5% of species tested." A third approach has been categorized as uncertainty analysis (Suter 1993). In this most complex case, the probability densities of all variables used in the functions to calculate exposure and effects are determined (Figure 19.3). These distributions are utilized in a Monte Carlo simulation that provide a distribution of risk values based on the entire range of potential outcomes, incorporating uncertainties of all variables. The ECOFRAM forum (1999) has recommended this last approach as one of the highest levels of risk-characterization efforts. A major advantage of this approach is the ability to explore different scenarios that can be used in environmental management decisions. Obviously, as more probability distributions and simulation studies are included in the risk-characterization component, data requirements and cost of risk assessments increase dramatically. The forum concluded that more development is needed in the design of tests, monitoring of environmental variables, and understanding ecological processes in order for full implementation of probabilistic risk assessment. Finally, while these techniques may provide a more realistic representation of the systems under study, they may also highlight the uncertainty encountered in specifying these models and the variability in the organisms under study.

Conclusions

Quantitative models are used throughout the ERA process. As the science of ERA continues to develop, quantitative models will play an increasingly important role in the protection and maintenance of environmental quality. The use of advanced-level bioaccumulation models, the use of regression-based endpoint estimates in effects assessment, and the use of probabilistic models in risk characterization all provide more accurate evaluations of ecological risk. However, these methods are computation- and data-intense, and thus are more expensive and time-consuming compared to traditional, default methods. Probabilistic methods of risk characterization present perhaps the largest challenge to date in the ecological risk assessment process. Many techniques are available, but the currently available basic ecotoxicological testing approaches produce data that are not always useful (Newman 1998). Numerous forums have been convened to discuss formalizing these methods, and investigators are now designing risk assessments and collecting data with probabilistic methods in mind. Regardless of the methods chosen, qualitative judgments must still be made in reference to biologically significant effects and acceptable levels of risk in populations and communities.

References

Bailer, A.J., and D.A. Dankovic. 1997. An introduction to the use of physiologically based pharmacokinetic models in risk assessment. *Statistical Methods in Medical Research* 6: 341–358.

Bailer, A.J., R.T. Elmore, B.J. Shumate, and J.T. Oris. 1999. Simulation study of characteristics of statistical estimators of inhibition concentration. *Environmental Toxicology and Chemistry* 19: 3068–3073.

Bailer, A.J., M.R. Hughes, D.L. Denton, and J.T. Oris. 2000a. An empirical comparison of effective concentration estimators for evaluating aquatic toxicity test responses. *Environmental Toxicology and Chemistry* 19: 141–150.

Bailer, A.J., and J.T. Oris. 1993. Modeling reproductive toxicity in Ceriodaphnia tests. *Environmental Toxicology and Chemistry* 12: 787–791.

————. 1997. Estimating inhibition concentrations for different response scales using generalized linear models. *Environmental Toxicology and Chemistry* 16: 1554–1559.

————. 1998. Incorporating hormesis in the routine testing of hazards. *Human and Experimental Toxicology* 17: 247–250.

Bailer, A.J., S. Walker, and K.J. Venis. 2000b. Statistical methods for estimating and comparing bioconcentration factors. *Environmental Toxicology and Chemistry* 19: 2338–2340.

Barber, M.C., L.A. Suarez, and R.R. Lassiter. 1991. Modeling bioaccumulation of organic pollutants in fish with an application to PCB's in Lake Ontario salmon. *Canadian Journal of Fisheries and Aquatic Science* 48: 318–337.

Barnthouse, L.W., D.L. DeAngelis, R.H. Gardner, R.V. O'Neill, G.W. Suter, and D.S. Vaughan. 1982. Methodology for environmental risk analysis. ORNL/TM-8167. Oak Ridge, TN: Oak Ridge National Laboratory.

Branstrator, D.K., G. Cabana, A. Mazumder, and J.B. Rasmussen. 2000. Measuring life-history omnivory in the opossum shrimp, *Mysis relicta*, with stable nitrogen isotopes. *Limnology and Oceanography* 45: 463–467.

Bruce, R.D., and D.J. Versteeg. 1992. A statistical procedure for modeling continuous toxicity data. *Environmental Toxicology and Chemistry* 11: 1485–1494.

Burns, L.A. 1997. *Exposure Analysis Modeling System: User's Guide for EXAMS II*, version 2.97.5. Ecosystems Research Division, Environmental Research Laboratory, Office of Research and Development, U.S. Environmental Protection Agency: Athens, GA.

Cabana, G., and J.B. Rasmussen. 1994. Modeling food chain structure and contaminant bioaccumulation using stable nitrogen isotopes. *Nature (London)* 372: 255–257.

Chapman, P.M., R.S. Caldwell, and P.F. Chapman. 1996. A warning: NOECs are inappropriate for regulatory use. *Environmental Toxicology and Chemistry* 15: 77–79.

Cousins, I.T., and D. Mackay. 2001. Strategies for including vegetation compartments in multimedia models. *Chemosphere* 44: 643–654.

Crane, M.C., and M.C. Newman. 2000. What level of effect is a no observed effect? *Environmental Toxicology and Chemistry* 18: 516–519.

Crump, K.S. 1984. A new method for determining allowable daily intake. *Fundamental and Applied Toxicology* 4: 854–871.

Diamond, M.L., D. Mackay, D.J. Poulton, and F.A. Stride. 1994. Development of a mass balance model of the fate of 17 chemicals in the Bay of Quinte. *Journal of Great Lakes Research* 20: 643–666.

Ecological committee on FIFRA risk assessment methods (ECOFRAM). 1999. *Report of the Aquatic Workshop.* U.S. EPA, Office of Pesticide Programs: Washington, DC.

EPA. 1996. *Proposed guidelines for ecological risk assessment. EPA/630/R-95/002B.* Washington, DC: US Environmental Protection Agency, Risk Assessment Forum, Office of Research and Development.

Gibaldi, M., and D. Perrier. 1982. *Pharmacokinetics*. 2d edition. New York: Marcel Dekker.

Hall, L.W., and R.D. Anderson. 1999. A deterministic ecological risk assessment for copper in European saltwater environments. *Marine Pollution Bulletin* 38: 207–218.

Hamer, M. 2000. Ecological risk assessment for agricultural pesticides. *Journal of Environmental Monitoring* 2: 104N–109N.

Hill, R.A., P.M. Chapman, G.S. Mann, and G.S. Lawrence. 2000. Level of detail in ecological risk assessments. *Marine Pollution Bulletin* 40: 471–477.

Hughes, M.R., A.J. Bailer, and D.L. Denton. 2001. Toxicant- and response-specific comparisons of statistical methods for estimating effective concentrations. *Environmental toxicology and Chemistry* 20: 1374–1380.

Hung, H., and D. Mackay. 1997. A novel and simple model of the uptake of organic chemicals by vegetation from air and soil. *Chemosphere* 35: 959–977.

Iannuzzi, T.J., and D.F. Ludwig. 2000. The role of food web models in the environmental management of bioaccumulative chemicals. *Soil and Sediment Contamination* 9: 181–195.

Knoben, R.A.E., M.A. Beek, and A.M. Durand. 1998. Application of species sensitivity distributions as ecological risk assessment tool for water management. *Journal of Hazardous Materials* 61: 203–207.

Koelmans, A.A., A. Van Der Heijde, L.M. Knijff, and R.H. Aalderink. 2001. Integrated modeling of eutrophication and organic contaminant fate and effects in aquatic ecosystems. A review. *Water Research* 35: 3517–3536.

Mackay, D. 1982. Correlation of bioconcentration factors. *Environmental Science and Technology* 16: 274–278.

Mackay, D., and A. Fraser 2000. Bioaccumulation of persistent organic chemicals: Mechanisms and models. *Environmental Pollution* 110: 375–391.

Mackay, D., and S. Paterson. 1982. Fugacity revisited. The fugacity approach to environmental transport. *Environmental Science and Technology* 116: 654A–660A.

Meylan, W.M., P.H. Howard, R.S. Boethling, D. Aronson, H. Printup, and S. Gouchie. 1999. Improved method for estimating bioconcentration/bioaccum-ulation factor from octanol/water partition coefficient. *Environmental Toxicology and Chemistry* 18: 664–672.

Morrison, H.A., F.A.P.C. Gobas, R. Lazar, D.M. Whittle, and G.D. Haffner. 1998. Projected changes to the trophodynamics of PCBs in the Western Lake Erie ecosystem attributed to the presence of zebra mussels (*Dreissena polymorpha*). *Environmental Science and Technology* 32: 3862–3867.

Morrison, H.A., D.M. Whittle, C.D. Metcalfe, and A.J. Niimi. 1999. Application of a food web bioaccumulation model for the prediction of polychlorinated biphenyl, dioxin, and furan congener concentrations in Lake Ontario aquatic biota. *Canadian Journal of Fisheries and Aquatic Sciences* 56: 1389–1400.

Murray, K., and M. Claassen. 1999. An interpretation and evaluation of the US Environmental Protection Agency ecological risk assessment guidelines. *Water SA* 25: 513–518.

Neeley, W.B., D.R. Branson, and G.E. Blau. 1974. Partition coefficient to measure bioconcentration potential of organic chemicals in fish. *Environmental Science and Technology* 8: 1113–1115.

Newman, M.C. 1995. *Quantitative Methods in Aquatic Ecotoxicology.* Boca Raton, FL: CRC Press.

———. 1998. *Fundamentals of Ecotoxicology.* Chelsea, MI: Ann Arbor Press.

Newman, M.C., D.R. Ownby, L.C.A. Mezin, D.C. Powell, T.R.L. Christensen, S.B. Lerberg, and B.A. Anderson. 1999. Applying species-sensitivity distributions in ecological risk assessment: Assumptions of distribution type and sufficient numbers of species. *Environmental Toxicology and Chemistry* 19: 508–515.

Norberg-King, T. 1993. *A Linear Interpolation Method for Sublethal Toxicity: The Inhibition Concentration (ICp) Approach.* National Effluent Toxicity Assessment Center Technical Report 03-93. U.S. EPA Environmental Research Laboratory: Duluth, MN.

NRC (National Research Council, U.S.). 1983. *Risk Assessment in the Federal Government: Managing the Process.* National Research Council Committee on the Institutional Means for Assessment of Risks to Public Health. Washington, DC: National Academy Press.

Oris, J.T., and A.J. Bailer. 1993. Statistical analysis of the *Ceriodaphnia* toxicity test: Sample size determination for reproductive effects. *Environmental Toxicology and Chemistry* 12: 85–90.

Parker, R.D., H. Nelson, and R.D. Jones. 1995. GENEEC: A screening model for pesticide environmental exposure assessment. Pp. 485–490 in C. Heatwole, editor. *Water Quality Modeling: Proceedings of the International Symposium.* St. Joseph, MI: American Society of Agricultural Engineers.

Sharpe, S., and D. Mackay. 2000. A framework for evaluating bioaccumulation in food webs. *Environmental Science and Technology* 34: 2373–2379.

Shukla, R., Q. Wang, F. Fulk, C. Deng, and D. Denton. 2000. Bioequivalence approach for whole effluent toxicity testing. *Environmental Toxicology and Chemistry* 19: 169–174.

Solomon, K., J. Giesy, and P. Jones. 2000. Probabilistic risk assessment of agrochemicals in the environment. *Crop Protection* 19: 649–655.

Stebbing, A.R.D. 1982. Hormesis: The stimulation of growth by low levels of inhibitors. *The Science of the Total Environment* 22: 213–234.

Suter, G.W. 1993. *Ecological Risk Assessment.* Chelsea, MI: Lewis Publishers.

Suter, G.W., L.W. Barnthouse, R.A. Efroymson, and H. Jager. 1999. Ecological risk assessment in a large river reservoir. 2. Fish community. *Environmental Toxicology and Chemistry* 18: 589–598.

U.S. Environmental Protection Agency (EPA). 1992. *Framework for Ecological Risk Assessment.* EPA/630/R-92/001. U.S. EPA, Risk Assessment Forum, Office of Research and Development: Washington, DC.

Vander Zanden, M.J., and J.B. Rasmussen. 1996. A trophic position model of pelagic food webs: Impact on contaminant bioaccumulation in lake trout. *Ecological Monographs* 66: 451–477.

Veith, G.D., D.L. Defoe, and B.V. Bergstedt. 1979. Measuring and estimating the bioconcentration factor of chemicals in fish *Pimephales promelas*. *Journal of the Fisheries Research Board of Canada* 36: 1040–1048.

Wang, S.C.D., and E.P. Smith. 2000. Adjusting for mortality effects in chronic toxicity testing: Mixture model approach. *Environmental Toxicology and Chemistry* 19: 204–209.

Watras, C.J., R.C. Back, S. Halvorsen, R.J.M. Hudson, K.A. Morrison, and S.P. Wente. 1998. Bioaccumulation of mercury in pelagic freshwater food webs. *Science of the Total Environment* 219: 183–208.

Weber, C.I., W.H. Peltier, T.J. Norberg-King, W.B. Horning, F.A. Kessler, J.R. Menkedick, T.W. Neiheisel, P.A. Lewis, D. Klemm, Q.H. Pickering, E.L. Robinson, J.M. Lazorchak, L.J. Wymer, and R.W. Freyberg. 1989. *Short-Term Methods for Estimating the Chronic Toxicity of Effluents and Receiving Waters to Freshwater Organisms.* 2d edition. EPA/600/4-89/001A. Cincinnati, OH: U.S. Environmental Protection Agency.

Wenning, R., D. Dodge, B. Peck, K. Shearer, W. Luksemburg, A. Della Sala, and R. Scazzola. 2000. Screening-level ecological risk assessment of polychlorinated dibenzo-p-dioxins and dibenzofurans in sediments and aquatic biota from the Venice Lagoon, Italy. *Chemosphere* 40: 1179–1187.

Yanagawa, T., and M. Kikuchi. 2001. Statistical issues on the determination of the no-observed-adverse-effect level in toxicology. *Environmetrics* 12: 319–326.

20

Effects of Plant Invaders on Nutrient Cycling: Using Models to Explore the Link between Invasion and Development of Species Effects

Carla M. D'Antonio and Jeffrey D. Corbin

Summary

Over the past decade, ecosystem ecologists have become increasingly interested in the effects of individual plant species on ecosystem processes. Several recent studies have demonstrated that rates of nitrogen mineralization are significantly different under plots with different plant composition and have correlated differences with plant traits such as C:N ratio or other measures of litter quality. However, these studies have not evaluated how rapidly changes in plant composition translate into changes in ecosystem N availability or cycling and the conditions under which individual species can change ecosystem nutrient fluxes. Exotic plant invasions offer an opportunity to evaluate how species effects develop as species move into communities where they have no prior history. We believe that they also offer an opportunity to explore how different and how abundant a species has to become to have a measurable impact on a key ecosystem process.

In this chapter, we review the literature on the effects of non-native plant invaders on nitrogen cycling with the goal of trying to identify when and where invaders should be expected to increase or decrease N cycling. We believe that the speed and magnitude of invader effects should vary as a function of how different invaders are from residents in particular traits, and how resistant the resident community and soils are to change. We also believe that the application of a variety of modeling approaches holds great promise in understanding interactions between invader traits and environmental conditions and that this application should be explored more completely. Because the database for this review is not large, we focus on identifying gaps in our knowledge that could be filled by carefully directed empirical work and quantitative modeling.

Introduction

Over the two past decades, the accelerated pace of species losses from and additions to natural ecosystems has fostered interest in the role that individual species play in affecting the structure and functioning of ecosystems. While we know a great deal about the role that certain animal species can play in influencing patterns of primary production and energy flow (e.g. keystone species; see Paine 1966; Estes and Palmisano 1974; Power et al. 1996), there has been little effort until recently (Vitousek 1990) to identify effects of individual plant species on such ecosystem properties as energy flux and the cycling of nutrients and water. Identifying plant-species effects in natural communities is challenging because of the intermingling of roots and shoots and the relatively slow rates of change that characterize plant succession in many regions. However, with the rapid spread of nonnative species across many landscapes, it has become more straightforward to study pathways through which plant species affect ecosystem processes in natural systems as highlighted by Vitousek (1990).

Simultaneous with the rising interest among ecosystem ecologists in understanding species effects, many population and community ecologists have focused on understanding controls over invasion of native communities by nonindigenous species (Mack et al. 2000). Research has focused on (1) identifying species traits associated with successful invaders or invasion potential and (2) identifying characteristics of communities that make them invasion resistant or susceptible. Although it has been difficult to find reliable generalities with regard to the first topic, Rejmanek and Richardson (1996) and Reichard and Hamilton (1997), examining woody species introduced into either South Africa or North America, respectively, found that species traits such as seed mass, juvenile period, and the ability to reproduce vegetatively were the most important traits for predicting successful invasion. Several investigators have also found that disturbance or alterations to disturbance regimes correlate strongly with increased community susceptibility to invasion (Hobbs and Huenneke 1992; Mack and D'Antonio 1998; D'Antonio et al. 1999).

Few ecologists have attempted to combine the study of controls over invasion with the study of when and how invasive species affect ecosystem functioning. Yet, such a synthesis is necessary if we are truly to develop the ability to predict the impacts of potential invaders. The often rapid increase in abundance of nonindigenous species in landscapes that encompass a mosaic of habitats offers the opportunity to investigate both questions regarding which sorts of species traits lead to ecosystem-level changes and the connection between abundance and ecosystem process change. The impact of a particular species is going to be a function both of traits of the invader and traits of the ecosystem being entered. Invader traits include life-history characters such as number of seeds produced, age at first reproduction, longevity, dispersal, and a suite of physiological and morphological traits, including how fast an individual takes up nutrients, how it allocates those nutrients, the phenology of uptake and quality and timing of litter production, the amount of carbon exuded from its roots, and the effect of an individual on heat flux, microclimate, and physical habitat structure. The net effect of these traits can be thought of as the effect of an individual or group of individuals on the surrounding biotic and abiotic environment. Traits of the target environment include climate, soil fertility, soil organic matter (SOM) pool size, hydrology, disturbance regime, and biotic interactions such as herbivory or mutualisms. Properties of the ecosystem affect the expression of the life history, physiology, and morphology of the invader and provide the template upon which the invader's individual effects will be played out.

Nitrogen addition has been demonstrated to increase plant production across a wide range of ecosystems (Vitousek and Howarth 1991) and to alter plant composition (e.g., Bobbink 1991; Wedin and Tilman 1996), including the promotion of exotic species invasions (e.g., Huenneke et al. 1990; Vinton and Burke 1995; Maron and Connors 1996). In addition, numerous investigators have demonstrated that species composition is sensitive to relatively small changes in rates of N cycling (e.g., Aerts and Berendse 1988; McLendon and Redente 1991; Tilman and Wedin 1991; Paschke et al. 2000; and many others). Because of the fundamental role that nitrogen plays in structuring plant communities and controlling primary productivity, linkages between invader abundance, species traits, and ecosystem N cycling are especially promising areas in which to combine studies of invasibility with invading species' effects. Thus, we focus here on evaluating the linkage between species invasions and rates of ecosystem N cycling.

In this chapter, we review what is known about when and how invasive plant species affect nutrient cycling, emphasizing details of the species and the recipient ecosystems that are important for predicting plant species effects. Overall, there is not a large literature on invader effects on nitrogen cycling, and we believe that this area could benefit from further empirical work. We also believe that mechanistic ecosystem models have been substantially underutilized in the study of species invasions. Such models hold great promise for increasing scientists' understanding of the importance of various species traits and ecosystem properties on the net effects an invader will have in its new environment. In addition, quantitative ecosystem modeling could be used to explore relationships between invader abundance and impact and to test the sensitivity of ecosystem N cycling to variation in species composition while manipulating or holding constant edaphic and climatic variables that are difficult to manipulate experimentally. Finally, modeling could help provide insight into the relationship between N limitation, invasion, and the persistence of species effects and could also help to direct future empirical work in this area. We discuss examples in the context of biological invasions and the factors known to influence invasion success, but we believe that the points we argue could also be applied to "native" invading or colonizing species. A comparison of traits of exotic invaders with native colonizers concluded that the two groups of species were indistinguishable (Thompson et al. 1995), and Levine and D'Antonio (1999) argued that native and exotic plant species respond similarly to the same ecological processes. Thus, the patterns and the applications of a modeling approach that we describe are not necessarily limited to a restrictive definition of species invasions.

Framework: What Do We Need to Know to Predict Species Effects on N Cycling?

In order to predict whether an invading species is likely to affect ecosystem N cycling, we need to know (1) the ability of the species to become abundant, (2) species traits likely to be important to ecosystem N cycling, (3) the degree of difference between the invader and resident species for those traits, and (4) resistance of both the community and abiotic environment to changes imposed by the invader. In order to illustrate how these questions interact, we could consider, for example, that leaf litter quality is a key control over ecosystem N cycling (2) and that an invader has a different lignin:N ratio from the average native litter (3). In this case, we might ask how much litter of the invader must enter the system (1) to shift the system to a new state where rates of N availability are substantially altered (4). The likelihood that the invader will have an effect will depend on how different it is from residents, how rapidly it becomes abundant, how quickly resident species drop out of the community as the invader increases, and how greatly the invader alters the microenvironment (Figure 20.1). Other features of the environment such as soil texture, soil organic matter pool size, availability of essential nutrients, and climate will also determine whether the invader alters nutrient turnover.

Our choice of leaf litter quality in the above example is important, because it is a species trait especially likely to give rise to ecosystem effects on N cycling (e.g., Pastor et al. 1984; Gower and Son 1992; Scott and Binkley 1997). Most of the literature dealing with plant-species effects on nitrogen cycling indeed focuses on litter decomposition. Through use of litter bags, tethered leaves, or lab incubations we have come to understand a great deal about how tissue quality influences rates of nutrient release from decomposing litter (e.g., Melillo, et al. 1982; Pastor et al. 1984; Aber et al. 1990; Taylor et al. 1991; Gower and Son 1992; Perez-Harguindeguy et al. 2000). Litter features such as the percentages of lignin, the lignin:N ratio, and the C:N ratio explain a large amount of variation among species in decomposition rates both within sites (e.g., Melillo et al. 1982; Taylor et al. 1991) and across ecosystems (Berg et al. 2000; Perez-Harguindeguy et al. 2000). Leaf litter quality has also been shown to correlate with soil N mineralization across a range of forested ecosystems (Gower and Son 1992; Scott and Binkley 1997) and in some grass-dominated systems (Wedin and Tilman 1991; Vinton and Burke 1995). Many of these studies have involved planted mono- versus mixed-culture plots or stands (but see Vinton and Burke 1995). While they have been instructive in providing predictive indices of the direction of species effects, they have not addressed the extent to

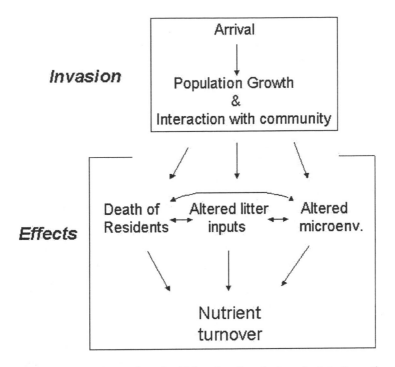

Figure 20.1. Some pathways through which an invading plant species (whether native or nonnative) could alter rates of ecosystem nitrogen turnover. Invaders might also alter nutrient cycling by altering disturbance regimes.

which species will have to become abundant in order for an impact to be measurable or how climate and soils interact with invasion to influence the rate of change in ecosystem processes as an invader becomes abundant on the landscape. In addition, almost no one has examined the extent to which species effects are driven directly by litter quality rather than by species interactions or species effects on soil moisture and temperature (Figure 20.1).

In the review that follows, we address what is known about invader impacts on ecosystem nitrogen cycling and divide examples into plants with symbiotic N fixation and those without. For each group we ask, What do we know about how likely they are to become abundant? How do their key traits relative to N cycling compare to those of natives, and are there any obvious features of the environment that appear to modulate impact? We also try to identify important unanswered questions and those that might best be explored with quantitative models.

Empirical Results

Impacts of Nitrogen-Fixers on Nitrogen Availability and Cycling

Nitrogen-fixing invaders are common on lists of exotic invaders and are considered to be overrepresented among the world's most problematic invaders of natural areas (Daehler 1998). They have been shown to decrease native species diversity directly and promote other invaders by leaving behind N-enriched soil (Maron and Connors 1996, Adler et al. 1998, others). Despite their suspected importance, there have been no systematic studies of factors influencing where and when they are likely to become abundant. We know that they have the potential to alter ecosystem N cycling because of their ability to fix atmospheric N. The expected mechanism by which N-fixing invaders increase total soil N pools and speed up N transformation rates is that N-rich litter from N fixers decomposes faster than litter from non-N-fixing natives, resulting in faster rates of N cycling such as gross and net N mineralization rates, nitrification rates, and the availability of soil inorganic N (Vitousek and Walker 1989; Hart et al. 1997; Maron and Jeffries 1999). Access to atmospheric N presumably also provides the potential for higher total litter N inputs. Nonetheless, we do not know the extent to which the impact of N fixers depends on their absolute and relative amounts of litter produced, soil factors, or climate in the region of invasion and/or their ability to displace resident species.

The ability of an exotic N-fixer to alter soil N availability has been most clearly demonstrated in young (less than 100 years old) volcanic soils in Hawaii (Vitousek et al. 1987; Vitousek and Walker 1989). In a system where N availability strongly limits vegetative productivity (Vitousek et al. 1993), where phosphorus levels are reasonably high, and where the native flora does not undertake symbioses with nitrogen-fixing bacteria, *Myrica faya* has proven to be an effective colonizer of primary succession sites. (First reported in Volcanoes National Park in 1961, the population increased to cover 12,200 ha by 1985 [Whiteaker and Gardner 1985, cited in Vitousek and Walker 1989]). Vitousek and colleagues (Vitousek et al. 1987; Vitousek and Walker 1989) found large differences in N fixation and leaf chemistry between indigenous non-N-fixing species and this N-fixing invader. At sites where *Myrica* was abundant, the exotic fixed 18.5 kg N ha^{-1} yr^{-1} through its actinorrhizal association, as compared to 0.17 kg N ha^{-1} yr^{-1} fixed by other known biological sources. The difference in access to N by *Myrica* relative to that of the dominant native tree *Metrosideros polymorpha* was reflected in the chemistry of each tree's senescent leaves: *Myrica* had significantly higher litter N concentrations than *Metrosideros* (1.33% vs. 0.56%), while the lignin:N ratio was lower (25.3 vs. 37.5). Decomposition of *Myrica* litter was found to be faster and to release more N than *Metrosideros* litter, as well. These differences between the tree species were clearly detectable in soil N availability and N cycling rates; total soil N, inorganic N pool sizes (ammonium and nitrate), and potential net N mineralization were all significantly higher under *Myrica* individuals than under *Metrosideros* individuals.

The impacts of the N-fixing exotic *Myrica faya* on soil N content and N cycling rates in Hawaii are consistent with the effects of other N-fixing species in their native ranges. Maron and Jeffries (1999) found that soils in live and dead patches of the N-fixing shrub *Lupinus arboreus* (bush lupine) in coastal California had higher total soil N content, inorganic N pool sizes, and potential net N mineralization rates than soils free of lupine individuals. Soil C:N ratios were lower in dead lupine patches than where there was no lupine. Similarly, soils where N-fixing *Alnus rubra* (red alder) trees were planted into mixed coniferous plantations had higher total soil N, microbial CO_2 evolution, net N mineralization and nitrification rates, and gross N mineralization and nitrification rates than soils without *Alnus* (Binkley et al. 1992; Hart et al. 1997).

The impact of N-fixing invaders on soil N and N cycling is likely to be dependent on the absolute abundance of the invader and/or its abundance relative to native species. For example, the data discussed above from Vitousek and Walker (1989) is from sites where *M. faya* was abundant. They also, however, examined the effect of *Myrica* in a site where it was not yet abundant and found that it had very little impact there, presumably because its inputs into the soil were still low. Hence, at least some minimal threshold of abundance must occur to detect a species effect, particularly if one is sampling soil processes randomly across a community. However, absolute abundance alone may not be the key to impact. *Myrica* is abundant in a nearby wet forest site where native species are more abundant than in either of the Vitousek and Walker (1989) sites. Its effects there have not been measured; but because its inputs into the community litter pool will be a smaller fraction of total annual litter inputs, and because SOM pools are probably already much larger there since they increase with precipitation in Hawaii (Schuur et al. 2001), we might predict that *Myrica* should have much less of an effect on N cycling in these wet sites than in the Vitousek and Walker (1989) sites.

Even when we control for climate, ecosystem impacts of invading N fixers are not always closely correlated with biomass. Haubensak and Parker (unpublished) examined the effect of the N-fixing shrub *Cytisus scoparius* (Scotch broom) on ecosystem N accumulation across a range of plots of varying invader biomass within the same climate zone in western Washington (Figure 20.2). Surprisingly they did not find a strong relationship between invader biomass and total soil N across a wide range of invader abundance. Soil C:N, however, decreased significantly with increasing broom biomass. Perhaps species effects on some soil N properties (e.g., total N) are not easily predicted but other properties (e.g., C:N ratio) may change more predictably with invader abundance.

In addition to abundance, both the leaf chemistry of the invading N fixer and soil properties should influence N turnover. Comparing two ecosystems of varying soil type in South Africa, each with a different species of exotic *Acacia*, Witkowski (1991) and Stock et al. (1995) found differences in the response of

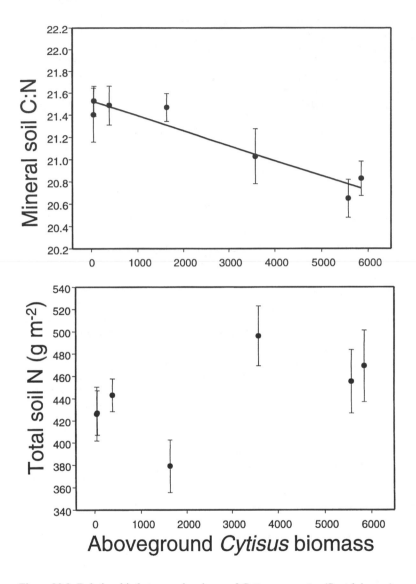

Figure 20.2. Relationship between abundance of *Cytisus scoparius* (Scotch broom) and soil C:N ration and total N pools in soils from glacial outwash prairies in eastern Washington, USA. Points represent means of 18 samples locations for each biomass value. Bars represent ± 1 standard error. (Data provided by Karen Haubensak from her dissertation research, University of California, Berkeley.)

soil N cycling to invasion. In both the calcareous strandveld ecosystems invaded by *A. cyclops* and the acidic fynbos ecosystems invaded by *A. saligna*, invasion resulted in an increase in total soil N content and decreases in litter C:N ratio. Decomposing leaf litter of both invading *Acacia* spp. in their respective ecosystems released significantly more N than indigenous leaf litter over a two-year period. The difference between N released by native species and the invader was greatest in the strandveld ecosystem. The rate of decomposition of *A. cyclops* litter was also significantly faster than that of *A. saligna*, and it released greater amounts of N (Witkowski 1991). Stock et al. (1995) found that only at the relatively fertile strandveld site were soil net N mineralization rates and resin-extractable N concentrations greater under *A. cyclops* individuals than in uninvaded areas. By contrast, at the relatively infertile fynbos site, there was no significant difference in net N mineralization or resin-extractable N in soils invaded by *A. saligna* compared to uninvaded soils.

In the South African example, the authors selected invaded fynbos and strandveld sites of similar age so as to avoid age and abundance as factors that might complicate the interpretation of species effects. But in their study it is not possible to separate the importance of litter chemistry (*A. cyclops* decomposes faster and releases more N than *A. saligna*) from soil characteristics in trying to understand the impact of invading species on ecosystem N transformations since species type varied with soil conditions. As with the Stock et al. (1995) study, Hart et al. (1997) found that *Alnus rubra* increased gross and net N mineralization rates to a greater extent in fertile soils than in infertile soils but the reasons why have not been explored. Rhoades et al. (2001) found the opposite when comparing *Alnus crispus* effects on mineral N pools in three Alaskan ecosystems: effects were greatest in the most infertile system. The degree of nutrient limitation of the vegetation, including the invader, and of microbes responsible for N transformations are both likely important in determining whether the introduction of an N fixer will alter N cycling, but there have been no studies examining this relationship. Quantitative models might help to explore the importance of ecosystem nutrient status and soil texture or chemistry in determining the rate of development of impact of an N fixer. The nutrient status of the recipient ecosystem should interact with the ability of the invader to become abundant as well as to have an impact.

Invasion of Non–N Fixers Into Natural Communities: Disturbance, Species Traits, and the Potential for Species Effects

Species traits associated with becoming a successful invader are frequently those associated with fast growth, particularly through the juvenile phase (e.g., Rejmanek 1996; Rejmanek and Richardson 1996). Fast growth, in turn, should be associated with relatively lower leaf construction costs, higher leaf tissue N, and lower N use efficiency: all of which are traits that should affect plant litter quality and hence ecosystem N cycling. In comparisons of traits of native and non-native woody species in Hawaii, Pattison et al. (1998) and Baruch and

Goldstein (1999) indeed found lower leaf construction costs and higher leaf tissue N in nonnative species compared to natives. Likewise Durand and Goldstein (2001) found that invasive nonnative tree ferns had higher N contents and shorter leaf life spans than leaves of native tree ferns. Baruch and Gomez (1996) also observed lower leaf construction costs in introduced compared to native grasses in Venezuela. These studies support the hypothesis that invasive nonnative species tend to have traits that should be associated with faster rates of N turnover.

Invasive nonindigenous species generally increase in abundance with disturbance (Hobbs and Huenneke 1992; D'Antonio et al. 1999), and N fertilization (Huenneke et al. 1990; Vinton and Burke 1995; Maron and Connors 1996) or rely upon a single disturbance event to provide a window of opportunity for establishment (Davis et al. 2000). Davis et al. (2000) present what they consider to be a general model to explain when invasion should occur in a given site, and their model incorporates the effects of fertilization and disturbance (Figure 20.3). They suggest that invasion should occur when the gross resource supply exceeds resource uptake by the resident community, whether by virtue of increases in gross resource supply through N fertilization or deposition (arrow B), of the disturbance reducing the uptake (arrow C), or through a combination of the two (arrow D). Invaders that get into a site because of an increase in the gross supply of N (arrow B, e.g., N deposition, fertilization, inputs by N fixers, etc.) should have low root allocation, high specific leaf area, and low N use efficiency relative to residents. These are traits that should favor faster cycling of N. Traits of invaders that rely upon a reduction in resource uptake, such as might occur with disturbance or stress, are possibly less predictable (Figure 20.3). Such invaders may also be fast-growing in order to monopolize resources before the resource pulse associated with disturbance has passed. However, in order to persist after the resource pulse has passed, invaders need to invest in organs that allow persistence, and this allocation may result in traits that do not promote fast cycling of nutrients. Whether disturbance should promote invaders that accelerate N cycling is thus dependent upon a host of issues besides just the identity of the invading species, including the amount of resource released by the disturbance, which in turn, depends on the nature of the disturbance and the fertility of the ecosystem in question. Disturbances in degraded sites where production has become limited due to successive losses of nutrient capital from the system or in sites where productivity is limited by some other "stress" factor, are not likely to promote invaders with fast growth/rapid N-cycling characteristics. In these systems, disturbance, by disrupting the priority effect of the resident species, may allow for establishment of an invader, but this invader may be more efficient at utilizing resources than residents, thereby slowing N cycling, or it may be similar to former residents and have no effect on N cycling. In an environment with chronic grazing pressure, successful invaders might be expected to contain high levels of antiherbivore compounds, some of which, like polyphenols, should slow nutrient cycling.

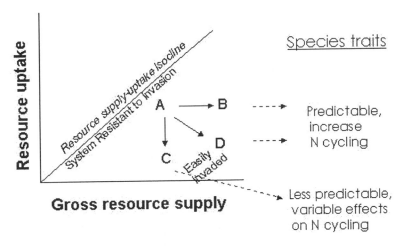

Figure 20.3. Relationship between gross resource supply and resource uptake as it influences biological invasion. A community's susceptibility to invasion increases as resource availability increases. Resource availability increases due to a pulse (or sustained increase) in resource supply (arrow A to B), a decline in resource uptake such as might occur when disturbance eliminates resident individuals (arrow A to C) or some combination of the two as when a disturbance stimulates nitrogen mineralization while at the same time it has reduced uptake. We propose that species traits that should affect nutrient cycling are most predictable under scenario A → B when invaders should exhibit traits that accelerate N cycling. This should also be true for scenario A → D although the degree of change from the resident community may not be as great depending on features of the disturbance and the soil. Traits that affect nutrient cycling are least predictable under scenario A→ C. (Modified from Davis et al. 2000.)

When Do Non-N-Fixing Invaders Affect N Cycling?

Our simple reasoning in the previous section implies that in relatively fertile systems, invaders that get into the sites because of disturbance or N addition should have traits that will bring about more rapid N cycling. Assuming that the invaders persist on the site for many years and that the direct positive effect of disturbance on resource release disappears within a short time, we should be able to measure increased N cycling in invaded versus comparable uninvaded sites several years after the disturbance event as a result of invader traits. Very few studies have made such measurements or have separated the direct effect of disturbance or N addition from the trait-based effect of the invader. Vinton and Burke (1995) demonstrated increased rates of nutrient cycling in the soil beneath introduced annual grasses compared to that under native perennial species in short-grass steppe sites in Colorado. The annual grasses invaded these sites after the addition of fertilizer more than 20 years before their study and have continued to dominate the sites. The litter-quality characteristics of the invaders are associated with faster N cycling, and their persistence suggests a positive feedback between tissue quality and N cycling that was triggered by the original nutrient addition. By contrast, Svejcar and Sheley (2001) found no consistent differences in the rates of N cycling between disturbed desert sites invaded by the annual grass *Bromus tectorum* and nearby undisturbed sites dominated by native perennial species. It is possible that in the low-productivity Great Basin sites of Svejcar and Sheley (2001) there was no significant pulse of resources released by the original disturbance, and that although the invader has now dominated the sites for more than 40 years, its dominance has little to do with N-cycling traits. Svejcar and Sheley (2001) conclude that species effects will not develop if invaders and natives are similar in total productivity and tissue chemistry even if they are different in overall life form.

In a fertile Hawaiian rainforest where disturbances to the forest canopy promote invasion, Scowcroft (1997) reported consistently elevated rates of decomposition and N release from litter of the invading vine *Passiflora mollisima*, compared to decomposition and N release from native litter (including that of a N-fixing tree). Interestingly, mixing *P. mollisima* litter with native litter, as might occur during an invasion, did not increase decomposition rates of native litter. This invader relies on openings in the forest canopy to establish itself and then apparently can climb to the canopy of established native trees and persist for decades. Scowcroft did not measure the rates of N mineralization in the soils under either the invaded or uninvaded stands, so it is not yet clear that the invader has led to increased soil N mineralization at a scale that could affect forest productivity or composition. However, the invaders' tissue quality and high rates of leaf turnover do suggest that over time, N cycling should increase following invasion. Similarly, disturbances in the alluvial forests of Argentina promote invasion by the tree *Ligustrum lucidum*. The litter of *Ligustrum* turns over 2.5 times as rapidly as the litter of native species, providing the potential to increase rates of soil N cycling in invaded sites (Descanio et al. 1994). Likewise, Kourtev et al. (1998) have found increased rates of leaf-litter decomposition and increased N availability in soils under stands of the invading shrub *Berberis thunbergii* in mesic deciduous forests of New Jersey. This understory invader is most abundant where the forest canopy is somewhat open (Kourtev et al. 1998), suggesting that it relies on elevated light availability to invade the forest understory. While their research generally supports the prediction that an invader that capitalizes on increased resource availability (in this case light) to get into a relatively fertile system should accelerate N cycling, the authors do not unambiguously separate site characteristics from species traits in determining the causes of altered N cycling in invaded sites. They suggest that higher decomposition and N mineralization are the result of earthworm activity associated with leaf litter of the invader rather than directly due to leaf-litter chemistry.

Several studies have reported on the effect of invaders on soil nutrients in sites where it was not clear that disturbance or a pulse in resource supply played a role in promoting invasion. There are no consistent trends in the outcomes of these studies. For example, Asner and Beatty (1996) compared N availability under stands of the African pasture grass *Melinis minutiflora* with availability of N under native shrubland on Molokai'i Island Hawaii, in a preserve where there was no evidence that disturbance had promoted *Melinis'* invasion. *Melinis* patches were consistently associated with elevated NH_4 availability, as measured by accumulation on ion-exchange resins. By contrast, Mack et al. (2001) found no effect of invasion by introduced pasture grasses (primarily *Schizachyrium condensatum* but also *Melinis minutiflora*) on annual net N mineralization in a young woodland on Hawaii Island that contained many of the same native woody plant species as the Asner and Beatty (1996) study. While the Mack et al. study was conducted on a much younger substrate, it is difficult to compare results because the metrics of N cycling used were very different. Asner and Beatty measured N availability only using ion-exchange resins across two separate one-month time periods, whereas Mack et al. used repeated intact core, in situ incubations throughout the year and estimated the annual net N mineralized. Since many of the same grass invaders occur across a range of soil ages and/or climate zones in the Hawaiian Islands (Smith 1985), it would be interesting to select one invader and one climate zone and see how invader impacts vary with soil age and nutrient status using standardized metrics of N cycling.

In another study of grass invasions, where again it was not clear that disturbance played a role in invasion, Williams (see Williams and Baruch 2000) reported no correlation between soil C:N ratios and cover of the African grass *Eragrostis lehmanniana* in desert grasslands of Arizona. However, he did find that cover of herbaceous dicots was negatively correlated with soil C:N ratios and that cover of herbaceous dicots declined with time since invasion by *E. lehmanniana*, a finding supported by others (Bock et al. 1986). He suggests that herbaceous dicot abundance may be the important link between nutrient cycling changes and *E. lehmanniana* invasion. This raises an important issue regarding pathways through which invading species may cause ecosystem process change: if the loss of residents takes many years, then changes in soil processes will likely be difficult to measure on timescales of dissertations or NSF funding.

Several studies report potential impacts of invaders on N cycling where the invader's abundance appears to be the result of recurrent or severe disturbance that has caused almost complete loss of native species. For example, Johnson and Wedin (1997) compared rates of N cycling and soil nutrient pools in grasslands dominated by an invasive Africa grass, *Hyparrhenia rufa*, with those of an intact, dry tropical forest in Costa Rica. They found that the grasslands had slower rates of N cycling and lower C stocks than soils in the woodland from which these grasslands were derived. However, it is not clear whether this reduction was due to the loss of forest species through clearing and recurrent fire or more directly to invasion by *H. rufa*. In contrast, Mack et al. (2001) found that grasslands formed by recurrent fire in Hawaiian woodlands had faster rates of N cycling and greater net N availability than the woodlands from which they were derived. They attributed this increase to the loss of native species with fire as well as to changes in microclimate and tissue quality associated with the conversion from woodland to grassland. In both cases, fire was responsible for the land-cover conversion, but the outcomes were the opposite of each other, perhaps because of differences in the nature of the forests from which these grasslands were derived. However, in both of these studies, elimination of the native species is likely to be an important reason why differences between invaded and uninvaded sites were large.

Saggar et al. (1999) examined the effect of an invading perennial herb, *Hieracium pilosella*, on soil nutrient status in a heavily grazed, tussock grassland in New Zealand. They note that the native vegetation had been "depleted" due to persistent sheep grazing and that this depletion is related to *Hieracium* invasion. *Hieracium* is known to have relatively high levels of polyphenols in its leaves (Makepeace et al. 1985). In these "depleted" soils they found that *Hieracium* increased organic C and N accumulation but decreased net N mineralized (in laboratory assays). They note, however, that in this sort of high alpine environment many species, including *Hieracium*, may take up organic N from polyphenols. Hence the pattern of increased total N accumulation under *Hieracium* may represent a positive feedback whereby *Hieracium* promotes its own success by locally concentrating N from the surrounding area into a form to which it has access (organic N). This example hints that to demonstrate species effects on N cycling, we must know what metric of N is important for plant uptake. In sites where plants have access to organic N, proposed relationships between plant traits and species effects on N cycling may have to be rethought. Species that have litter chemistry that slows decomposition may still have positive effects on N availability if plant-available N includes leached amino acids and polyphenols. Also, in sites where grazing is a major source of disturbance or a chronic stress, it may select for plant traits (e.g., high levels of polyphenols) that slow N mineralization from litter and SOM.

Synthesis and the Role of Modeling

Despite the rising interest in the importance and effects of individual species on ecosystem processes, our review confirmed that our knowledge about controls over when invading species should have effects on N cycling is fragmentary. The best evidence of direct plant-species effects on N transformations comes from studies of N fixers invading low fertility sites with little resident cover (e.g., Vitousek et al. 1987; Vitousek and Walker 1989; Rhoades et al. 2001). In addition, in the case of N addition or disturbance in relatively fertile systems, evidence suggests that invaders exhibit traits that should and do accelerate the cycling of N. Yet, even in the case of N fixers, we understand very little about the range of conditions over which invaders will have effects particularly when they invade systems where residents are abundant.

Because N fixers are common invaders of native ecosystems, are often considered to be ecologically damaging for resident species (Daehler 1998, D'Antonio and Haubensak 1998), and are frequently the target of removal efforts, it is important to try to understand where they will have substantial effects on soil processes. Plant species that have associations with N-fixing bacteria vary enormously in the amount of leaf litter they produce and the amount of nodulation that they undergo. For example, in coastal California, French (*Genista monspessulana*) and Scotch broom (*Cytisus scoparius*) are both common shrubby invaders of coastal prairie ecosystems. Individuals of the former produce a large amount of leaf litter, while individuals of the latter have photosynthetic stems and produce very little leaf litter. As a result, their effects on soil processes might differ because of variation in amount of organic matter input despite being relatively similar in overall life form. Also, a single species can vary in its allocation to stems, leaves, and nodules across ecosystems. For example, Haubensak (pers. comm.) has noted that Scotch broom plants produce more leaves and leaf litter in Washington state than in California. Hence, to predict the effect of an N fixer, we need to know how tissue allocation and decomposition will change with environment and how resident species and soil texture and chemistry will respond to the invader inputs.

In the case of non-N-fixing invaders entering relatively undisturbed settings or settings where disturbance does not greatly increase the availability of resources, it is difficult to make predictions about when invaders will have impacts. If invaders increase the productivity of the system by accessing otherwise underutilized resources (e.g. deep-rooted trees invading a grassland), they might ultimately increase the rates of N cycling. However, if they have higher N use efficiency than residents, they might slow the cycling of nutrients by tying up nutrients in recalcitrant litter. Likewise, invaders entering degraded or stressful environments will likely have traits that either slow or have no effect on N cycling. Additional experimental work is needed across a broader range of environments to begin to develop more meaningful generalities in this area.

Applications of Models in Understanding the Impacts of Invaders on Ecosystems

We believe that mechanistic models of soil biogeochemistry and plant production can be important tools for understanding invaders' impacts, particularly if these are modified to include species dynamics. We have argued that we must better understand the interaction between plant traits, such as litter lignin:N ratios, and ecosystem characteristics, such as nutrient availability, disturbance history, soil structure, and climatic conditions, if we are to predict invaders' influences on nutrient cycling. Models that allow researchers to examine the interactions between these factors should be explored as a way of increasing our understanding of species impacts and as a way of guiding future experimental tests of the impacts of invaders on ecosystems.

Some of the ecosystem properties that might influence the development of species effects are difficult to manipulate experimentally. Hence, models provide a means of gaining insight into the relationship between a particular variable and a species effect. For example, soil texture can control many aspects of soil organic matter accumulation and turnover (e.g., Burke et al. 1989), and it is an important controller of ecosystem biogeochemistry in many quantitative models (Parton et al. 1987, 1988; Fan et al. 1998). Yet, it is very difficult to manipulate in the field. Most empirical studies that consider its role in ecosystem nutrient cycling compare processes in soils of different textures or across gradients where more than one factor may be changing simultaneously (e.g., Burke et al. 1989; Neill et al. 1997; Koutika et al. 1999; Hook and Burke 2000). Ecosystem models might allow testing the sensitivity of soil N processes to changes in species traits across a gradient of soil textures that would not be possible to find or create in a controlled experimental field setting.

One popular model, CENTURY (Parton et al. 1987, 1988), can be used as a tool in this area. CENTURY uses multiple compartments for SOM and nutrients such as N and P. Carbon in above- and belowground plant residues is partitioned into either structural or metabolic decomposition pools as a function of the lignin:N ratio in the residue. With increases in the ratio, more of the residue is partitioned to the structural pools, which have much slower decay rates than the metabolic pools. Decomposing litter pools flow into one of the three SOM pools (slow, passive, and active) via a microbial pool. N in plant residue follows the flow of C, and the amount of N entering each of the soil pools varies as linear functions of the size of the mineral N pool. In other words, as soil mineral N increases, the C:N ratios of the slow, passive, and active N pools decrease, reflecting the greater N content of each of the pools relative to C.

The decomposition of each of the SOM and N pools is a function of climatic conditions such as soil temperature and precipitation, soil texture as it influences leaching losses, organic matter, or soil water content, and disturbances such as fire or atmospheric N deposition that alter soil N and SOM content. Plant productivity is modeled separately for grassland/crop ecosystems, forest ecosystems, and savannah ecosystems and is a function of the genetic potential of the vegetation, temperature, soil moisture, ecosystem nutrient status, and shading.

CENTURY can be parameterized to predict the impact on ecosystem N cycling of an invading species that alters litter quantity or chemistry. Most simply, the model could be run for an ecosystem with a background quantity of litter input and litter lignin:N ratio ("uninvaded") and compared to a series of similar ecosystems in which an invader has increased the quantity of litter, altered the litter chemistry, or both ("invaded"). This might provide insight into how different an invader has to be from the resident species to cause a measurable impact. Of more theoretical value, however, is the use of models to understand how the impact of an invader on soil N cycling varies with such ecosystem characteristics as climate, soil nutrient availability, or certain disturbances. For example, CENTURY can be used to investigate whether the difference in the litter lignin:N ratios between the invaded and uninvaded ecosystems must be larger where soil nutrient availability is low or where SOM pools are large before ecosystem N cycling is affected. Such modeling exercises can lead to predictions that guide experimental studies of the mechanisms of invader impacts.

Most current models of ecosystem C and N dynamics do not include the community dynamics that may play a role in the degree to which invading species alter ecosystem N cycling. Relative abundances of community members, which could be used to examine the importance of invader abundance, are not explicitly included in biogeochemical models such as CENTURY, which limits our ability to examine how ecosystem biogeochemistry changes during transient phases of community change. Furthermore, the theoretical model of community invasion presented by Davis et al. (2000) views the likelihood of invader success as a function of competition among community members for unused resources such as N. Future advances in model design should attempt to link the cycling of C and N in ecosystem compartments with changes in the composition of plant communities in an effort to more realistically predict the likelihood of both invader success and impact. For example, the relative abundances of resident and invading species, each possessing different litter inputs, litter lignin:N ratios, and nutrient uptake rates, could be explicitly included in CENTURY as input variables at each time step. In this way, the community composition could be allowed to vary more realistically leading to a more accurate understanding of the interaction between invader abundance, nutrient availability, and ecosystem N cycling. Bachelet et al.'s (2001) recent global vegetation model, which links vegetation change (in her case associated with climate change) to CENTURY biogeochemistry, may be a useful approach for incorporating the transient dynamics associated with plant invasions.

While models such as CENTURY might be relatively easily adapted to the study of species composition change and its subsequent effects on N cycling, they also face the limitation that the mode of action of the invader is controlled solely by litter chemistry (specifically lignin:N). Because species may affect N cycling through other pathways (e.g., Figure 20.1), predictions or insights from CENTURY may be misleading in some settings. Models that incorporate effects on other factors that potentially control N cycling, such as microclimate, microbial and faunal composition, disturbance regime, and labile C exudation from live plant roots, might provide additional insights into conditions under which invaders have impact.

Acknowledgments. The authors would like to acknowledge the contributions of the D'Antonio lab group to discussions that helped this manuscript take shape and in particular discussion with Eric Berlow. C. D'Antonio would also like to thank C. Canham and participants in Cary Conference IX for valuable feedback, I. Burke for inspiration, and an anonymous reviewer for useful comments. Karen Haubensak graciously provided the data for Figure 20.2 and NSF DEB 9910008 provided salary support to J. Corbin.

References

Aber, J.D., J.M. Melillo, and C.A. McClaugherty. 1990. Predicting long-term patterns of mass loss, nitrogen dynamics and soil organic matter formation from initial fine litter chemistry in temperate forest ecosystems. *Canadian Journal of Botany* 68: 2201–2208.

Adler, P., C.M. D'Antonio, and J.T. Tunison. 1998. Understory succession following a dieback of *Myrica faya* in Hawaii Volcanoes National Park. *Pacific Science* 52: 69–78.

Aerts, R., and F. Berendse. 1988. The effect of increased nutrient availability on vegetation dynamics in wet heathlands. *Vegetatio* 76: 63–69.

Asner, G.P., and S.W. Beatty. 1996. Effects of an African grass invasion on Hawaiian shrubland nitrogen biogeochemistry. *Plant and Soil* 186: 205–211.

Bachelet, D., J.M. Lenihan, C. Daly, R. Neilson, D. Ojima, and W.J. Parton. 2001. MC1: A dynamic vegetation model for estimating the distribution of vegetation and associated carbon, nutrients and water. Technical documentation. Version 1.0. USDA Forest Service PNW Res. Station Gen. Tech. Rep. PNW-GTR 508: 1–95.

Baruch, Z., and G. Goldstein. 1999. Leaf construction cost, nutrient concentration, and net CO_2 assimilation of native and invasive species in Hawaii. *Oecologia* 121: 183–192.

Baruch, Z., and J.A. Gomez. 1996. Dynamics of energy and nutrient concentration and construction cost in a native and two alien C4 grasses from two neotropical savannas. *Plant and Soil* 181: 175–184.

Berg, B., M-B. Johansson, and V. Meentemeyer. 2000. Litter decomposition in a transect of Norway spruce forests: Substrate quality and climate control. *Canadian Journal of Forest Research* 30: 1136–1147.

Binkley, D., P. Sollins, R. Bell, D. Sachs, and D. Myrold. 1992. Biogeochemistry of adjacent conifer and alder-conifer stands. *Ecology* 73: 2022–2033.

Bobbink, R. 1991. Effects of nutrient enrichment in Dutch chalk grasslands. *Journal of Applied Ecology* 28: 28–41.

Bock, C.E., J.H. Bock, K.L. Jepson, and J.C. Ortega. 1986. Ecological effects of planting African lovegrasses in Arizona. *National Geographic Research* 2: 456–463.

Burke, I.C., C.M. Yonker, W.J. Parton, D.V. Cole, K. Flach, and D.S. Schimel. 1989. Texture, climate and cultivation effects on soil organic matter content in U.S. grassland soils. *Soil Science Society of America Journal* 53: 800–805.

Daehler, C. 1998. The taxonomic distribution of invasive angiosperm plants: Ecological insights and comparison to agricultural weeds. *Biological Conservation* 84: 167–180.

D'Antonio, C.M., T.L. Dudley, and M. Mack. 1999. Disturbance and biological invasions: direct effects and feedbacks. Pp. 413–452 in L.R. Walker, editor. *Ecosystems of Disturbed Ground*. New York: Elsevier.

D'Antonio, C.M., and K.A. Haubensak. 1998. Community and ecosystem impacts of introduced species in California. *Fremontia* 26: 13–18.

Davis, M.A., J.P. Grime, and K. Thompson. 2000. Fluctuating resources in plant communities: A general theory of invasibility. *Journal of Ecology* 88: 528–534.

Descanio, L.M., M.D. Barrera, and J.L. Frangi. 1994. Biomass structure and dry matter dynamics of subtropical alluvial and exotic *Ligustrum* forests at the Rio de la Plata, Argentina. *Vegetatio* 115: 61–76.

Durand, L.Z., and G. Goldstein. 2001. Photosynthesis, photoinhibition, and nitrogen use efficiency in native and invasive tree ferns in Hawaii. *Oecologia* 126: 345–354.

Estes, J.A., and J.F. Palmisano. 1974. Sea otters: Their role in structuring nearshore communities. *Science* 185: 1058–1060.

Fan, W., J.C. Randolph, and J.L. Ehmann. 1998. Regional estimation of nitrogen mineralization in forest ecosystems using Geographic Information Systems. *Ecological Applications* 8: 734–747.

Gower, S.T., and Y. Son. 1992. Differences in soil and leaf litterfall nitrogen dynamics for five forest plantations. *Soil Science Society of America Journal* 56: 1959–1966.

Hart, S.C., D. Binkley, and D.A. Perry. 1997. Influence of red alder on soil nitrogen transformations in two conifer forests of contrasting productivity. *Soil Biology and Biochemistry* 29: 1111–1123.

Hobbs, R.J., and L. Huenneke. 1992. Disturbance, diversity and invasion: Implications for conservation. *Conservation Biology* 6: 324–337.

Hook, P.B., and I.C. Burke. 2000. Biogeochemistry in a shortgrass landscape: Control by topography, soil texture, and microclimate. *Ecology* 81: 2686–2703.

Huenneke, L.F., S.P. Hamburg, R. Koide, H.A. Mooney, and P.M. Vitousek. 1990. Effects of soil resources on plant invasion and community structure in California serpentine grassland. *Ecology* 71: 478–491.

Johnson, N.C., and D.A. Wedin. 1997. Soil carbon, nutrients, and mycorrhizae during conversion of dry tropical forest to grassland. *Ecological Applications* 7: 171–182.

Kourtev, P.S., J.G. Ehrenfeld, and W.Z. Huang. 1998. Effects of exotic plant species on soil properties in hardwood forests of New Jersey. *Water, Air and Soil Pollution* 105: 493–501.

Koutika, L.-S., T. Chone, F. Andreux, G. Burtin, and C.C. Cerri. 1999. Factors influencing carbon decomposition of topsoils from the Brazilian Amazon Basin. *Biology and Fertility of Soils* 28: 436–438.

Levine, J.M., and C.M. D'Antonio. 1999. Elton revisited: A review of evidence linking diversity and invasibility. *Oikos* 87: 15–26.

Mack, M.C., and C.M. D'Antonio. 1998. Impacts of biological invasions on disturbance regimes. *Trends in Ecology and Evolution* 13: 195–198.

Mack, M.C., C.M. D'Antonio, and R. Ley. 2001. Alteration of ecosystem nitrogen dynamics by exotic plants: A case study of C4 grasses in Hawaii. *Ecological Applications* 11: 1323–1335.

Mack, R.N., D. Simberloff, W.M. Lonsdale, H. Evans, and M. Clout. 2000. Biotic invasions: Causes, Epidemiology, global consequences and control. *Ecological Applications* 10: 689–710.

Makepeace, W., A.T. Dobson, and D. Scott. 1985. Interference phenomena due to mouse-ear and king devil hawkweed. *New Zealand Journal of Botany* 23: 79–90.

Maron, J.L., and P.G. Connors. 1996. A native nitrogen-fixing shrub facilitates weed invasion. *Oecologia* 105: 302–312.

Maron, J.L., and R.L. Jefferies. 1999. Bush lupine mortality, altered resource availability, and alternative vegetation states. *Ecology* 80: 443–454.

McLendon, T., and E.F. Redente. 1991. Nitrogen and phosphorus effects on secondary succession dynamics on a semi-arid sagebrush site. *Ecology* 72: 2016–2024.

Melillo, J.M., J.D. Aber, and J.F. Muratore. 1982. Nitrogen and lignin control of hardwood leaf litter decomposition dynamics. *Ecology* 63: 621–626.

Neill, C., M.C. Piccolo, C.C. Cerri, P.A. Steudler, J.M. Melillo, and M. Brito. 1997. Net nitrogen mineralization and net nitrification rates in soils following deforestation for pasture across the southwestern Brazilian Amazon Basin landscape. *Oecologia* 110: 243–252.

Paine, R.T. 1966. Food web complexity and species diversity. *American Naturalist* 100: 65–75.

Parton, W.J., S. Schimel, C.V. Pole, and D.S. Ojima. 1987. Analysis of factors controlling soil organic matter levels in the Great Plains grasslands. *Soil Science Society of America Journal* 51: 1173–1179.

Parton, W.J., W.B. Stewart, and C.V. Cole. 1988. Dynamics of C, N, P, and S in grassland soils: A model. *Biogeochemistry* 5: 109–131.

Paschke, M.W., T. McLendon, and E.F. Redente. 2000. Nitrogen availability and old-field succession in a shortgrass steppe. *Ecosystems* 3: 144–158.

Pastor, J., J.D. Aber, C.A. McClaugherty, and J.M. Melillo. 1984. Aboveground production and N and P cycling along a nitrogen mineralization gradient on Blackhawk Island, Wisconsin. *Ecology* 65: 256–268.

Pattison, R.R., G. Goldstein, and A. Ares. 1998. Growth, biomass allocation and photosynthesis of invasive and native Hawaiian rainforest species. *Oecologia* 117: 449–459.

Perez-Harguindeguy, N., S. Diaz, J.H.C. Cornelissen, F. Vendramini, M. Cabido, and A. Castellanos. 2000. Chemistry and toughness predict leaf litter decomposition rates over a wide spectrum of functional types and taxa in central Argentina. *Plant and Soil* 218: 21–30.

Power, M.E., D. Tilman, J.A. Estes, B. A. Menge, W. Bond, S.L. Mills, G. Daily, J.C. Castillo, J. Lubchenco, and R.T. Paine. 1996. Challenges in the quest of Keystones: Identifying keystone species is difficult, but essential to understanding how loss of species will affect ecosystems. *BioScience* 46: 609–620.

Reichard, S.H., and C.W. Hamilton. 1997. Predicting invasions of woody plants introduced into North America. *Conservation Biology* 11: 193–203.

Rejmanek, M. 1996. A theory of seed plant invasiveness: The first sketch. *Biological Conservation* 78: 171–181.

Rejmanek, M., and D.M. Richardson. 1996. What attributes make some plants invasive? *Ecology* 77: 1655–1661.

Rhoades, C., H. Oskarsson, D. Binkley, and B. Stottlemyer. 2001. Alder (*Alnus crispa*) effects on soils in ecosystems of the Agashashok River valley, northwest Alaska. *Ecoscience* 8: 89–95.

Saggar, S., P.D. McIntosh, C.B. Hedley, and H. Knicker. 1999. Changes in soil microbial biomass, metabolic quotient and organic matter turnover under *Hieracium* (*H. pilosella* L). *Biology and Fertility of Soils* 30: 232–238.

Schuur, E.A.G., O.A. Chadwick, and P.A. Matson. 2001. Carbon cycling and soil carbon storage in mesic to wet Hawaiian montane forests. *Ecology* 82: 3182–3192.

Scott, N.A., and D. Binkley. 1997. Foliage litter quality and annual net N mineralization: comparison across North American forest sites. *Oecologia* 111: 151–159.

Scowcroft, P.G. 1997. Mass and nutrient dynamics of decaying litter from *Passiflora mollisima* and selected native species in a Hawaiian montane rain forest. *Journal of Tropical Ecology* 13: 407–426.

Smith, C.W. 1985. Impact of alien plants on Hawaii's native biota. Pp. 180–250 in C.P. Stone and J.M. Scott, editors. *Hawaii's Terrestrial Ecosystems: Preservation and Management.* Honolulu: Cooperative National Park Resources Study Unit, University of Hawaii.

Stock, W.D., K.T. Wienand, and A.C. Baker. 1995. Impacts of invading N_2 fixing *Acacia* species on patterns of nutrient cycling in two Cape ecosystems: Evidence from soil incubation studies and ^{15}N natural abundance values. *Oecologia* 101: 375–382.

Svejcar, T., and R. Sheley. 2001. Nitrogen dynamics in perennial- and annual-dominated arid rangeland. *Journal of Arid Environments* 47(1): 33–46.

Taylor, B.R., C.E. Prescott, W.J.F. Parsons, and D. Parkinson. 1991. Substrate control of litter decomposition in four Rocky Mountain coniferous forests. *Canadian Journal of Botany* 60: 2242-2250.

Thompson, K., J.G. Hodgson, and T.C.G. Rich. 1995. Native and alien invasive plants: More of the same? *Ecography* 18: 390–402.

Tilman, D., and D. Wedin. 1991. Plant traits and resource reduction for five grasses growing on a nitrogen gradient. *Ecology* 72: 685–700.

Vinton, M.A., and I.C. Burke. 1995. Interactions between individual plant species and soil nutrient status in shortgrass steppe. *Ecology* 76: 1116–1133.

Vitousek, P.M. 1990. Biological invasions and ecosystems processes: Towards an integration of population biology and ecosystem studies. *Oikos* 57: 7–13.

Vitousek, P.M., and R.W. Howarth. 1991. Nitrogen limitation of land and sea: How can it occur? *Biogeochemistry* 13: 87–115.

Vitousek, P.M., and L.R. Walker. 1989. Biological invasion by *Myrica faya* in Hawaii: Plant demography, nitrogen fixation and ecosystem effects. *Ecological Monographs* 59: 247–265.

Vitousek, P.M., L.R. Walker, L.D. Whiteaker, and P.A. Matson. 1993. Nutrient limitations to plant growth during primary succession in Hawaii Volcanoes National Park. *Biogeochemistry* 23: 197–215.

Vitousek, P. M., L. Walker, L. Whiteaker, D. Mueller-Dombois, and P. Matson. 1987. Biological invasion by *Myrica faya* alters ecosystem development in Hawaii. *Science* 238: 802–804.

Wedin, D.A., and D. Tilman. 1996. Influence of nitrogen loading and species composition on the carbon balance of grasslands. *Science* 274: 1720–1723.

Williams, D.G., and Z. Baruch. 2000. African grass invasion in the Americas: ecosystem consequences and the role of ecophysiology. *Biological Invasions* 2: 123–140.

Witkowski, E.T.F. 1991. Effects of alien Acacias on nutrient cycling in coastal lowlands of the Cape Fynbos. *Journal of Applied Ecology* 28: 1–15.

21

Predicting the Ecosystem Effects of Climate Change

Harald K.M. Bugmann

Summary

The assessment of the ecological impacts of anthropogenic changes in atmospheric CO_2 and climate has become increasingly important for both the scientific community and policy makers. "Predictions" of the future fate and state of ecosystems are demanded, raising a number of new challenges for ecological modeling. In this chapter, definitions for the terms "prediction," "forecast," and "projection" are proposed, so as to reduce confusion within the scientific community and to avoid misunderstandings when scientists communicate with land managers and policy makers: "prediction" implies certainty, "forecast" implies likelihood, and "projection" implies possibility. Based on these definitions, our capability to predict the future climate is reviewed, because this is a precondition for predicting the *ecosystem effects* of such changes. It is concluded that we cannot predict the future climate due to a number of uncertainties.

Next, the state of research with respect to assessing the impacts of climate change on agriculture, on forest growth and biogeochemistry at the stand scale, and on forest succession is reviewed. In agriculture, predictability of the biophysical impacts of climate change is fairly high, but future agriculture is determined largely by processes other than biophysics, including management practices (e.g., irrigation, fertilizer use), new breeding technologies, and economic globalization. These processes strongly reduce predictability of future agriculture.

In forests, predictability is hindered by the temporal mismatch between the scales of observational/experimental data versus that of the target variables. Acclimation and adaptation phenomena, carbon allocation, and belowground dynamics are poorly understood yet remain central to projections of future forest growth and carbon fluxes. In forest succession, there are significant problems with respect to species-specific parameter estimation and with the temporal and spatial scales involved in evaluating the behavior of these models. In addition, forest succession, by definition, includes tree population dynamics (recruitment and mortality), which are highly stochastic, difficult to study, and not well understood. In conclusion, we cannot predict the ecosystem effects of climate change. Scientific research at the present time is dealing with *projections* of future ecosystem development. Assuming that we can fully specify the uncertainties involved in our analyses, we may be able to provide *ecological forecasts*, but this is a longer-term research goal.

Introduction

Prediction is difficult, especially about the future.—Yogi Berra

The publication of *Limits to Growth* (Meadows et al. 1972) strongly increased the awareness of environmental issues among the broad public, and the oil crisis in the early 1970s highlighted the unsustainable basis of modern economies. Air and water pollution problems became quite important in the same decade. Widespread concern developed over potential global climatic cooling, induced in the short to medium term (years to decades) by industrial and agricultural aerosols (Kellogg 1987). However, the work by Manabe and Wetherald (1975) focused attention on the longer-term (decades to centuries) climatic effects of rising CO_2. This study triggered scientific research activities to "predict" the impacts of increasing CO_2 and climate change on a variety of systems, including ecosystems, and at a variety of levels of organization, ranging from plant physiology to entire ecosystems.

Any such impact study must be based on one or several models of some sort, ranging from simple analogues taken from the present world, across static regression-based models, to complex, multidimensional mathematical models that require numerical simulation methods to arrive at a solution. The first ecological impact studies of anthropogenic climate change were performed in the early 1980s (e.g., Solomon et al. 1984). These studies certainly were timely, but they had to build upon relatively coarse methods and data sets regarding both climate drivers and the impact models themselves. For example, Kettunen et al. (1988) used a simple, empirical, static crop yield model coupled to an equally simple economic model to assess the impacts of a $2\times CO_2$ equilibrium climate scenario on crop yields and gross margin of barley and oats across Finland (Figure 21.1). In the case of forest resources, Solomon (1986) performed a sensitivity study with a succession model across twenty sites in eastern North America, simulating species composition and aboveground biomass under current conditions and into a $4\times CO_2$ world (Figure 21.1).

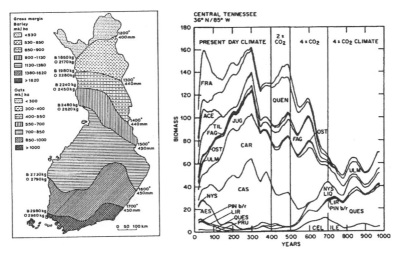

Figure 21.1. Examples of early impact assessments of climate change on ecosystems. *Left:* Harvest yield per hectare and gross margin for barley and oats in regions of Finland that are characterized by similar effective temperature sum of a 2× CO_2 scenario (from Kettunen et al 1988, reproduced with permission of Kluwer Academic Publishers). *Right:* Forest succession as simulated under the current climate (simulation years 0–400), under a change to a 2× CO_2 climate (years 400–500) and a 4× CO_2 climate (years 500–700), followed by a period of a constant 4× CO_2 climate (years 700–1000) for a climate division in central Tennessee, USA (from Solomon 1986, reproduced with permission of Springer Verlag).

Were these early studies "predictions" of future ecosystem dynamics? Probably most readers will intuitively agree that they were not, because the methods employed were not developed well enough for the results to qualify as "predictions." Given the magnitude of the environmental changes that we anticipate to take place in the twenty-first century and beyond, it is clear that future ecosystem dynamics are of concern not only for scientists, but also for land managers and policy makers, as evidenced by the Intergovernmental Panel on Climate Change (IPCC) and the Kyoto protocol, which aims to reduce the global anthropogenic carbon emissions. Quite understandably, land managers and policy makers today are asking for precise statements on the future state and fate of ecosystems under the influence of climatic change. This demand to *predict* the *ecosystem effects* of *climate change* poses new challenges for scientists, particularly regarding the robustness of their assessments and the state of the art of modeling in the physical and ecological sciences.

At the same time, the demand to predict the ecosystem effects of climate change raises questions regarding the precise meaning of these terms. What exactly is a "prediction," what are "ecosystem effects," and what is "climate change"? This may sound pedantic, but an array of different terms are being used widely in the literature to denote the trajectories that are obtained from models, and these terms are loosely defined at best. They include "prediction," "forecast," "projection," and various others. Inconsistency in definitions can lead to misunderstandings, particularly when scientists are communicating with land managers and policy makers. Therefore, the first aim of this chapter is to elaborate exact definitions of various terms as a basis for the further discussion.

If predicting the ecosystem effects of climate change is to be useful for land managers or policy makers, it requires that we can predict the future climate in the first place and use it as an input in impact analyses. Hence, the second aim of the paper is to briefly discuss whether we currently can predict the future climate.

Based on these considerations and starting from the assumption that we can predict the future climate indeed, I will review the problem of predicting the ecosystem effects of climate change, focusing on three types of ecological dynamics that operate on different spatial and temporal scales: agriculture, forest growth, and forest succession.

Terminology

In the scientific literature, the term "prediction" is used with widely different meanings. Some scientists use it to generally denote the outcome of a simulation (e.g., "model prediction"—the results obtained from a certain simulation experiment), by others as a synonym of "forecast" (e.g., similar to "weather forecast"), or as a weaker term compared to "forecast" ("the temporal predictions of models are more appropriately seen as scenarios, but there is growing interest for forecasts in a literal sense." (Canham et al. unpublished ms).

When linguistic confusion arises, the time has come to consult a dictionary. Based on the definitions in the *Merriam-Webster Dictionary* (2001), there is a clear distinction between the various terms, as compiled in Table 21.1.

It is possible to argue that a mathematical model *predicts* something, because the values of the model's output variables are a logical and *certain* consequence of the model assumptions, parameter values, and input variables. (Strictly speaking, this argument does not apply for stochastic models; however, it can be extended easily to include this model type.) Within the model world, this may be an appropriate use of language. However, there is uncertainty in the model assumptions, in the values of model parameters, and in the driving variables. As I will argue below, this is particularly true for models of long-term ecological processes (model structure and parameters) and for the case of global climate change (input variables). In addition, this uncertainty is often difficult to quantify at the level of the model parameters and input variables and typically impossible to quantify at the level of model assumptions.

Table 21.1. Terms, their definitions based on the compilation from the Merriam-Webster Dictionary (available on the Internet under http://www.m-w.com/netdict.htm), and some examples.

Term	Definition	Example
Prediction	Commonly denotes inference from facts or accepted laws of nature, implying *certainty*.	Astronomers predict an eclipse.
Forecast	Adds the implication of anticipating eventualities and differs from prediction in being concerned with *probabilities*.	Forecasting snow
Projection	An estimate of future *possibilities*.	
Scenario	An account or synopsis of a *possible* course of action or events.	

Thus, I would argue that no model could ever truly predict ecological phenomena because of the various kinds of uncertainties that are involved. At best, we may talk about forecasting if we put forth the optimistic assumption that all the uncertainties can be quantified in a probabilistic manner. Yet, it was only a few years ago that planning toward ecological forecasting was initiated as part of, among others, the Sustainable Biosphere Initiative of the Ecological Society of America, where ecological forecasting was defined as "the process of predicting the state of ecosystems, ecosystem services, and natural capital, with fully specified uncertainties, contingent on explicit scenarios for climate, land use, human population, technologies, and economic activity" (Clark et al. 2001).

Actually, calling model results "predictions" may even be dangerous when communicating with stakeholders such as land managers and policy makers. In such circumstances, the term "predict" may be interpreted in the context of the ecological dynamics that are being studied rather than in the context of a specific model. Hence, under such circumstances, "model predictions" quickly turn into "predictions of the future behavior of a specific ecosystem," and this would most likely be quite inappropriate.

Next, the meaning of "climatic change" needs to be narrowed down in the context of this contribution, because climate has changed and will continue to change on a multitude of time scales. Below, I will focus on anthropogenic climatic changes of the past few decades and into the future, with a time horizon of one hundred years or even longer, depending on the sector that will be considered.

Having defined the terms "prediction" and "climate change," we can now turn to analyzing the meaning of the "ecosystem effects" of climate change. There is a wide range of ecological phenomena that are affected by climate, ranging from physiological processes such as photosynthesis and respiration rates of plants, across tissue growth, rates of herbivory and mortality, to changes in the fluxes of trace gases between the land surface and the atmosphere. For land managers and policy makers, it is mainly the level of entire ecosystems, landscapes, and regions that is of interest. Therefore, I define the "ecosystem effects" of climate change as the following:

- Impacts on ecosystem function; examples of such functions are (1) the fluxes of carbon, nitrogen and water between the land surface and the atmosphere, (2) the water retention capacity, and (3) crop yields.

- Impacts on ecosystem structure; examples are (1) the number of trophic levels or the foodweb structure; (2) the number and sizes of the pools of carbon, nitrogen, and other elements; (3) biodiversity, from the genetic to the landscape level, and (4) species composition.

Predicting Future Climate

Atmospheric and climate sciences are often conceived of as branches of physics. Therefore, one might expect the predictability of the future climate to be high, because of the dominance of physical processes that should be easier to capture than biological and ecological processes. Upon closer inspection, it becomes evident that this is not really the case, due to three major reasons: (1) atmospheric modeling is facing very considerable scaling issues in both time and space, as evidenced by the problem of cloud parameterization in general circulation models, which has striking similarities with the problem of scaling process information in other fields, including ecology (Bugmann et al. 2000; Harvey 2000); (2) the climate system is more than the atmosphere, and it is thus not determined exclusively by pure physics, but rather by the complex interactions between physical components and the biota (Foley et al. 2000); and (3) the future state of the atmosphere is dependent on the trajectories of the anthropogenic emissions of trace gases and aerosols.

It is noteworthy that a fairly long cascade of processes is determining emission trajectories. The future human population size in the different regions of the world, which is difficult to estimate in itself (Lutz et al. 2001), is an inadequate predictor of future emissions without taking into account economic, societal, and technological developments that may occur across the coming decades. These "human dimensions of global change," however, are not predictable into the future if we adopt the definitions of Table 21.1.

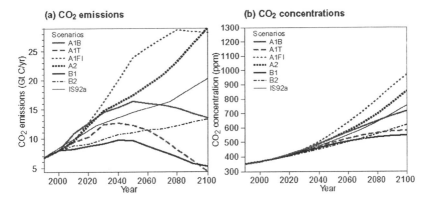

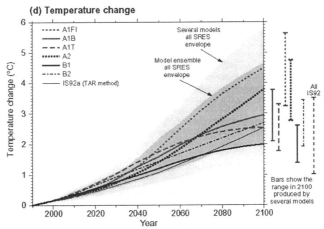

Figure 21.2. Anthropogenic emissions and atmospheric concentrations of CO_2 and the resulting changes of the global average surface temperature according to the various scenarios of the IPCC's Third Assessment Report (IPCC 2001, reproduced with permission of the IPCC).

From trace gas emissions, radiative forcing can be calculated relatively well, which leads to a signal of global climate change. The current generation of climate models is doing fairly well with respect to recovering the pattern of global average temperatures from 1860 A.D. to today (IPCC 2001), but the derivation of regional to local changes of the climate continues to constitute a major problem for the climate modeling community.

It is for all these reasons that the third assessment report of the IPCC (IPCC 2001; Figure 21.2) provided a fairly large number of global CO_2 emission sce-

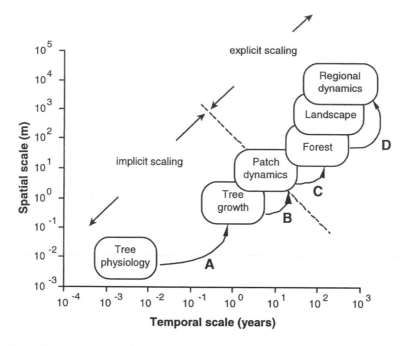

Figure 21.3. Temporal and spatial scales that are relevant for ecological dynamics, using forests as an example. Redrawn from Bugmann et al. (2000.)

narios that cover a wide range (5–30 Gt C/yr for the year 2100). These emissions give rise to atmospheric CO_2 concentrations between 500 and nearly 1,000 ppm in the year 2100, which in turn lead to temperature changes in the range 1.4–5.8 K. It is important to note that these trajectories of emissions, concentrations, and climatic parameters (Figure 21.3) have no probabilities attached. Yet, these scenarios are not equally probable, nor is a simple averaging procedure across all scenarios bound to yield the most likely future trajectory.

Thus, even if ecosystem models were available that are free of uncertainties at the level of model assumptions and parameter values, uncertainties in the climatic input data would remain that cannot currently be quantified, and we thus could neither predict nor forecast the ecosystem effects of climate change (Table 21.1).

In conclusion, the trajectories of future climate that are produced by state-of-the-art scientific methods are neither predictions nor forecasts, but projections (see Table 21.1). It would be a tremendous step forward if a potential fourth assessment report of the IPCC could begin attaching probabilities to the various scenarios, that is, if the transition to "climatic forecasting" could be achieved.

Predicting the Ecosystem Effects of Climate Change

Even in the absence of a capability to predict the future climate, it is useful to assess the state of the art with respect to our predictive capabilities of ecological dynamics. These dynamics span large ranges in space and time (Bugmann et al. 2000). For example, the enzymatic reactions of photosynthesis operate within fractions of a second; tree foliage development after bud break in spring takes a few days to weeks, while the growth of an individual plant may last years to centuries, and the dynamics of soil organic matter span millennia. The germination of a seed takes place at the scale of a square centimeter, a sunfleck moving over the forest floor covers a few square meters; a dominant tree in the forest canopy occupies 0.01–0.1 ha, and the quasi-equilibrium of a landscape may be reached on the scale of several hectares to square kilometers only, depending on the nature of the dominant disturbance agent (Shugart and Urban 1989).

Levin (1992) hypothesized that the central problem in ecology is that of pattern and scale, and that the various temporal, spatial, and organizational scales should be interfaced in order to understand the dynamics of ecosystems. Consequently, I will organize the following discussion around the typical scales in space and time on which selected ecological processes are operating (Figure 21.3). Based on a review of the literature and qualitative considerations regarding the match or mismatch between the scale of experimental and observational data and the scale of the model simulations, I will identify challenges for improving our predictive capabilities in three selected categories of systems: (1) agriculture; (2) forest growth and biogeochemistry at the stand scale; and (3) forest succession.

Agriculture

Predicting the ecosystem effects of climate change in agriculture (e.g., Parry et al. 1988a,b) revolves around target variables such as net primary productivity (NPP), net ecosystem productivity (NEP), and crop yield. The major agricultural crops are annual plants, which makes these systems amenable to experimental manipulations of environmental parameters, using growth chambers, open-top chambers or the free-air carbon dioxide enrichment (FACE) methodology (Gregory et al. 1999). In addition, past extreme years, such as the droughts in the United States in the 1930s and 1950s, may serve as (partial) analogues for future conditions (Rosenberg 1993), assuming that past and present cultivars will continue to be used in the future.

The behavior of agricultural models under changed environmental conditions can therefore be evaluated against observational or experimental data at a variety of scales: the scales of observational data and model results agree well (Figure 21.4). For example, the simulated course of photosynthesis at the leaf level can be compared to high-resolution measurements from FACE experiments (Figure 21.4 left), corroborating the adequacy of process formulations in

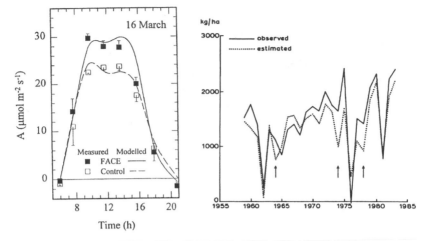

Figure 21.4. Testing of agricultural crop models. *Left:* Model evaluation at the physiological level: The diurnal course of the mean (±1 SD) measured leaf CO_2 uptake (symbols) and the predicted leaf CO_2 uptake (lines) are shown for 16 March 1993 in the Arizona Free-Air Carbon Dioxide Enrichment (FACE) Experiment. Predicted rates were calculated from a mechanistic model of photosynthesis. 'FACE' = elevated CO_2 partial pressure of ≈55 Pa; 'control' = current ambient CO_2 partial pressure of ≈37 Pa (from Garcia et al. 1998, reproduced with permission of Blackwell Academic Publishers). *Right:* Model evaluation at the level of the target variables (here, crop yield): Spring wheat yield model for North Karelia (from Kettunen et al. 1988, reprinted with permission of Kluwer Academic Publishers).

these models. At the same time, model behavior can be evaluated at the level of the target variables themselves, such as crop yield (Figure 21.4 right).

It appears that earlier methodological limitations on experimental approaches for studying the impacts of climate change on agricultural systems have largely been overcome (Mooney et al. 1999), so that the temporal scaling from experimental approaches to future real-world situations may not be a major issue any more. However, some concerns remain regarding spatial scaling, that is scaling from the limited number of situations that have been and can be studied experimentally, to responses across entire regions that include a wide variety of soil types, hydrological regimes, etc.

The above considerations were based on the implicit assumption that a number of factors and processes will remain constant or at least similar to their present conditions. It is in these areas that there are still grand challenges for assessing the ecosystem effects of climate change in agriculture. These challenges include the following

- Little is known about the magnitude and frequency of extreme climatic situations in the future, and historical agricultural data cover only a limited set of extreme events. Hence, exploring the effects of future climate variability and particularly the effects of extreme events constitutes a challenge in agricultural sciences.

- Herbivores and pathogens and their controls play a major role in current agriculture, and we cannot make the simplistic assumption that the range of herbivores and pathogens found today will remain the same under future atmospheric CO_2 concentrations, rates of nitrogen and sulphur deposition, and climatic parameters (Porter et al. 1991; Gregory et al. 1999). In addition, the relationship between existing herbivores/pathogens and agricultural plants may change in the future, with nonpests becoming pests, and vice versa. Our knowledge on such changes in trophic interactions is quite limited, but these factors cannot be ignored.

- Breeding technology has a very long history, and we cannot assume that current cultivars will be used in future agriculture. Recent methodological developments, including genetic engineering, make it likely that fifty years from now humankind will be using cultivars whose properties cannot be anticipated at the current time.

- Last but not least, biophysical constraints on agriculture are becoming less and less important. This is not only and not primarily because of the increased use of fertilization and irrigation in developed countries, but rather because agriculture has been subject to globalization over the past decades. Therefore, agricultural land use is increasingly determined by global economic processes. Biophysical constraints will continue to determine where certain crops *cannot* be grown, but whether they will actually be grown when biophysical factors are not severely limiting depends on technological, economic, and societal constraints.

In conclusion, due to the good match between observational and experimental data on the one hand and the modeled processes and the scale of the target variables on the other hand, predictability of the ecosystem effects of climate change is high in agriculture if we assume that current agricultural practices, cultivars, and trophic interactions will persist in the future. When we adopt a more realistic view by considering future changes in practices, cultivars, and trophic interactions, it becomes clear that even in the case of the relatively "simple" and highly managed agricultural systems, we cannot predict the ecosystem effects of climate change.

Forest Growth and Biogeochemistry at the Stand Scale

Major variables that are of interest in forest growth and biogeochemistry assessments (e.g., Dixon et al. 1990) include the flows of carbon (NPP, NEP) and water, aboveground and belowground carbon storage, standing wood volume, and harvestable timber. Such assessments typically cover up to one rotation period, which is up to a century. Due to this significantly longer time scale as compared to agriculture, historical analogues to a future climate are not readily available. Sometimes the space-for-time analogy can be used, that is the growth of stands comprising the same species can be compared across climatically different regions, but this has to be done with caution. As in the case of agriculture, these are likely to be partial analogues at best.

Ecosystem experimentation on forest growth and biogeochemistry is logistically more demanding than in the case of agriculture. Therefore, most experiments have typically been restricted to young trees, which raises the problem of extrapolating from juvenile growth responses to adult growth responses (Körner 1996). In most modeling studies, it is assumed that the long-term responses of trees to climatic change and CO_2 increase are identical to the responses obtained in short-term experiments. Long-term experimental or observational studies would be required to address the potential for acclimation and adaptation phenomena, but only few such studies exist.

For example, Hättenschwiler et al. (1997) examined growth responses of oaks in the only documented natural CO_2 fertilization experiment in CO_2 springs in Italy. They discovered that the positive growth response of the trees growing under naturally enhanced CO_2 levels was restricted to the first thirty years of tree life; older trees inside and outside the springs did not differ in their growth rates. More recently, Oren et al. (2001) found that in one of the FACE experiments with trees, the positive growth response to CO_2 enhancement was restricted to the sites that were fertilized with nutrients, and the growth response was much stronger during the first three years of the experiment than in the following four years. These examples clearly suggest that the extrapolation from short-term experiments, often conducted with irrigated and fertilized trees, to long-term ecosystem responses may be inappropriate. Therefore, a significant scaling problem continues to exist in assessments of forest growth responses and biogeochemistry to changing climatic conditions and atmospheric CO_2 concentrations because of the mismatch between the temporal scale of the ecosystem response and the scale at which most experiments and observations are done (Figure 21.3).

Still, a considerable amount of data are available to parameterize, for example, the growth function of the simpler, empirical forest models (e.g., Figure 21.5 left), and the same goes for the more mechanistic representations of ecological processes in complex forest growth models. At the level of the target variables, it is possible to examine simulated carbon and water fluxes against, for example, measured data from flux networks or FACE experiments at annual or shorter time scales. However, evaluating the behavior of forest growth models at the level of standing volume at the end of a rotation period under climate change, or with respect to time series of timber yield under changing climatic conditions, is hardly possible. High-quality, long-term forest growth data are available to evaluate such models under current climate (e.g., Figure 21.5 right), but this does not corroborate their applicability in a future climate.

Thus, I conclude that the uncertainties at the biophysical level are larger in the case of assessments of forest growth and biogeochemistry responses to changing climatic conditions and changing atmospheric CO_2 concentrations as compared to agricultural impact assessments. There are several challenges in forest growth and biogeochemistry modeling that relate to biophysical processes, including the following:

- Carbon allocation is a poorly understood process, and it is currently not possible to identify one modeling approach that would capture the essential features of allocation. Consequently, a wide range of approaches is being used for modeling carbon allocation in forest growth and biogeochemistry models. Carbon allocation is strongly dependent on the sink strength of the various plant organs, which is tied to the growth rate of the plant organs. Plant growth, however, is much less understood than net photosynthesis, that is, the process that provides the carbon source. There is increasing evidence that under many circumstances ecosystem carbon sequestration is not primarily determined by carbon acquisition (net photosynthesis), but rather by the carbon sink strength of plant organs (growth rates; Körner 1998).

- Belowground processes in forests are not well understood (Hendricks et al. 1993), but they are quite important for predicting the long-term effects of climate change on forest growth and biogeochemistry. The various types of roots (in the simplest case, fine roots vs. coarse roots) have quite different turnover times, but allocation of carbon to roots is treated coarsely in most forest growth models, simply because of a lack of good data and understanding regarding the associated processes (Cairns et al. 1997). Furthermore, vertical root distribution at larger spatial scales had for the most part been neglected until a few years ago (Jackson et al. 1996), and the functional role of roots with respect to competition for water and nutrients deserves further attention as well (Schulze et al. 1996).

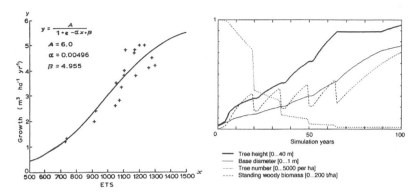

Figure 21.5. Parameter estimation and testing of forest growth models. *Left:* Calibration of the growth process formulation in a simple forest yield model based on the concept of the "effective temperature sum" using empirical data (from Kauppi and Posch 1988, reproduced with permission of Kluwer Academic Publishers). *Right:* Simulation results at the stand level from the TREEDYN model that can be compared against a host of growth-and-yield data from forest trail experiments (redrawn from Bossel 1994).

- Historically, forest growth modeling has been concerned with single-species, even-aged stands. In recent years, there has been a shift in the forest management practices of many countries towards multispecies and uneven-aged stands, for a variety of reasons including long-term sustainability and biodiversity issues. The simulation of multispecies, uneven-aged stands requires a conceptual shift in forest growth modeling, which presumably will narrow the gap between forest growth and forest succession models (see below).

Thus, even in the absence of challenges beyond the biophysical level, assessments of the impacts of climate change on forest growth and biogeochemistry face serious challenges, which seriously limit predictability in this area of science.

Forest Succession

In assessments of the impacts of climatic change on forest succession (Finegan 1984; Shugart 1984; Bugmann et al. 2001a), the target variables of interest include above- and/or belowground biomass, stand structure (e.g., diameter and height distributions of trees), and tree species composition. Given the generally long life span of trees and the fact that forest succession by definition deals with several tree generations, it is evident that time periods of a few to many centuries need to be considered. In addition, due to the role of large-scale disturbances in forest succession, patterns and processes need to be taken into account over large areas (Figure 21.3; Turner et al. 1993).

A considerable problem with current forest succession models (Bugmann 2001) is that there is limited information available for estimating the parameter values of a large number of tree species. The autecological properties of a few, commercially relevant species are well known, but few data are available concerning the vast majority of the tree species. Commercially important species tend to be those that are dominant in the current landscape, but we cannot assume that they will remain dominant in a future climate that has no recent analog. Therefore, species about which there is limited information today might be among the future dominants, and this poses a significant problem for the robustness of our assessments. Collecting the required data is a major undertaking, even for a limited set of species that occur at a given site (e.g., Pacala et al. 1993, 1996). However, such efforts show that it is possible to obtain these data, and it is evident that the forest succession modeling community has not focused enough on this problem. As with forest growth models, a related issue is the estimation of the autecological response of tree species to changes in environmental parameters, and basically the same scaling issues apply as in the case of forest growth models.

Evaluating the behavior of succession models under conditions of climatic change poses another set of problems. Past analog situations for future forest succession would have to come from pollen records, macrofossils, and similar proxies, which typically have a relatively coarse resolution in time (several decades to centuries) and contain only limited information on the exact species proportions, for example, with respect to basal area or biomass, which would be of primary interest. The paleoecological evaluation of succession models is important, however (Solomon et al. 1980; Solomon and Webb 1985; Lotter and Kienast 1992). For the more recent past, we lack century-long data on natural or near-natural forest dynamics, and where such data are available, it is difficult to provide an accurate, century-long reconstruction of the driving variables (namely, climate and disturbance regimes; Bugmann and Pfister 2000; Reynolds et al. 2001). Therefore, while any single model–data comparison study tends to remain inconclusive, it constitutes a crucial element for a rigorous model evaluation that must draw on a multitude of data, all of which have specific strengths, but also specific limitations (Bugmann 2001).

These two problem areas call for a stronger collaboration between "modelers" and "empiricists" who focus on successional processes. At the present time, forest succession modeling is facing very considerable scaling problems in time and space (Figure 21.3; Bugmann et al. 2000), which implies strong uncertainties with respect to the robustness of impact assessments. Under such circumstances, the role and value of model comparison exercises should not be underestimated (Rastetter 1996; Chapter 12). Obviously, pure model–model comparisons are of limited value, but when they are combined with model–data comparisons, they can be quite instructive (Chapter 12; Bugmann et al. 2001a).

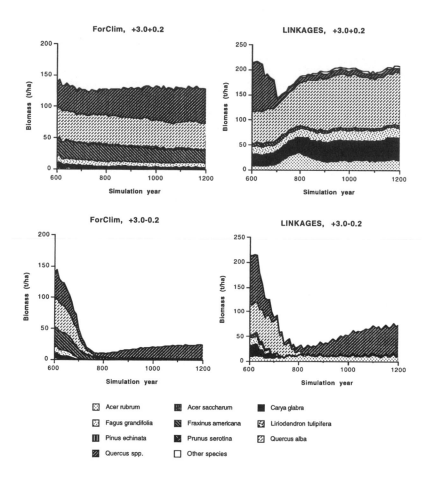

Figure 21.6. Results from a comparison of forest succession models conducted for the Walker Branch Watershed site. The simulation starts with a forest that is in equilibrium with the current climate. Between the simulation years 600 and 700, a linear change to a new climate takes place (top row: +3 °C and +20% precipitation; bottom row: +3 °C and –20% precipitation). The two models employed are FORCLIM (Bugmann 1996; Bugmann and Solomon 2000) and LINKAGES v2.0 (Wullschleger et al. Oak Ridge National Laboratory, unpublished). After the simulation year 700, the climate is kept constant at the new values (from Bugmann et al. 2001b).

For example, a comparison between two succession models and measured data on forest structure was performed for Walker Branch Watershed in Tennessee (USA) as a component of a larger study (Figure 21.6). The comparison between simulated rates of succession with data on forest stands that were 40–70 years old revealed specific deficiencies of the two models; the long-term successional behavior was compared to qualitative descriptions of the potential natural vegetation, and both models achieved a reasonable match, although there were considerable differences between the models with respect to the abundance of individual species (Figure 21.6). The models were then subjected to a series of hypothetical changes of the climate, starting from an equilibrium between the current climate and vegetation composition. These scenario runs exhibited very similar responses to changing soil dryness in both models; taking into account that the two models differ most in the modeling of soil moisture and its ecological effects, this convergence of model behavior corroborates our confidence in the usefulness of these models for assessments of the impacts of climatic change on forest succession—at least at the Walker Branch Watershed site and for the range of scenarios considered in this study.

It is not surprising that there are a number of challenges in forest succession modeling that strongly impede our predictive capabilities in this field. Since forest growth is an integral part of forest succession, the challenges in prediction of forest growth also apply here. In addition, there are two major issues that need to be addressed (for a more thorough treatment of the topic, see Reynolds et al. 2001):

- Recruitment and mortality are the central elements of tree population dynamics (Keane et al. 2001; Price et al. 2001). However, most of the research to date has focused on improving the formulation of the growth submodel in forest succession models (e.g., Bugmann 1996; Friend et al. 1997), which may have led to a certain imbalance in the representation of the various basic processes in current succession models (Bugmann 2001). For example, several simulation studies showed that different mortality algorithms could lead to drastically different simulation results (Bugmann 2001; Wyckoff and Clark 2002; Figure 21.7). In spite of methodological difficulties that are pervasive in research on recruitment and mortality processes of forest trees, these aspects should be emphasized in future research.

- The direct and indirect effects of increased atmospheric CO_2 are not easily built into most current succession models because of the still highly simplified representations of the associated processes, such as photosynthesis and transpiration (Bugmann 2001). "Hybrid" models, which include population dynamics from succession models

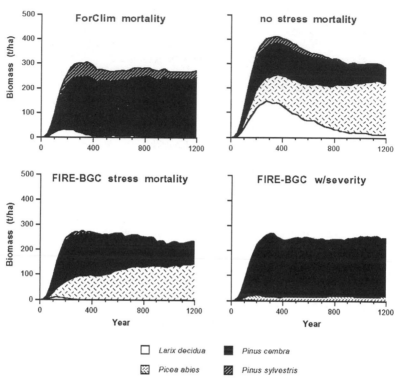

Figure 21.7. Sensitivity of a forest gap model to different assumptions regarding the formulation of stress-induced mortality rates (from Bugmann 2001).

and growth formulations from forest growth models, have been proposed to overcome these limitations, and it is inevitable that such approaches will be included in more succession models. It should be kept in mind, however, that in the absence of additional long-term studies on the effects of enhanced CO_2 on adult trees (Hättenschwiler et al. 1997), any simulation results from succession models that contain the growth response to enhanced CO_2 from short-term experimental investigations should be treated as sensitivity studies, not as predictions of the long-term behavior of forest ecosystems in a high-CO_2 world.

In view of these limitations and challenges for forest succession modeling, and taking into account the disparity of the scales of observations and experiments on the one hand and the scales of the simulation results obtained from these models on the other hand, it is clear that assessments of the impacts of climate change on forest succession constitute projections, rather than forecasts or predictions.

Synthesis and Conclusion

Human beings, together with all other creatures on earth, constantly face uncertainty. We have learned to live with uncertainty as long as we have some notion of its bounds. In particular, we use past experience to cope with uncertainty. Anthropogenic environmental changes, however, may be of an unprecedented magnitude that makes it unlikely that human beings, human societies, and the entire planet will continue to function as they did in the past (IPCC 2001). Given the magnitude and rate of anthropogenic environmental changes, we seek to reduce our uncertainty about future developments by "predicting," among others, the ecosystem effects of climate change. This goal is understandable, but there are many complicating issues with prediction in general and with predicting ecosystem effects in particular.

I hope to have shown in this chapter that prediction in the true sense is not possible. "Prediction" implies certainty; "forecasting" implies probability, whereas "projection" and "scenario" imply possibility. Even in the field of the physical climate itself, it is impossible to predict the future, let alone in the biological and ecological sciences. Hence, our discussion should focus on the value of projections, and on the next step that we may need to take, that is to move to ecological forecasts (Clark et al. 2001).

It is important to be careful about using appropriate terms, particularly when we deal with quantitative, mathematical models of ecological processes. In the strict sense, it is appropriate to talk about model predictions, because model output follows without any uncertainty from the structure of the model and the values of the model parameters and other boundary conditions. To a land manager or politician, a "model prediction" is likely to be the same as a "prediction" of the behavior of the system that is being studied (not of the model!), so that misunderstandings and misinterpretations are quite likely. Therefore, I suggest that we abandon the term "prediction" in the context of global change.

In this chapter, I considered agriculture, forest growth and biogeochemistry, and forest succession as examples of the problems that are associated with anticipating the ecosystem effects of climate change. It is the match (or mismatch) between the temporal and spatial scales at which observational or experimental data are available, and the scales at which the dominant processes in various ecosystems operate, that determines the degree of "predictability" of the ecosystem effects of climate change in these three areas. In this sense, we can state that "predictability" decreases when one moves from agriculture to forest growth and biogeochemistry and, finally, to forest succession.

However, this simple ranking is not entirely appropriate as soon as we address the three areas in more detail. This ranking was based on the predictability of biophysical processes. However, predictability of future states and processes in agriculture is exceedingly difficult because biophysics at the plant level alone is not the answer: the use of fertilizers and irrigation, changes in herbivores and pathogens, the development of new breeding technologies, including genetic engineering, and globalization will have large effects on agriculture in most areas of the world, which counteracts the predictability of biophysical processes in agriculture.

In the case of forest growth modeling, predictability is hindered strongly by the temporal mismatch between the scale of observational data and experiments on the one hand and the scale of the target variables on the other hand. Acclimation and adaptation phenomena of growth processes, carbon allocation, and belowground dynamics are poorly understood, but they are central to projections of future forest growth and carbon fluxes between the atmosphere and the terrestrial biosphere.

In forest succession modeling, there are significant problems with respect to parameter estimation for a large number of tree species and with the temporal and spatial scales involved in evaluating the behavior of these models. In addition to the issues mentioned with forest growth modeling, there arises a problem in forest succession modeling because of the need to encapsulate tree population dynamics, that is recruitment and mortality, which are highly stochastic, difficult to study, and therefore not well understood.

Finally, it is important to emphasize that climate change is not operating in isolation from other environmental changes, such as air pollution, which leads to increased nutrient deposition, acidification, and high ozone levels (DeHayes et al. 1999). For example, there is evidence of increased tree mortality rates since the 1950s that may be due to air pollution (Loucks 1998). The capacity to model pollutant effects at the ecosystem level is increasing rapidly (e.g., Retzlaff et al. 1997), but for a comprehensive assessment of the future development of forest resources, the combined effects of multiple environmental changes, not just climate change, need to be taken into account. Also from this point of view, it is fair to say that we cannot predict ecosystem dynamics. Scientific research at the present time is dealing with projections of future ecosystem development. Given that we can fully specify the uncertainties involved in our analyses, we may be able to provide ecological forecasts, but this is a longer-term research goal.

References

Bossel, H. 1994. TREEDYN3 Forest Simulation Model: Mathematical model, program documentation, and simulation results. *Reports of the Forschungszentrum Waldökosysteme, Series B* 35.

Bugmann, H. 1996. A simplified forest model to study species composition along climate gradients. *Ecology* 77: 2055–2074.

————. 2001. A review of forest gap models. *Climatic Change* 51: 259–305.

Bugmann, H., M. Lindner, P. Lasch, M. Flechsig, B. Ebert, and W. Cramer. 2000. Scaling issues in forest succession modeling. *Climatic Change* 44: 265–289.

Bugmann, H., and C. Pfister. 2000. Impacts of interannual climate variability on past and future forest composition. *Regional Environmental Change* 1: 112–125.

Bugmann, H., J.F. Reynolds, L.F. Pitelka, editors. 2001a. How much physiology is needed in forest gap models for simulating long-term vegetation response to global change? *Climatic Change* 51: 249–557.

Bugmann, H.K.M., and A.M. Solomon. 2000. Explaining forest biomass and species composition across multiple biogeographical regions. *Ecological Applications* 10: 95–114.

Bugmann, H.K.M., S.D. Wullschleger, D.T. Price, K. Ogle, D.F. Clark, and A.M. Solomon. 2001b. Comparing the performance of forest gap models in North America. *Climatic Change* 51: 349–388.

Cairns, M.A., S. Brown, E.H. Helmer, and G.A. Baumgardner. 1997. Root biomass allocation in the world's upland forests. *Oecologia* 111: 1–11.

Clark, J.S., and 16 co-authors. 2001. Ecological forecasts: An emerging imperative. *Science* 293: 657–660.

DeHayes, D.H., P.G. Schaberg, G.J. Hawley, and G.R. Strimbeck. 1999. Acid rain impacts on calcium nutrition and forest health. *BioScience* 49: 789–800.

Dixon, R.K., R.S. Meldahl, G.A. Ruark, and W.G. Warren, editors. 1990. *Process Modeling of Forest Growth Responses to Environmental Stress.* Portland: Timber Press.

Finegan, B. 1984. Forest succession. *Nature* 312: 109–114.

Foley, J.A., S. Levis, M.H. Costa, W. Cramer, and D. Pollard. 2000. Incorporating dynamic vegetation cover within global climate models. *Ecological Applications* 10: 1620–1632.

Friend, A.D., A.K. Stevens, R.G. Knox, and M.G.R. Cannell. 1997. A process-based, terrestrial biosphere model of ecosystem dynamics (HYBRID v3.0). *Ecological Modelling* 95: 249–287.

Garcia, R.L., S.P. Long, G.W. Wall, C.P. Osborne, B.A. Kimball, G.Y. Nie, P.J. Pinter, R.L. LaMorte, and F. Wechsung. 1998. Photosynthesis and conductance of spring-wheat leaves: Field response to continuous free-air atmospheric CO_2 enrichment. *Plant, Cell, and Environment* 21: 659–669.

Gregory, P.J., J.S. Ingram, B. Campbell, J. Goudriaan, L.A. Hunt, J.J. Landsberg, S. Linder, M. Stafford Smith, R.W. Sutherst, and C. Valentin. 1999. Managed production systems. Pp. 229–270 in B.H. Walker, W.L. Steffen, J. Canadell, and J.S.I. Ingram, editors. *Global Change and the Terrestrial Biosphere: Implications for Natural and Managed Ecosystems—A Synthesis of GCTE and Related Research.* IGBP Book Series No. 4, Cambridge: Cambridge University Press.

Grossman, S., T. Kartschall, B.A. Kimball, D.J. Hunsaker, R.L. LaMorte, R.L. Garcia, G.W. Wall, and P.J. Pinter. 1995. Simulated responses of energy and water fluxes to ambient atmosphere and free-air carbon dioxide enrichment in wheat. *Journal of Biogeography* 22: 601–609.

Harvey, L.D.D. 2000. Upscaling in global change research. *Climatic Change* 44: 225–263.

Hättenschwiler, S., F. Miglietta, A. Raschi, and C. Körner. 1997. Thirty years of in situ tree growth under elevated CO_2: A model for future forest responses? *Global Change Biology* 3: 463–471.

Hendricks, J.J., K.J. Nadelhoffer, and J.D. Aber. 1993. Assessing the role of fine roots in carbon and nutrient cycling. *Trends in Ecology and Evolution* 8: 174–178.

IPCC (Intergovernmental Panel on Climate Change). 2001. Summary for policy makers: A report of Working Group I of the Intergovernmental Panel on Climate Change. WMO/UNEP, Geneva, Switzerland. (available online at http://www.ipcc.ch/).

Jackson, R.B., J. Canadell, J.R. Ehleringer, H.A. Mooney, O.E. Sala, E.D. Schulze. 1996. A global analysis of root distributions for terrestrial biomes. *Oecologia* 108: 389–411.

Kauppi, P., and M. Posch. 1988. A case study of the effects of CO_2-induced climatic warming on forest growth and the forest sector. A. Productivity reactions of northern boreal forests. Pp. 183–195 in M.L. Parry, T.R. Carter, and N.T. Konijn, editors. *Assessments in Cool Temperate and Cold Regions. Vol. 1 of The Impact of Climatic Variations on Agriculture.* Dordrecht: Kluwer.

Keane, R.E., M. Austin, C. Field, A. Huth, M.J. Lexer, D. Peters, A. Solomon, and P. Wyckoff. 2001. Tree mortality in gap models: Application to climate change. *Climatic Change* 51: 509–540.

Kellogg, W.W. 1987. Man's impact on climate: the evolution of an awareness. *Climatic Change* 10: 113–136.

Kettunen, L., J. Mukula, V. Pohjonen, O. Rantanen, and U. Varjo. 1988. The effects of climatic variations on agriculture in Finland. Pp. 511–614 in M.L. Parry, T.R. Carter, and N.T. Konijn, editors. *Assessments in Cool Temperate and Cold Regions. Vol. 1 of The Impact of Climatic Variations on Agriculture.* Dordrecht: Kluwer.

Körner, C. 1996. The response of complex multispecies systems to elevated CO_2. Pp. in 20–42 in B. Walker and W. Steffen, editors. *Global Change and Terrestrial Ecosystems.* Cambridge: Cambridge University Press.

———. 1998. A re-assessment of high elevation treeline positions and their explanation. *Oecologia* 115: 445–459.

Levin, S.A. 1992. The problem of pattern and scale in ecology. *Ecology* 73: 1943–1967.

Lotter, A., and F. Kienast. 1992. Validation of a forest succession model by means of annually laminated sediments. In M. Saarnisto and A. Kahra, editors. Proceedings of the INQUA workshop on laminated sediments. Lammi, Finland. June 4–6, 1990. *Geological Survey of Finland Special Paper Series* 14: 25–31.

Loucks, O.L. 1998. In changing forests, a search for answers. Pp. 85-97 in H. Ayers, J. Hager, and C. Little, editors. *An Appalachian Tragedy: Air Pollution and Tree Death in the Eastern Forests of North America*. San Francisco: Sierra Club Books.

Lutz, W., W. Sanderson, and S. Scherbov. 2001. The end of world population growth. *Nature* 412: 543–545.

Manabe, S., and R.T. Wetherald. 1975. The effect of doubling the CO_2-concentration on the climate of a general circulation model. *Journal of Atmospheric Science* 32: 3–15.

Meadows, D.H., et al. 1972. *The Limits to Growth: A Report for the Club of Rome's Project on the Predicament of Mankind*. London: Earth Island.

Merriam-Webster Dictionary. 2001. Available on the Internet under the following URL: http://www.m-w.com/netdict.htm.

Mooney, H.A., J. Canadell, F.S. Chapin III, J.R. Ehleringer, C.H. Körner, R.E. McMurtrie, W.J. Parton, L.F. Pitelka, and E.-D. Shulze. 1999. Ecosystem physiology responses to global change. Pp. 141–189 in B.H. Walker, W.L. Steffen, J. Canadell, and J.S.I. Ingram, editors. *Global Change and the Terrestrial Biosphere: Implications for Natural and Managed Ecosystems: A Synthesis of GCTE and Related Research*. IGBP Book Series No. 4. Cambridge: Cambridge University Press.

Oren, R., D.S. Ellsworth, K.H. Johnsen, N. Phillips, B.E. Ewers, C. Maier, K.V.R. Schäfer, H. McCarthy, G. Hendrey, S.G. McNulty, and G.G. Katul. 2001. Soil fertility limits carbon sequestration by forest ecosystems in a CO_2-enriched atmosphere. *Nature* 411: 469–472.

Pacala, S.W., C.D. Canham, and J.A. Silander, Jr. 1993. Forest models defined by field measurements. I. The design of a northeastern forest simulator. *Canadian Journal of Forest Research* 23: 1980–1988.

Pacala, S.W., C.D. Canham, J. Saponara, J.A. Silander, R.K. Kobe, and E. Ribbens. 1996. Forest models defined by field measurements: Estimation, error analysis and dynamics. *Ecological Monographs* 66: 1–43.

Parry, M.L., T.R. Carter, and N.T. Konijn, editors. 1988a. *Assessments in Cool Temperate and Cold Regions. Vol. 1 of The Impact of Climatic Variations on Agriculture*. Dordrecht: Kluwer.

Parry, M.L., T.R. Carter, and N.T. Konijn, editors. 1988b. *Assessments in Semi-arid Regions. Vol. 2 of The Impact of Climatic Variations on Agriculture*. Dordrecht: Kluwer.

Porter, J.H., M.L. Parry, and T.R. Carter. 1991. The potential effects of climatic change on agricultural insect pests. *Agricultural and Forest Meteorology* 57: 221–240.

Price, D.T., N.E. Zimmermann, P.J. van der Meer, M.J. Lexer, P. Leadley, I.T.M. Jorritsma, J. Schaber, D.F. Clark, P. Lasch, S. McNulty, J. Wu, and

B. Smith. 2001. Regeneration in gap models: Priority issues for studying forest responses to global change. *Climatic Change* 51: 475–508.

Rastetter, E.B. 1996. Validating models of ecosystem response to global change. *BioScience* 46: 190–198.

Retzlaff, W.A., D.A. Weinstein, J.A. Laurence, and B. Gollands. 1997. Simulating the growth of a 160-year-old sugar maple (*Acer saccharum*) tree with and without ozone exposure using the TREEGRO model. *Canadian Journal of Forest Research* 27: 783–789.

Reynolds, J.F., H. Bugmann, and L.F. Pitelka. 2001. How much physiology is needed in forest gap models for simulating long-term vegetation response to global change? Challenges, limitations, and potentials. *Climatic Change* 51: 541–557.

Rosenberg, N.J., editor. 1993. Towards an integrated impact assessment of climate change: The MINK (Missouri-Iowa-Nebraska-Kansas) study. *Climatic Change* 24 (special issue).

Schulze, E.D., H.A. Mooney, O.E. Sala, E. Jobbagy, N. Buchmann, G. Bauer, J. Canadell, R.B. Jackson, J. Loreti, M. Oesterheld, and J.R. Ehleringer. 1996. Rooting depth, water availability, and vegetation cover along an aridity gradient in Patagonia. *Oecologia* 108: 503–511.

Shugart, H.H. 1984. *A Theory of Forest Dynamics. The Ecological Implications of Forest Succession Models*. New York: Springer. 278 pp.

Shugart, H.H., and D.L. Urban. 1989. Factors affecting the relative abundances of tree species. Pp. 249–273 in P.J. Grubb and J.B. Whittaker, editors. *Toward a More Exact Ecology*. 30th Symposium of the British Ecological Society. Oxford: Blackwell.

Solomon, A.M. 1986. Transient-response of forests to CO_2-induced climate change: Simulation modeling experiments in eastern North America. *Oecologia* 58: 567–579.

Solomon, A.M., H.R. Delcourt, D.C. West, and T.J. Blasing. 1980. Testing a simulation model for reconstruction of prehistoric forest-stand dynamics. *Quaternary Research* 14: 275–293.

Solomon, A.M., M.L. Tharp, D.C. West, G.E. Taylor, J.W. Webb, and J.L. Trimble. 1984. Response of unmanaged forests to CO_2-induced climate change: Available information, initial tests and data requirements. DOE/NBB-0053, National Technical Information Service, U.S. Dept. Comm. Springfield, Virginia.

Solomon, A.M., and T. Webb. 1985. Computer-aided reconstruction of late-quaternary landscape dynamics. *Annual Review of Ecology and Systematics* 16: 63–84.

Turner, M.G., W.H. Romme, R.H. Gardner, R.V. O'Neill, and T.K. Kratz. 1993. A revised concept of landscape equilibrium: Disturbance and stability on scaled landscapes. *Landscape Ecology* 8: 213–227.

Wyckoff, P.H., and J.S. Clark. 2002. The relationship between growth and mortality for seven co-occurring tree species in the southern Appalachian Mountains. *Journal of Ecology* 90: 604–615.

Part IV

The Future of Modeling in Ecosystem Science

22

The Role of Modeling in Undergraduate Education

Holly A. Ewing, Kathleen Hogan, Felicia Keesing,
Harald K.M. Bugmann, Alan R. Berkowitz, Louis J.
Gross, James T. Oris, and Justin P. Wright

Introduction

Models—from box-and-arrow diagrams to statistical tests and equations—are ubiquitous in science. Indeed, it is the dynamic interplay between models and empirical studies, the repeated process of posing and revising conceptual models, on which the scientific method rests. As is clear from the other chapters in this volume, models are one of the most basic tools for scientific experimentation, synthesis, and prediction. Yet, the teaching of modeling is relegated to the corners of the undergraduate curriculum. We believe that the purposes and the practices of modeling have enormous educational potential but that this potential is currently largely unmet.

This is not to say that models and modeling are underrepresented in the classroom. Orbiting electrons, spiraling ladders of DNA, hamburger-bun ribosomes, logistic growth curves, and predator-prey oscillations are part of nearly every introductory biology class, including those taken by nonscience majors. But most students do not recognize any of these as models, and even fewer have thought about the assumptions behind them or the ways in which models clarify or constrain our thinking about the way the world works.

Our basic premise is that undergraduates at all levels would benefit from learning more about modeling. We will argue that modeling—especially conceptual and mathematical modeling—could be incorporated more explicitly into undergraduate science classes to help students strengthen their quantitative and critical thinking skills, as well as to see the roles models play in science and society. We will present a framework for thinking about modeling as an integral part of an undergraduate science curriculum. Our comments are offered simply to suggest a framework rather than a prescription for what should be done with any given group of students.

We organize our suggestions as responses to four key questions: (1) Why should we teach modeling? (2) What core messages are important to convey about modeling? (3) What do students bring to the learning of modeling? and

(4) What concepts, skills, and tools should be emphasized in courses at different levels? While we often write specifically with respect to teaching ecology in introductory through advanced courses, our statements and recommendations could apply to teaching any field of undergraduate science.

Why Should We Teach Modeling?

Students use models extensively and yet shy away from anything called a model, so the primary goals of teaching modeling are to make their importance and role in science explicit and to help students become more comfortable and skilled at using, critiquing, and building them. The specific rationale for teaching modeling in any given class will vary depending on the teacher's objectives and the course content. Here we propose four compelling reasons for teaching modeling that are applicable across the curriculum.

To Foster Understanding and Appreciation of How Models Are Used in Science and Society

Fundamental at all levels, and perhaps most importantly for nonmajors, is to convey an appreciation of the general role of models in science and everyday life. Models are central to the process of science. The recognition that every hypothesis is a model and that the testing and revision of those models is the foundation of the scientific method would go a long way toward demystifying models and modeling. Beyond this foundation, models are frequently used in science to clarify thinking about the structure of a system, questions for investigation, and factors influencing the future trajectory of a system. Detailed consideration of these uses can be found in other chapters in this volume. In society at large, predictive ecological models—for example, those dealing with fish stocks or global climatic changes—are frequently used as a basis for shaping environmental policy, and it is this use that most frequently places models in news stories. Students who have some familiarity with models and understand that they are simplifications of reality with underlying assumptions are more likely to ask critical questions about the basis and interpretations of science-based policies which they encounter in the media. Such questions are the foundation of personal and corporate responsibility and have the potential to inform the legislative process.

To Enhance Quantitative Literacy

To the greatest degree possible, we would like both science and nonscience majors to become comfortable with the quantitative side of science and modeling. While some people are comfortable with and interested in mathematical abstractions of natural phenomena, many people become flustered, overwhelmed, or simply inattentive when a discussion or piece of writing turns to

numbers and equations. Because both scientific ideas and environmental policies are often compared based on quantitative outcomes, ease with and an ability to evaluate quantitative arguments is essential for scientists and nonscientists alike. This points to the obvious need for some ease (or at least familiarity)
with mathematics in order to become either an informed citizen or a scientist
conversant in modeling. With greater quantitative skills and familiarity with
modeling, students have a new lens through which to see the world, more tools
available to them, and less fear about attempting to interpret the work of other
scientists.

We urge the inclusion of quantitative models or other kinds of quantitative
training within the curriculum to whatever extent possible given the skills of the
students. However, most educators who teach models and modeling either build
models themselves or have found the concepts of science more exciting or accessible through exploration of mathematical models. Thus, both the need for
quantitative skills and the predisposition of many teachers has led to significant
emphasis on mathematics in courses that deal with modeling. Gauging how
much emphasis to place on the quantitative can be difficult, particularly in the
face of differences among students in mathematical background and thinking
styles. We encourage teachers to consider course objectives, desired learning
outcomes, student background, and the "meta" messages they want to convey
about modeling when deciding how much and what kind of quantitative reasoning to include in any given course having a modeling component.

To Enrich and Facilitate the Learning of Ecological Concepts

Ecological concepts are complex and multidimensional. Constructing simple
models such as box-and-arrow diagrams can help students synthesize information and consolidate their understanding about complex ecological concepts
while making their thinking visible to an instructor who can then intervene to
clarify misconceptions. However, since learning ecology also requires understanding the inherently dynamic nature of ecological systems, static representa-
tional tools are of limited use.

Describing and predicting how complex systems behave necessitates going
beyond structural depictions of cause/effect relationships among multiple factors to show the net effects of all of those variables interacting simultaneously.
This is possible for students to do with computer-based modeling and simulation tools. Students do learn more about system behaviors when they build or
use dynamic models than they do by creating static depictions of system relationships (Kurtz dos Santos and Ogborn 1994; Stratford 1997). Also, using
modeling software, such as Stella, enhances students' understanding of systems
concepts such as feedback, variation, and complex causality (Mandinach and
Cline 1996). Thus, computer modeling can help students represent, describe,
and analyze ecological systems, test their conceptualizations and assumptions
about systems, and learn about system behaviors and emergent properties.

To Educate Scientists to Embrace, Use, Understand, Modify, and Create Models

For students who go on to become science majors, a focus on models and modeling provides an opportunity to delve more deeply into the role of models. While answering both theoretical and applied research questions, students are pushed toward greater familiarity and comfort with mathematical abstractions of phenomena and are allowed to take an active role in designing or creating models. Students who are science majors are likely to encounter multiple ways in which models can be used in science, so making modeling an explicit theme of their studies should help them become competent in analyzing diverse types of models. Providing students the opportunity to build and analyze models for answering their own research questions gives them an invaluable tool. Hands-on construction will help students more fully develop their skills in quantitative abstraction and their comprehension of the ways in which models inform, constrain, and advance the process of science.

What Core Messages Are Important to Convey about Modeling?

Instead of coming away from experiences with models with the core message that modeling is hard or irrelevant, as some students do, we would like students to take away several more substantive ideas about the role and nature of modeling in science. The first concerns the centrality of models, particularly conceptual models, to the scientific process. For students with a focus on ecology, we would like to convey the role that models have played in the development of many widely accepted theories in ecology. But beyond these, we would also like students to come away with deeper ideas about the way we think about and relate to the natural world.

All students should learn that models are abstractions of reality that, of necessity, make a number of simplifying assumptions (Chapter 2). These assumptions can lead to greater understanding of how natural systems work, but they can also constrain our understanding in critical ways. Understanding the many kinds and ubiquity of models can make it more clear how models form the foundation of scientific discourse and thus the basis of knowledge claims. Because scientists see the world from different perspectives, they sometimes disagree in their simplifications of reality. This provides an opportunity to illustrate the contested nature of knowledge and to show how science can move ahead through disagreement, modification of models, and eventually agreement on a new paradigm. The discussion and revision of models through collaboration and exchange between data and models should clearly illustrate the collaborative nature of science. If the role of models in the conceptual foundation of science becomes clear, so should the artificiality of the distinction between "modelers" and "empiricists." Indeed, this artificial distinction is misleading and detrimental to science.

With respect to the use of models, we must educate students to see models as one type of tool to be used in approaching scientific problems. Whether or not they have the mathematical aptitude to construct mathematical models of their own, they should recognize the strengths and weaknesses of using models to address particular types of questions. They should be willing to modify or discard a model should it become clear that the hypothesis or assumptions behind the model are not supported by data. Students should also see that each model is tied to a specific context and knowledge construction purpose, so the GEM (Global Everything Model) is an empty, if not impossible, pursuit. For students who have had hands-on experience building models, we would like them to have the confidence and self-efficacy to cast the complexity of the world in simplified, quantitative terms. And with that, we would like them to develop a good measure of humility in recognizing the limits of our ability to simplify and understand the natural world.

The following stories illustrate many of the core ideas about modeling that we have mentioned. We present these two examples, one based on a conceptual model and the other on a mathematical model, to demonstrate how analysis of specific cases can convey to students important messages about how models both guide and constrain research.

Keller (1983) describes an excellent example of how a conceptual model can constrain scientific thinking in her story of the work of plant geneticist Barbara McClintock. In 1948, McClintock observed corn-kernel color patterns that could only be explained by genes moving within the genome. When she reported on this work in 1951, other scientists responded with everything from confusion to hostility, perhaps in part because of the complexity of her thinking or the context in which she cast the work (Comfort 2001). McClintock's "jumping genes" model countered the traditional view (model) of genes as having fixed locations on chromosomes. Her ideas were radical both because they allowed that genes were not in fixed locations on chromosomes and because she proposed that their movement was important in controlling development. The central model (if not dogma) of molecular genetics, just coming into its own in the 1950s and 1960s, held that information moved from the nucleus to the cytoplasm of a cell in a deterministic way with little consideration of the developmental context. In the eyes of her colleagues, McClintock's propositions were so far outside the conceptual models of the time as to be readily set aside. To McClintock, on the other hand, the evidence of her experiments led to a new vision of what was possible within a cell, a new line of research, and new possibilities in evolution. Though there was acceptance of movable genes in corn by the mid-1950s (Comfort 2001), it was not until molecular biologists began discovering the same phenomenon in other organisms two decades later that the mobility of genes became widely accepted. In 1983, McClintock won the Nobel Prize for her work on transposable elements. While some of her ideas about genetic control of development have been disproved, her more radical vision integrating genetics, development, and evolution is only now being understood (Comfort 2001). McClintock was quietly persistent in her development and exploration of many conceptual models, and her story shows both how models may be poorly understood, rejected, revised, or accepted and that science benefits from the consideration of multiple (even radical) conceptual models.

While the constraint that conceptual models can have on scientific thinking is clear in the previous example, it is often less clear to students how mathematical models can clarify and constrain a particular problem. Unpacking a particular example might best illustrate how a model can be used to show students the interplay between models and data, and the way in which modelers can simplify a problem, but must make assumptions about the system. An example relevant to students with experience in calculus would be to consider a simple model of pollen deposition on a lake surface (modified from Sugita 1993):

$$y_{i,\,lake}(z, R) \equiv \frac{\pi R}{2} P_i X_i(z) \int_{z-R}^{z+R} g_i(x)dx \qquad (22.1)$$

where each individual of species i located distance z from the center of a lake with radius R is contributing pollen to the lake. P_i is the pollen productivity for species i per unit quantity of the plant (such as basal area, biomass, or leaf mass). $X_i(z)$ is the mean abundance of species i at radius z from the center of the lake. $g_i(x)$ is a function of the fall speed of the pollen type and describes proportion of pollen deposited at distance x relative to the total pollen released at the source (from Chamberlain 1975 and Prentice 1988; cited in Sugita 1993, p. 239). The integral in this equation sums the pollen contribution to the lake for all individuals of species i located distance z from the center of the lake.

In this model, pollen deposition depends on the pollen production of each species, the rate at which the pollen type falls to the ground, the abundance of each species, and the distance from the lake being considered. The model approximates the pollen contribution to the lake based upon a ratio of the area of the lake to the area of the concentric ring between a distance $z - R$ and a distance $z + R$ from each plant. The parameters for the model and estimates of the appropriate form for the functions $X_i(z)$ and $g_i(x)$ can all be obtained in practice from data, so the model could guide empirical work by directing which data to collect. Observations of pollen assemblages and associated vegetation can provide a test of the model and an answer to the question of whether the model adequately describes the phenomenon. This illustrates the give and take between data and model.

The assumptions behind the model are a little less transparent than the list of parameters suggests, and discussion of them is important for understanding the constraints of the model. That each species is assigned a pollen productivity and a pollen fall speed that determines the amount of pollen deposited on the lake should be clear. Less apparent are the assumptions that the wind blows equally from all directions, that the lake is circular, that there is an approximation used for the fraction of pollen landing in the lake, that vegetation is homogeneous, and that pollen transported through the trunk space, in streams, and by rainfall is neglected. These assumptions are clear only from what terms are missing from the model, rather than by those included. The absence of terms that might seem reasonable to include and the identical treatment of all individuals of a species are a source of discussion. Students may want to consider whether a single value for pollen productivity or fall speed for all individuals of any species is adequate. Or, they might want the processes that occur in the real world to be better represented in the model. But how would these be included? Would such a model be tractable? Considering these questions would illustrate the trade-off between realism and simplicity.

What Do Students Bring to the Learning of Modeling?

Student background is usually considered in terms of the prerequisites students need for a course, but we would like to expand that view. We note the importance of students' preparation in quantitative reasoning through other classes. But we also suggest that there are other aspects of student background, including their experiences with technology, their ecological misconceptions, and their beliefs about the nature of knowledge and learning that strongly influence what (or even whether) students learn about models and modeling.

For classes in quantitative modeling or classes in which consideration of quantitative models is a significant component, students' mathematical background is of particular concern. In general, the earlier in a sequence of science classes a student is, the more likely a student is to be uncomfortable with mathematical abstractions, either because they have not had the mathematical prerequisites or because they lack experience or comfort with abstract thinking. To some extent, many teachers of advanced science courses rely on instructors of math and statistics classes both to provide the necessary background and to "weed out" students who are uncomfortable with quantitative thinking from science classes. Even students with the mathematical prerequisites may lack the ability to transfer those skills to a modeling context.

The tendency of biology faculty to expect students to have developed appropriate quantitative training in their math and statistics classes should be tempered by the reality that these classes are typically not designed to provide the diversity of concepts that arise in many biological contexts. Few universities provide any first-year math courses that do more than basic calculus. Yet much of ecological theory involves a breadth of topics from discrete dynamics (insect population models), differential equations (community models), linear algebra (structured population models), and probability (population genetics models) that students cannot easily learn without taking many math courses beyond calculus. Early examples of how faculty attempted to bridge this gap within individual classes can be found in Watt (1965), Patten (1966), and Innis (1971). At the curriculum level, we recommend a two-pronged approach: (1) encourage mathematics faculty to provide a broader course sequence for biology students that emphasizes applications which can motivate these students; and (2) encourage biology faculty to incorporate quantitative ideas throughout their courses so that students realize that math is an integral part of doing biology, rather than something isolated in quantitative courses (Gross 2000). With respect to students in existing science courses, appreciation of their difficulty and provision of practice opportunities in quantitative reasoning may help alleviate this problem.

Aspects of students' background other than their mathematical competency are also relevant to their ability to understand, analyze, and build models. These include a variety of experiences with the tools and technology we use to teach modeling, especially computers. For example, many students are accustomed to navigating within computer simulation games and programs. These experiences may help them rapidly learn to navigate within modeling simulation programs. However, it also means that they probably have some deep-seated assumptions about what simulations are and how they work. When ecological concepts are presented in the same guise, they may be targets for student comprehension (that there are rules behind the game, and in this case the rules are models of ecological phenomena) or for student ignorance (things just happen, the game really represents the world accurately, and there have been no subjective decisions about how the world is to be represented).

Students' perspectives on how ecological systems work may also influence their ability to learn about models and modeling. For example, both precollege and undergraduate students often think of food chains rather than food webs (Munson 1994). On the surface, this might be a simple misconception to correct substitute "web" for "chain" and make sure students can give an example. But at a deeper level, this exemplifies the tendency precollege students have to identify only one-way, one-step, linear interactions within ecological systems rather than recognizing two-way, multistep, and reciprocal relationships (Hogan 2000)—a tendency that persists among students at the undergraduate level (Green 1997; White 2000). Such views could make it difficult for students to understand or effectively analyze a model with many interacting components *even if they have the appropriate mathematical skills.* Obviously, a linear view of ecological interactions would also limit any attempt a student might make at modeling a complex system.

Students also bring with them a range of theories and beliefs about the nature of knowledge, knowing, and learning, which together constitute their epistemological perspectives. Undergraduate students' epistemological perspectives on knowledge range from absolutist views of knowledge as unchanging dogma to relativistic views of knowledge as being rootless and flexible to more centrist views of the grounded yet evolving nature of knowledge and scientific claims (Hofer and Pintrich 1997). Since epistemic concerns focus on the nature and grounds of knowledge, particularly its limits and validity, they can potentially have an important influence on how students evaluate models and whether they even value the enterprise of modeling as a legitimate approach to building scientific understanding.

The degree of students' epistemological understanding of models as intellectual tools potentially influences their ability to use them as such (Grosslight et al. 1991). While scientists recognize that models are a caricature of reality that can never capture every subtlety of a system and often aim to build the simplest representation of a system that can best explain what drives a particular behavior, students tend to want to create an exhaustive representation of real-world complexities (Hogan and Thomas 2001). So directly addressing students' conceptions about the nature and purpose of models could help them appreciate the value of simplifying reality in order to understand it.

Students also have different approaches to learning science that stem in part from their perspectives on the nature of knowledge and learning. Some students approach modeling tasks with an exploratory mindset, eager to manipulate models and learn through making discoveries, while others are more concerned with doing exactly the right things to obtain correct answers (Hogan and Thomas 2001). Students have different degrees of comfort and familiarity with ill-defined, open problems that are typical of science and which differ markedly from the well-defined, closed problems that typify many academic tasks.

Only an exploration of student background and assumptions will tell a teacher what students bring to the learning of modeling, and all of these dimensions of students' experiences and cognition are relevant in the classroom. To the extent possible, this should be recognized and explicitly addressed—for example through adjusting teaching methods and educational objectives—to help students make efficient progress in learning about and through modeling.

What Modeling Concepts, Skills, and Tools Should Be Emphasized in Courses at Different Levels?

Once a teacher has decided to include models in a class, she or he is faced with designing the syllabus. In the following sections, we provide examples of the issues and topics we believe to be important, including some things that could be taught but are often overlooked in introductory, intermediate, and advanced undergraduate science classes. These suggestions are presented to help teachers set educational objectives for their classes. At all levels, labs and in-lecture interactive problems may be the best way to emphasize the dynamic nature of modeling. In general, we would expect the proportion of the class spent in labs to increase as students move toward analysis and construction of models. There are many possible approaches to teaching modeling, including contextualizing abstraction within real problems, using excellent software packages, linking empirical information collected by students with models, role playing for scenario exploration, and team- or project-based instruction. We encourage the exploration of these and other teaching techniques at all levels but cannot discuss them all in detail.

Introductory Science Classes (For Majors and Nonmajors)

In order to accomplish the primary goal of teaching modeling—having all students develop an understanding that models are simplifications of complex phenomena and that they are underlain by simplifying assumptions—we believe it is important for instruction in models to begin in the earliest college-level science courses. Approaching modeling through physical or conceptual models (rather than mathematical ones) may provide an appropriate level of introduction for first- or second-year students. Fortunately, historical examples of the ways in which models have constrained or advanced science are now often included in college textbooks. For example, in biology texts, Barbara McClintock's research on jumping genes and James Watson and Francis Crick's scramble to determine the structure of DNA are often described in historical context. Examples such as these might provide relatively seamless integration of introductory aspects of modeling into the existing curriculum. However, these historical discussions rarely, if ever, explicitly indicate the aspects of these events that were based on conceptual or physical models; doing that remains the responsibility of the instructor.

Beyond understanding the purpose of models in general, introductory students can also benefit from learning how to use mathematical models. While most students should be able to handle simple mathematical models based on high school math, many introductory students have difficulty with abstraction. As a consequence, simulation models may be effective tools for introducing students to the use of models. Excellent educational simulations (e.g., Bio-Quest, Biology Labs OnLine, EcoBeaker) are now widely available and have been used in numerous courses. However, while these simulations are obviously useful for imparting course content (e.g., simulating evolutionary time scales, experimenting with DNA), the degree to which simulations are useful for teaching modeling is less clear. Instructors should consider emphasizing the modeling aspects of simulations by incorporating exercises such as having students list the assumptions of the models they are using or vary the model's assumptions and compare results across runs.

While our recommendations to use physical, conceptual, and simulation models emphasize nonquantitative approaches to teaching modeling at this level, we feel that instructors will have done students a disservice if they do not also urge them to develop their skills in math and abstraction. Exercises that require students to visualize data, analyze data sets, and use algebraic relationships provide a foundation for more mathematical approaches to modeling and have a number of other fundamental applications.

Intermediate Classes

For intermediate-level classes (e.g., ecology for majors), we suggest deeper exploration of the construction and use of both conceptual and quantitative models. Quantitative models such as exponential and logistic growth and

Lotka-Volterra predator–prey models often comprise a substantial part of the standard ecological curriculum, and conceptual models of biomes, food webs, and refugia are just as pervasive. Nevertheless, instruction frequently focuses on definitions or model results rather than on assumptions, construction, and use of models. We suggest that students at this level be explicitly introduced to the assumptions behind models from both quantitative and qualitative perspectives. For example, when teaching an exponential model of growth (e.g., population growth— $N_t = N_o e^{Kt}$) or decay (e.g., radioactive decay, light penetration, elimination of toxins— $N_t = N_o e^{-Kt}$), consideration should be given not just to the shape of the curve, but also to assumptions about and implications for the process under consideration; examination of the negative exponential is a good way to demonstrate the persistence of substances such as mercury or radioactive chemicals, since N_t will never equal zero. Students who can scrutinize mathematical formulas have access to what the model builder assumed about the world when she or he chose to represent a phenomenon in those mathematical terms or constructs—an area of potentially fruitful discussion both about ecological principles and the conduct of science. Such analysis requires that students have greater facility with math and abstraction, but it should be clear from the examples of exponential and pollen-deposition models how powerful such skills can be at opening avenues of understanding.

Simulation models can be valuable for students at this stage to explore model behavior, especially equilibria and parameter sensitivity, and in some cases, the impact of structural changes to the model. When more extensive consideration of models is desired, stability analysis and model calibration, verification, and "validation" may also be appropriate.

Quantitative and ecological ideas can be linked through the use of projects —individual, group, and class. Individual projects allow students the flexibility to choose a particular area of the course and develop an understanding of quantitative approaches in that area. Group and class projects capitalize on differences in student backgrounds, allowing students to learn from each other and develop their ability to share skills in a team—an experience that will serve them well in future endeavors. Incorporating quantitative methods into such projects can be accomplished in many ways: carrying out a literature review on models used to address particular problems, reviewing a software product that is designed for ecological research, implementing a mathematical model and analyzing it, comparing various field sampling schemes, or even building a simple, spreadsheet-based model. Such projects, done so that they are not simply a class add-on, can enhance student investment in the course, provide hands-on experience with the multidisciplinary nature of science, and foster an understanding that students can learn new quantitative skills on their own, outside a formal quantitative class.

However, even when students are able to understand and use mathematics in modeling, they will not necessarily find mathematical abstractions of natural phenomena a compelling way to study ecology. Therefore, extensive emphasis on these abstractions in intermediate-level classes may not enhance students' investment in either ecology or modeling. Providing an anchor to the concrete world, both in terms of the phenomena being modeled and in terms of the utility of models as tools for building understanding within and beyond the field of ecology, is essential.

Advanced Undergraduate Classes, Especially Those with a Modeling Emphasis

Incorporating a modeling emphasis into advanced undergraduate or beginning graduate ecology courses should provide an opportunity for students to uncover the role of models in framing research questions. To familiarize students with the interplay between data and models, students could be urged to conceptualize a research question or a simple ecological system as a conceptual model (e.g., a box-and-arrow diagram). Such a conceptual model is an excellent foundation both for discussion of what kinds of things should be included in a model and how models are revised as a result of data collection. Discussion should make clear that models are hypotheses that can and should be tested and revised. Given a firm understanding of the assumptions behind models and the role of models in science, the move to a mathematical form may be less mysterious. When students move to a mathematical model, consideration of options for parameterizing it may also help remind students about the assumptions behind the model and their malleability. Students can implement their models with existing software (e.g., Stella) or develop their own algorithms in spreadsheets or programming languages with which they have or can gain facility. Again, group projects may be especially useful in bringing together students with different skills and allowing students access to a greater range of tools than each would have in an individual project. Advanced students can also be expected to analyze a greater variety of models, including individual- and event-based models, models with explicit spatial structure, and models based on partial differential equations.

The trade-offs between realism and feasibility can be a major emphasis when students both construct and analyze models. Essential questions about what gets included in a model—both how and why—should be an explicit theme of modeling discussions. Students who have constructed, as well as analyzed, models should be able to see the tradeoffs among realism, accuracy, and precision of models (Levins 1966). That is, complexity, realism, and reliability of models are not always closely correlated. At this level, discussions of ecological processes can be dealt with from both conceptual and mathematical perspectives. Discussing real-world, conceptual, and mathematical trade-offs in the context of constructing models will help students build an understanding of complex linkages within the natural world, between data and models, and between conceptual and mathematical descriptions of the world.

A Note on Assessment

"What is the right answer?" is one of the most common questions professors hear from students, reflecting the instincts that students have developed through years of experiencing educational assessments that emphasize evaluation of the products more often than the process of learning and thinking. Yet the dynamic exchange between data and models, the role of models in the process of science, the analysis of problems, and the construction of models that we advocate emphasizing in science courses center on modeling processes, practices, and uses. This situation calls for creative approaches to assessing students' modeling expertise in undergraduate courses. We encourage teachers to think broadly about how to prompt students to demonstrate facility with process and analysis, whether through explication and justification of modeling approaches or discussion of policy implications of model outcomes. The challenge is how to capture students' thought processes in a formal way as a means of evaluating their performance. Other authors have dealt with this issue (e.g., Mintzes et al. 2000), so we encourage instructors to explore the literature on alternative assessment methods to expand their repertoire for evaluating what and how students learn through modeling.

Conclusion

A thread of models—their utility and potential pitfalls—could and should be woven into the fabric of undergraduate science education. At the most fundamental level, all students need to understand that the purpose of models is to simplify and clarify more complex systems. Further, students need to understand that building models requires making simplifying assumptions about the important features of the system being modeled. More advanced students need to gain facility with manipulating, analyzing, and perhaps even constructing models to further develop their understanding both of the systems under study and of the nature, uses, benefits, and potential constraints of models in general. In addition to developing "model literacy," hands-on work with models can reinforce essential mathematical concepts within the context of real-world problems and can encourage thought about the conduct of science. And perhaps most importantly, appropriate training in modeling could help develop versatile and fearless scientists, undaunted by models and math, and critical thinkers, capable of considering and evaluating the assumptions behind scientific discourse.

Toward these ends, we believe it is necessary to evaluate not only the ways we teach our individual classes, but also the ways we incorporate models into the curriculum for both science and nonscience students. While the latter is a significant challenge, developing appropriate model literacy within individual courses may be much more immediately attainable. Careful consideration of both the goals of teaching modeling at each level and the background that students bring to class should guide these efforts. We further recommend that teaching methods and assessment techniques should reflect not only the concepts, skills, and tools we want students to learn, but also the deeper, general messages about modeling that we want them to embrace.

We have advocated making modeling an integral part of an undergraduate education in science and have offered a framework for thinking about how to incorporate modeling into existing courses. However, achieving our goals requires significant effort beyond individual classrooms, and thus we recommend the following next steps:

- More explicit treatment of models in introductory textbooks

- The development, evaluation, and dissemination of specific examples of using models at different educational levels, e.g., through a targeted curriculum development effort, through solicited articles in appropriate journals, or through workshops and/or symposia at appropriate meetings

- An expansion of the Ecological Society of America's *Teaching Issues and Experiments in Ecology* (TIEE) efforts to include educational uses of models in the undergraduate curriculum

- An effort at the National Center for Ecological Analysis and Synthesis (NCEAS) aimed at infusing ecological modeling into ecological education, exploring both the intellectual and pragmatic issues involved

- Greater research in teaching and learning to explore issues such as the nature of student conceptual and skill development with respect to modeling, the relationship between teachers' own disposition towards modeling and their use of modeling in teaching, and barriers to and mechanisms for inclusion of modeling within curricula, instructional programs, and degrees.

References

Chamberlain, A.C. 1975. The movement of particles in plant communities. *Vegetation and the Atmosphere* 1: 155–203 New York: Academic Press, *cited in* Sugita, S. 1993. A model of pollen source area for an entire lake surface. *Quaternary Research* 39: 239–244.

Comfort, N.C. 2001. *The Tangled Field: Barbara McClintock's Search for the Patterns of Genetic Control.* Cambridge, MA: Harvard University Press.

Green, D.W. 1997. Explaining and envisaging an ecological phenomenon. *British Journal of Psychology* 88: 199–217.

Gross, L.J. 2000. Education for a biocomplex future. *Science* 288(5467): 807.

Grosslight, L., C. Unger, E. Jay, and C.L. Smith. 1991. Understanding models and their use in science: Conceptions of middle and high school students and experts. *Journal of Research in Science Teaching* 2: 799–822.

Hofer, B., and P. Pintrich. 1997. The development of epistemological theories: Beliefs about knowledge and knowing and their relation to learning. *Review of Educational Research* 67: 88–140.

Hogan, K. 2000. Assessing students' systems reasoning in ecology. *Journal of Biological Education* 35(1): 22–28.

Hogan, K., and D. Thomas. 2001. Cognitive comparisons of students' systems modeling in ecology. *Journal of Science Education and Technology* 10(4): 319–345.

Innis, G. 1971. An experimental undergraduate course in systems ecology. *BioScience* 21: 283–284.

Keller, E.F. 1983. *A Feeling for the Organism: The Life and Work of Barbara McClintock.* San Francisco: W.H. Freeman.

Kurtz dos Santos, A.C., and J. Ogborn. 1994. Sixth form students' ability to engage in computational modeling. *Journal of Computer Assisted Learning* 10: 182–200.

Levins, R. 1966. The strategy of model building in population biology. *American Scientist* 54: 421–431.

Mandinach, E.B., and H.F. Cline. 1996. Classroom dynamics: The impact of a technology-based curriculum innovation on teaching and learning. *Journal of Educational Computing Research* 14: 83–102.

Mintzes, J.J., J.H. Wandersee, and J.D. Novak, editors. 2000. *Assessing Science Understanding: A Human Constructivist View.* New York: Academic Press.

Munson, B.H. 1994. Ecological misconceptions. *Journal of Environmental Education* 24(4): 30–34.

Patten, B.C. 1966. Systems ecology: A course sequence in mathematical ecology. *BioScience* 16: 593 598.

Prentice, I.C. 1988. Records of vegetation in time and space: the principles of pollen analysis. Pp. 17–42 in B. Huntley and T. Webb, III, editors. *Vegetation History.* Dordrecht: Kluwer, *cited in* Sugita, S. 1993. A model of pollen source area for an entire lake surface. *Quaternary Research* 39: 239–244.

Stratford, S.J. 1997. A review of computer-based model research in precollege classrooms. *Journal of Computers in Mathematics and Science Teaching* 16: 3–23.

Sugita, S. 1993. A model of pollen source area for an entire lake surface. *Quaternary Research* 39: 239–244.

Watt, K.E.F. 1965. An experimental graduate training program in biomathematics. *BioScience* 15: 777–780.

White, P.A. 2000. Naïve analysis of food web dynamics: A study of causal judgment about complex physical systems. *Cognitive Science* 24: 605–650.

23

Increasing Modeling Savvy: Strategies to Advance Quantitative Modeling Skills for Professionals within Ecology

Kathryn L. Cottingham, Darren L. Bade, Zoe G. Cardon, Carla M. D'Antonio, C. Lisa Dent, Stuart E.G. Findlay, William K. Lauenroth, Kathleen M. LoGiudice, Robert S. Stelzer, David L. Strayer

Summary

To increase the contributions of modeling to the science of ecology, we propose that the ecological community would benefit from (1) greater clarification of the potential roles of quantitative models in ecology, (2) widespread proficiency in the construction and evaluation of simple quantitative models, and (3) assistance in making more effective use of complex quantitative models. This chapter describes a variety of strategies for improving modeling skills so that these goals can be achieved. Our suggestions fall into three general categories: convincing nonmodelers what modeling can do, helping interested ecologists build relatively simple models, and increasing the effective use of existing, complex models. Each of these goals is addressed in turn. Briefly, the first goal might be met by developing a statement on what models can and cannot do, supplemented with case studies; improving the dialog between modelers and empiricists; and introducing models as a research tool early in the educational process. Second, to help ecologists learn to develop simple models, we propose mechanisms to help new modelers identify what tools are available, what skills are needed to use these tools, and where to get help. We also support the development of short-term workshops to introduce common models and model-building techniques and longer-term apprenticeships for more in-depth training. Third, to help ecologists use complex models more effectively, we emphasize the importance of clear and up-to-date documentation, recognizing that funding must be explicitly allocated to support the time and infrastructure necessary for this task. Finally, all of the activities proposed above would be greatly facilitated by the creation of a national or international center to develop quantitative modeling skills and models for the ecological community.

Introduction

We met during Cary Conference IX to brainstorm strategies for advancing quantitative modeling skills among professional ecologists. Our initial premises were that (1) quantitative modeling is an important tool that should be used more broadly than at present (Chapter 3) and (2) given the proper motivation, training, and assistance, all ecologists can learn to use modeling as a tool. Before we tackled our primary goal—determining what the proper motivation, training, and assistance might be—we spent some time defining "modeling skills," identifying two focal modeling approaches, and classifying nonmodelers into groups based on the reasons why they don't use modeling in their own research.

What Are Modeling Skills?

We defined "modeling skills" quite broadly: our list includes not only math and computing skills, but also understanding what quantitative models can do, when such models are useful, and which models might be appropriate for a particular question. Thus, we seek to advance four types of modeling skills:

- *Appreciation for the substantive role of quantitative models in ecological research.* We would like for all ecologists to understand both the benefits and the limitations of models and to be able to answer the following questions: What can models do? What can't models do? How have models been used in the past? How might models be used in the future?

- *The ability to recognize when modeling could be used to answer a research question.* When confronted with an ecological or environmental problem, ecologists should be able to think "Aha, let's see what I can find out with a model" in the same way that they might think about finding an answer through an experiment, cross-system comparison, or a literature survey. Thus, we would like to develop not only the technical skills involved in modeling, but also the inclination and intuition to use those skills.

- *Awareness of the options available to answer particular questions.* The bewildering array of approaches to quantitative modeling is often a stumbling block to getting started in modeling. We would like to see more attention paid to helping new modelers navigate through the potential approaches by developing a taxonomy of model types and a list of contacts (publications, web sites, people) that could assist with model development.

- *Sufficient mathematical and computing skills to understand how models work and how to use model results.* For example, all ecologists should understand the algebra involved in finding equilibria and computing mass balances, as well as calculus through differential equations. The required level of computing sophistication will vary with model complexity, but at a minimum, ecologists need to be able to program formulas and make graphs in a spreadsheet package such as Excel or interact with graphical user interfaces such as Stella or SimuLink. Researchers who want to build their own models are likely to need additional skills in differential equations, linear algebra, numerical methods of solving differential equations, probability theory, and programming in a matrix-based (MatLab, Gauss) or programming (Visual Basic, C, C++, Java) language.

Identifying Focal Model Types

After defining "modeling skills," we identified a few specific modeling approaches around which to focus the rest of our discussion. There is a broad continuum of model complexity that ranges from simple mechanistic models (e.g., allometry, photosynthesis-irradiance curves, functional responses) to one-box population models to two- or three-box models to highly complex models with many boxes and large data needs. We singled out two particular types of models along this continuum as worthy of particular attention: simple, targeted models and complex, "big scary" models.

By simple targeted models, we mean models with relatively simple mathematical structure that are designed to answer a particular research question. Examples of such models include a stage-based matrix model for an endangered bird species parameterized from field data (Sillett 2000) and a two-species predator–prey model for wolves and ungulates (Eberhardt and Peterson 1999). With experience, these models can be developed relatively quickly, for little expense, and can be easily confronted with field data (Hilborn and Mangel 1997). Some researchers call these "throw away" or "disposable" models because they are unlikely to be reused in their original structure, although parts may be cannibalized for later projects. However, this name belies the power of simple models as a research tool, especially early in a research project (Chapters 4 and 6). We believe that all ecologists should to be able to build and use this sort of model.

"Big scary" models that are highly sophisticated and tend to be put to multiple uses lie at the end of the complexity continuum. Very few researchers are likely to build such models on their own or to completely understand the computer programs that run them. Rather, such models are community resources that are built and maintained by one research team and used by many others. Because of their complexity and flexibility, "big scary" models can be used to stimulate research, generate new questions and hypotheses, and to iterate between models and field studies. Examples include the CENTURY model for terrestrial ecosystems (Parton et al. 1987) and the DYRESM model for freshwater lakes and reservoirs (Imberger and Patterson 1981). Others will be developed, and one of our goals is to increase support for development of, and access to, such models.

Why Many Researchers Do Not Use Quantitative Models

We identified three groups of professionals within ecology who are not currently engaged in modeling:

1. Those with a strong aversion to modeling

2. Those who are interested in modeling and have questions amenable to relatively simple modeling approaches, but who are uncertain about how to begin

3. Those who are interested in modeling but have questions sufficiently complex to require using "big scary" models

Our subsequent discussions sought to develop strategies for improving modeling skills within each of these groups, as well as within the community at large. We summarize the highlights of this discussion in the next three sections, one for each group of nonmodelers.

Strategies to Promote the Use of Modeling, Particularly among Skeptics

Convincing people that quantitative modeling is a valuable tool for advancing ecological understanding continues to be a big hurdle. We propose four strategies to increase ecologists' appreciation for what modeling can do.

Develop a concise statement on what models can do, and what they can't, to create realistic expectations. We believe that models play a number of valuable roles in ecology. For example, quantitative models can serve as statements of quantitative hypotheses and as tools to discriminate among competing hypotheses. Models can also be used to generate ideas to test in the field, to direct field research (e.g., compare experimental designs, prioritize field measurements) and to make sense of field data. In addition, models can explore management decisions by predicting what future ecosystem behavior might be like under different decision-making scenarios. However, we need to be wary of overconfidence: models are only as good as the data and understanding that goes into them, and models cannot be used to make precise quantitative forecasts far into the future (Chapters 2 and 7). We need to do a better job of conveying both the strengths and weaknesses of modeling throughout all stages of ecological training.

Develop a list of case studies that demonstrate how modeling has advanced ecological understanding. We believe that a compilation of productive models would play a key role in convincing both ecologists and the public of the key role of modeling in ecosystem science, especially if this document were written in a highly accessible style such as that used in the *Issues in Ecology* series (e.g., Daily et al. 1997; Vitousek et al. 1997). Ideally, such a document would include a broad array of examples with different degrees of model complexity, application to real-world problems, and stages at which models entered into the study. Some excellent examples for such a document could come from many of the chapters in this volume. A specific example would be the modeling studies of the endangered Kemp's ridley sea turtle (*Lepidochelys kempi*), which have predicted the impact of turtle excluder devices (TEDs; Crowder et al. 1994) and illustrated the futility of expensive headstart programs to reduce mortality in the egg and hatchling stages (Heppell et al. 1996). The sea turtle case study also provides an excellent example of a situation where modeling prior to the implementation of the headstart program might have saved millions of dollars and focused attention on the true problem—high mortality rates among large turtles.

Improve the dialog between modelers and empiricists. We believe that an improved dialog between model creators, model users, and data collectors would enhance our science. This dialog could be facilitated by increased attention to model transparency, honesty about model limitations, better documentation of field data, and training of people who are familiar with the pros and cons of both theoretical and empirical approaches to ecology. It is also essential that we explicitly recognize that an improved dialog will require money to ensure full participation. For example, grants could include funding for empiricists to document and make data generally available and for modelers to better document their models and train nonmodelers in using them. At present, the dialog is underfunded—empiricists and modelers are generally expected to do these jobs for free, which probably impedes rather than fosters links between them.

Introduce the role of models as a research tool early in the educational process. Introducing modeling as a tool in our undergraduate courses (Chapter 22) would have tremendous benefits for our science. In particular, students should develop an early appreciation for the diversity of tools that can be used to answer ecological questions, as well as the background needed to choose the tool(s) most appropriate for a particular question without a bias for any one approach. Early exposure to modeling might also stir greater interest and enthusiasm in developing a strong quantitative background during the undergraduate years.

Strategies to Help Ecologists Develop Simple Quantitative Models

Once ecologists appreciate and strongly support the role of modeling in eco-logical research, the next step is to get them actively involved in modeling. Several of us commented on being stymied once we recognized the need for modeling in our own research. Several steps might be taken by the ecological community and its funding agencies to facilitate developing a relatively simple model targeted at a specific ecological question.

- *Provide formal help in identifying what tools are available, what skills are needed to use these tools, and who might be available for consultation or collaboration on the project.* Many new modelers find the diversity of modeling options overwhelming, and assistance navigating through these options would be incredibly valuable. For example, the ecological community could develop a taxonomy of available modeling approaches together with a simple expert system to help new modelers choose an appropriate strategy and then get started in developing a model. An alternative to the expert system approach would be to disseminate the model taxonomy through workshops, articles in the *Bulletin of the Ecological Society of America*, and on-line bibliographies of useful resources. Several re-cent "how to" papers and books might prove valuable in compiling this taxonomy, including Starfield et al. (1990), Jackson et al. (2000), and Peck (2000).

  The community could also develop a Web-based guide to contacts who have particular expertise. Identifying contacts is likely to be relatively straightforward for simple models and considerably more difficult as the level of mathematical and/or computing sophistica-tion increases. We recognize that there could be a mismatch between supply (the relatively small pool of experienced modelers) and the potentially high demand for assistance. This discrepancy will need to be addressed, but hopefully the problem will be short-lived as more ecologists become familiar with modeling.

- *Offer training workshops and short courses that introduce common model structures and model-building techniques.* We liked the idea of hands-on training programs, both workshops and short courses, but did not spend a great deal of time developing specific recom-mendations. The Ecological Society of America could take the lead here, either by coordinating short workshops with its annual meeting (as is done by the North American Benthological Society and the American Fisheries Society) or by sponsoring in-service training workshops.

- *Continue developing useful, intuitive tools such as Excel, Stella, SimuLink, and EcoBeaker.* Programs like these make it possible to build simple models without knowing a programming language. We strongly support the continued development of spreadsheet and graphical tools that make modeling accessible to all ecologists.

- *Promote formal programs that pair people with quantitative skills to those in demand of such skills.* Apprenticeships are an important form of training, and additional programs to financially support such partnerships would be very worthwhile. Programs already in place include the National Science Foundation's Ecological Circuitry Collaboratory to bring graduate students from mathematics and computer science into ecology (see http://www.ecostudies.org/cc/) and the Postdoctoral Fellowships in Biological Informatics to develop the quantitative skills of postdoctoral researchers.

- *Encourage the National Science Foundation to consider funding Integrative Graduate Education and Research Training (IGERT) programs to promote the development of skills in both modeling and empiricism during graduate training.*

Strategies to Help Ecologists Develop and Use Complex Models

Although simple modeling approaches are quite powerful, not all ecological questions are amenable to such approaches—some questions require big, complex models. We would like to propose two strategies for improving our ability to develop and use the types of complex models that become community resources for multiple research questions.

Complex models that are made available as resources for the ecological community need to have clear and up-to-date documentation. At present, there are a number of complex, quantitative models that are available as community resources (e.g., BIOME-BGC, TEM). However, use of such models can be hindered by incomplete or obsolete documentation (Chapter 13). Ideally, modeling teams should provide accurate information about the sources of equations and parameters, the output from sensitivity and uncertainty analyses, and any known limitations of the model's structure or function. Sufficient information must be available to allow responsible use of publicly available code (Chapter 13).

Develop a national or international center to develop modeling skills and models for the ecological community. We strongly believe that a facility that emphasizes quantitative modeling would have tremendous benefits for both the ecological community and society. We foresee a place that serves many needs, including offering workshops to develop the modeling skills of both novice and experienced modelers; providing a clearinghouse of models for use in learning and research; maintaining the code and documentation for these models; assisting with finding and accessing the empirical data needed to develop, parameterize, calibrate, and test models; and arranging apprenticeships with experienced modelers. We expect that experienced modelers available for consultation and possibly long-term collaborations would play a key role in the functioning of such a center, similar to the role that postdoctoral researchers currently play at the National Center for Ecological Analysis and Synthesis (NCEAS). This facility could take many physical forms, including an expanded role for modeling and modelers at NCEAS, an expanded vision of the proposed National Center for Ecological Forecasting, a brand-new place, or even a virtual center. Such a place might be patterned after the International Institute for Applied Systems Analysis (IIASA) in Vienna, which conducts interdisciplinary scientific studies on environmental, economic, technological and social issues in the context of human dimensions of global change (see http://www.iiasa. ac.at/docs/what-is-iiasa.html for more information).

Conclusions

We believe that quantitative modeling plays a central role in advancing ecological science, and that increased attention to modeling as a research tool could have a tremendous impact on our field. As such, we propose three goals for the ecological community: to clarify the potential roles of models in ecology, to learn to construct and evaluate simple models, and to use complex quantitative models more effectively. We strongly believe that the ecological community can meet these goals if we make them a priority and commit the financial resources to support them.

Acknowledgments. We thank Lou Gross for discussions during the Cary Conference, Dean Urban for introducing the phrase "savvy modelers" during his summary comments at the meeting, and Kass Hogan for providing references on modeling tools. David Peart provided a helpful review of the manuscript.

References

Crowder, L.B., D.T. Crouse, S.S. Heppell, and T.H. Martin. 1994. Predicting the impact of turtle excluder devices on loggerhead sea turtle populations. *Ecological Applications* 4: 437–445.

Daily, G.C., S. Alexander, P.R. Ehrlich, L. Goulder, J. Lubchenco, P.A. Matson, H.A. Mooney, S. Postel, S.H. Scheider, D. Tilman, and G.M. Woodwell. 1997. *Ecosystem Services: Benefits Supplied to Human Societies by Natural Ecosystems.* Vol. 2 of Issues in Ecology.

Eberhardt, L.L., and R.O. Peterson. 1999. Predicting the wolf-grey equilibrium point. *Canadian Journal Of Zoology* 77: 494–498.

Heppell, S.S., L.B Crowder, and D.T. Crouse. 1996. Models to evaluate head-starting as a management tool for long-lived turtles. *Ecological Applications* 6: 556–565.

Hilborn, R., and M. Mangel. 1997. *The Ecological Detective: Confronting Models with Data.* New York: Cambridge University Press.

Imberger, J., and J.C. Patterson. 1981. A dynamic reservoir simulation model: DYRESM 5. Pp. 310–361 in K.W. Thornton, B.L. Kimmel, and F.E. Payne, editors. *Reservoir Limnology: Ecological Perspectives.* New York: Wiley.

Jackson, L.J., A.S. Trebitz, and K.L. Cottingham. 2000. A simulation primer for ecological modeling. *Bioscience* 50: 694–706.

Parton, W.J., D.S. Schimel, C.V. Cole, and D.S. Ojima. 1987. Analysis of factors controlling soil organic matter levels in Great Plains grasslands. *Soil Science Society of America Journal* 51: 1173–1179.

Peck, S.L. 2000. A tutorial for understanding ecological modeling papers for the nonmodeler. *American Entomologist* 46: 40–49.

Sillett, T.S. 2000. Long-term demographic trends, population limitation, and the regulation of abundance in a migratory songbird. Ph.D. dissertation. Dartmouth College, Hanover, NH.

Starfield, A.M., K.A. Smith, and A.L. Bleloch. 1990. *How to Model It: Problem Solving for the Computer Age.* New York: McGraw-Hill.

Vitousek, P.M., J. Aber, R.W. Howarth, G.E. Likens, P.A. Matson, D.W. Schindler, W.H. Schlesinger, and G.D. Tilman. 1997. *Human Alteration of the Global Nitrogen Cycle: Causes And Consequences.* Vol. 1 of Issues in Ecology.

24

The Limits to Models in Ecology

Carlos M. Duarte, Jeffrey S. Amthor, Donald L. DeAngelis,
Linda A. Joyce, Roxane J. Maranger, Michael L. Pace,
John Pastor, and Steven W. Running

Summary

Models are convenient tools to summarize, organize, and synthesize knowledge or data in forms allowing the formulation of quantitative, probabilistic or Bayesian statements about possible or future states of the modeled entity. Modeling has a long tradition in the earth sciences, where the capacity to predict ecologically relevant phenomena is ancient (e.g., motion of planets and stars). Since then, models have been developed to examine phenomena at many levels of complexity, from physiological systems and individual organisms to whole ecosystems and the globe.

The demand for reliable predictions and, therefore, models is rapidly rising as environmental issues become a prominent concern of society. In addition, the enormous technological capacity to generate and share data creates a considerable pressure to assimilate these data into coherent syntheses, which are typically provided by models. Yet, modeling still encompasses a very modest fraction of the ecological literature, and modeling skills are remarkably sparse among ecologists (Chapter 3). The growing demand for models is in contrast with their limited contribution to the ecological literature, which suggests that either there are serious constraints to the development of models or that there are limitations in the conceptualization and/or acquisition of the elements required for model construction, both of which result in the requirements for products from models not being met at present.

Models are apparently not being as widely used as expected given the present demand. Hence, the identification of bottlenecks for the development of models in ecology and of the limitations of their application are important goals. This chapter reports on the

conclusions of a discussion group at Cary Conference IX that addressed these issues.

Background

Ecological models can be classified in a number of ways. One of the most useful is the distinction between single-level descriptive (or empirical) models and
hierarchical, or multilevel, explanatory (or mechanistic) models (de Wit 1970,
1993; Loomis et al. 1979). An example of a single-level descriptive model is a
regression equation relating annual net primary production (NPP) or crop yield
to annual precipitation and/or temperature. When used within the range of precipitation and temperature included in the formulation of the regression equa-
tion(s), such a model may be rather accurate for interpolative prediction. It does
not, however, "explain" the operation of the systems, and the model may fail
when applied to conditions outside the environmental envelope used for parameter estimation or when applied to a different ecosystem. Explanatory models often include at least two levels of biological or ecological organization,
using knowledge at one level of organization (e.g., biological organs) to simulate behavior at the next higher level of organization (e.g., organisms), although
other factors may come into play. Information at the lower levels may be empirical or descriptive information (Loomis et al. 1979) that helps explain behavior at the level of the organism. Of course, in explanatory ecological models,
knowledge gaps arise and simplifications are inevitable.

Modeling terrestrial NPP provides a robust example of the spectrum of modeling possible in ecology (Figure 24.1). The highest modeling complexity is
found with mechanistic carbon-flux models for plot-level applications that require intensive data for execution (Rastetter et al. 1997), and yet more sophisticated mechanistic models are becoming now available (Amthor et al. 2001).
These models compute photosynthesis-minus-respiration balances of leaves and
model the partitioning of photosynthate to plant growth. Somewhat more generalized models, such as FOREST-BGC, do not treat each plant explicitly but
compute the integrated carbon, water, and nutrient biogeochemistry of a landscape (Running 1994). Recent attempts to distill these complex NPP models to
use more simplified input data, such as PnET-II and 3-PGS, have been impres-
sively successful for forests (Law et al. 2000; Aber and Melillo 2001). Other
models have explicitly considered treatment of NPP by different vegetation
biomes from a common logical framework (VEMAP 1995). These multi-biome
NPP models are becoming important to policy issues regarding terrestrial carbon source and sink dynamics (Schimel et al. 2000). A new generation of NPP
models uses satellite data for input and a simple light-conversion efficiency
factor to compute NPP from absorbed photosynthetically active radiation. Use
of satellite data for primary input data has allowed broad mapping of NPP from
regional up to global scales (Coops and Waring 2001; Running et al 2000).

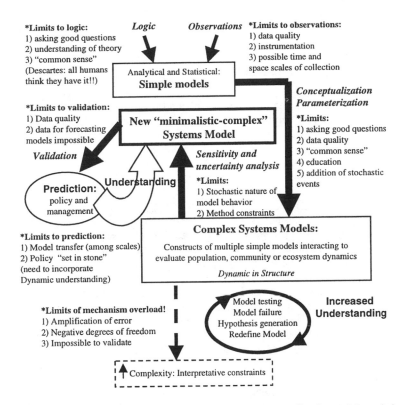

*Limits to logic:
1) asking good questions
2) understanding of theory
3) "common sense"
(Descartes: all humans
think they have it!!)

Logic   *Observations*

*Limits to observations:
1) data quality
2) instrumentation
3) possible time and
space scales of collection

Analytical and Statistical:
Simple models

Conceptualization
Parameterization

*Limits to validation:
1) Data quality
2) data for forecasting
models impossible

New "minimalistic-complex"
Systems Model

*Limits:
1) asking good questions
2) data quality
3) "common sense"
4) education
5) addition of stochastic
events

Validation

Sensitivity and
uncertainty analysis

Prediction:
policy and
management

Under**standing**

*Limits:
1) Stochastic nature of
model behavior
2) Method constraints

*Limits to prediction:
1) Model transfer (among scales)
2) Policy "set in stone"
(need to incorporate
Dynamic understanding)

Complex Systems Models:
Constructs of multiple simple models interacting to
evaluate population, community or ecosystem dynamics
Dynamic in Structure

*Limits of mechanism overload!
1) Amplification of error
2) Negative degrees of freedom
3) Impossible to validate

Model testing
Model failure
Hypothesis generation
Redefine Model

Increased
Understanding

Complexity: Interpretative constraints

Figure 24.1. A schematic overview of systems model conceptualization and formulation,
driven by understanding, with the ultimate objective of prediction (or projection). Simple
analytical and statistical models formulated using logic or theory and empirically derived
observations are compiled into larger complex systems models. Model parameters are
rigorously tested using a variety of techniques resulting in a feedback loop that increases
parameter and model understanding and identifies those parameters that are the most
ecologically relevant in order to create "minimalistic-complex" system models, which are
potentially useful for prediction. Solid arrows outline this process. The large white arrow
linking prediction to model development and reevaluation via understanding emphasizes
that model building is a dynamic process benefiting from both elements. The broken
arrow denotes the "danger" of increasing model complexity and the limitations associated with doing so.

This dramatic simplification from full photosynthesis – respiration balances
to a simple light use efficiency model to compute a common variable, NPP,
exemplifies the range of logic used in ecosystem modeling (Figure 24.1). No
single model can be optimally used throughout all scales (Figure 24.1), which

span from the individual leaf to the global biosphere (Waring and Running 1998).

Limits to the Development of Models

Bottlenecks impairing the development of models seem to be multiple and to encompass limitations in the scientific community (Chapter 23). Improvement will require actions at the capacity-building stage and the removal of techno-logical bottlenecks and limitations in the data available to build, drive, and validate the models.

Capacity Building

University curricula in biology are notoriously poor in providing students with modeling skills, which probably derives from a general neglect of solid mathe-matical training in biology programs, at both undergraduate and graduate lev-els. Consequently, modeling is largely a self-taught craft. This imposes severe limitations in the recruitment of modelers to the community, which accounts for the relatively small number of ecological modelers. Poor training also limits the understanding of the data requirements for model construction by experimental ecologists, resulting in insufficient coordination between data acquisition and the requirements of the models. The ineffective communication between mod-elers and experimental ecologists has other important consequences, such as the present tendency of papers on ecological models to be relegated to particular journals. By increasing their isolation, this tendency further enforces the insuf-ficient communication between modelers and experimental ecologists.

Ecological modeling will also benefit from establishing firm interdiscipli-nary links, which will allow capitalizing on developments in other fields. Such developments include recent advances in computing science and technology, such as optimized algorithms for parallel computing. In addition, recent devel-opments in new modeling approaches, such as dynamic modeling of complex systems in mathematics and theoretical physics, could be conveyed faster to the community of ecological modelers if platforms to foster interdisciplinary links between these different disciplines were better established. The opportunity to interact with ecological modelers could also benefit mathematicians and theo-retical physicists, who are in a continuous search for complex systems to serve as test benches for their new developments. In addition, human intervention in ecological processes is now widespread at all scales, requiring this element to be factored into model formulations. Doing so will require greater collaboration between social scientists and ecologists to incorporate human influence in eco-logical models (National Assessment Synthesis Team 2000).

Technological Developments

Technological bottlenecks appear to be minor at present because computing power has increased so rapidly that the gains made since ecological simulation began more than thirty years ago are no less than staggering. Only the larger, global models seem to be constrained by access to adequate computing facilities. In addition, software developments have provided modeling platforms and analytical tools that now render model construction and analysis a relatively simple task. Quantitative, robust approaches to decide on the optimal size of ecological models are now becoming available (Chapter 8) and guiding model development. As a result, technological bottlenecks cannot, in general, be held responsible for the insufficient development of models in ecology, although increased networking between supercomputer facilities could improve their availability to the scientific community.

Appropriate Observational Basis

The empirical basis for developing and testing models is generally insufficient and may well be the ultimate bottleneck for the development of mechanistic models. Even available data are usually poorly fit to model needs, for they may lack the required spatial and temporal coverage or they may poorly encompass the range of gradients that must be covered by the model. This limitation may be somewhat alleviated by developing modeling approaches flexible enough to incorporate and combine a variety of data, including data at various hierarchical levels (e.g., population, class within a population, and individual), and addressing tactical questions that do not require complete data descriptions of a system (Chapter 5). Nevertheless, data limitation will remain a chronic problem because, despite the activity of many ingenious field ecologists, collection of some types of data is by its nature difficult and expensive. The data sets available to test models are, therefore, limited, leading to growing concern that most models may be validated against a few, common data sets, a practice that—while allowing for model comparison—involves the danger that these few data sets become the "world" the models re-create. Moreover, there are also unduly long time lags between data acquisition and when these data are made available for model construction and validation; such delays preclude the development of on-line models, which are becoming available in other disciplines (e.g., operational oceanography). The inadequacy of ecological data for modeling purposes largely stems from insufficient appreciation of the requirements for model construction, again calling for greater interaction between modelers and experimental ecologists in education programs and during the formulation of research programs. Results from present experimental research focuses heavily on statistical analysis of end points rather than explanation of the processes yielding those end points, whereas knowledge of processes forms the basis of explanatory (or extrapolative) models. In addition, experiments are rarely useful as a base for modeling because of the emphasis on contrasts and ANOVA-directed designs instead of on gradient and process designs (Chapter 12).

Limits to the Achievements of Models

The bottlenecks identified above are substantial but alone cannot explain the limited application of modeling approaches to ecological problems. We suggest that there must be other, external limits to what models can achieve. An appreciation of these limits requires, however, an identification of the goals of models as a prerequisite to identifying the circumstances that may preclude the achievement of these goals.

The Goals of Models

Models are often used as heuristic tools to organize existing knowledge, identify gaps, formulate hypotheses, and design experiments. Models can also be used as analytical tools to increase our understanding about the relative importance and interplay of the various processes involved in the control of populations, communities, or ecosystems and to use this understanding to examine their behavior at scales that extend beyond direct observation. Societal needs impose increasing pressure for ecological models to help inform management decisions as to the likely consequences of alternative management options and future scenarios of the status of populations, communities, or ecosystems (Chapter 7). Indeed, heavy human pressures upon the earth's resources are leading to major changes in the functions of the earth system and loss of the biodiversity it contains. The need for ecological models by society in the twenty-first century will become a benchmark upon which the robustness of ecological knowledge will be assessed. This demand is largely articulated through the societal request for predictions, which must be supplied along with the logical rationalization of the model's mechanistic basis that provides reassurance on the reliability and generality of the model's principles.

The role of models as tools in formulating hypotheses is a noncontroversial—although insufficiently explored—use. In contrast, conflicts between the goals of understanding versus prediction have been the subject of recurrent debates (e.g., Peters 1986, 1991; Lehman 1986; Pickett et al. 1994; Pace 2001). These debates mirror those in other branches of science, where in some cases mechanistic models have been claimed to be overemphasized to the detriment of progress (Greene 2001). We contend, however, that prediction and understanding are mutually supportive components of scientific progress and that ecological modeling would greatly benefit from an improved integration of both these goals, which should, if possible, progress in parallel (Figure 24.1). Even the simplest statistical models are used to test hypotheses (Peters 1991) through the mechanistic reasoning that generally guides the selection of candidate predictors, thereby leading to increased understanding (Pace 2001). On the other hand, models developed to test the reliability of our understanding generally do so by comparison between model output and observations, thereby assessing the predictive power of the models. We, therefore, contend that the discussion on the priority of prediction versus understanding in model formulation is largely futile. Whether the models are empirical, statistical models that make few assumptions on the underlying mechanisms or are mechanism-rich cannot be considered, a priori, to offer any particular advantage when predictions are sought, for "whether the cat is black or white does not matter, as long as it catches the mice" (Mao Tse-tung).

The key issue is, therefore, whether the model works, that is, whether it is conducive to the formulation of reliable, validated predictions. For instance, Chinese observers were able to accurately predict tides more than two thousand years ago, despite the fact that they understood the process to result from the breathing of the sea (Cartwright 2001). More than two millennia later, we have been able to improve somewhat our capacity to predict tides; however, despite what we believe is an adequate understanding of the tide phenomenon, the prediction is still based on site-specific, empirically fitted curves (Cartwright 2001).

The Complementary Goals of Prediction and Understanding

Whereas predictive capacity is an objective trait amenable to quantitative test (Chapter 8), the degree of understanding achieved through modeling is more difficult to assess in any specific way. Although understanding is not a prerequisite to the achievement of predictive power, modeling approaches that offer no possibility to test hypotheses and, therefore, gain understanding are regarded as suspect and should not be used as sources of predictions. Hence, both scientists and society are not satisfied with projections of past dynamics into the future that provide little or no understanding, but, rather they require predictions, which must be derived or be consistent with theory (Pace 2001). This requirement provides an indication of the importance scientists and humans in general assign to understanding, as reflected in the quote: "I believe many will discover in themselves a longing for mechanical explanation which has all the tenancy of original sin." (Bridgman 1927, 27).

In addition, prediction and understanding are linked through effective feedback processes, for predictions are derived from theory and, at the same time, their success or failure serves to improve theory. Predictions can be derived from simple statistical models, analytical models, or complicated simulation models. The growing complexity of the models may provide a greater sense of understanding if successful predictions can be derived, but often this complexity leads to reduced predictive power and/or various interpretive constraints (Figure 24.1). Such loss of predictive power is clearly illustrated in the comparison of the predictive record of statistical versus mechanistic models of El Niño–La Niña events (Chapter 4). Probably the most studied and implemented use of statistical (regression) models in ecology involves "prediction" of crop yield based on environmental conditions (e.g., mean air temperature during the growing season or total precipitation). However, simple statistical models are not formulated from a random trial-and-error approach. Rather, the candidate independent variables that achieve predictive power are deduced from theory, such as sea-surface temperature anomalies and atmospheric pressure locations in the case of statistical models used for El Niño–La Niña prediction (Chapter 4), or evapotranspiration in models of terrestrial NPP (Leith 1975).

Mechanistic models are not necessarily bounded in their predictive capacity, because the predictive capacity of NPP models has increased significantly since the initial formulations. However, the requirements of mechanistic models do exceed those of statistical descriptive models by virtue of the simple fact that they contain (and therefore need as input) more elaborated information. Indeed, the more realistic an explanatory model is sought to be, the more information is needed to parameterize the model and initialize its driving variables. "If these (data) are unavailable, then the regression model may still be our best option" (Penning de Vries 1983, 128). A common pitfall of statistical models is, however, oversimplification ("everything should be made as simple as possible, but not one bit simpler" [A. Einstein]). Applying simple models to different locations, or even to different years, often fails to yield accurate predictions unless the environment is relatively uniform (Penning de Vries 1983).

Mechanistic constraints can also be imposed upon empirical models (e.g., potential quantum use efficiency based on underlying biophysics can be used to constrain an empirical light-response curve for leaf photosynthesis), thereby using understanding to formulate them. Indeed, most models are neither purely statistical nor mechanistic models but, rather are hybrids between the two, usually containing empirical relationships or parameters, linked according to theory. Freshwater eutrophication models are among the most widely applied in the management context (e.g., Dillon and Rigler 1975; Vollenweider 1976; Smith 1998). The model generally applied is a semiempirical one (Vollenweider 1976), combining regression-derived relationships between chlorophyll a concentration and total phosphorus concentrations, with a simple mass-balance approach to predict the phosphorus concentration in lake waters. Mass-balance approaches, which are essentially empirical models guided by simple, fundamental principles, are widely used in ecological modeling (Chapter 15). The search for the optimal compromise in the complexity and the empirical versus mechanistic components of models must be guided by parsimonious criteria and is a key milestone in the successful development of models (Chapter 8). Indeed complex models should be rigorously evaluated to be made as "minimalistically complex" as possible (Figure 24.1). This objective would allow for a greater focus on a more specific or relevant mechanistic understanding of a given process and would provide a more useful tool to policy makers and managers with a greater chance of being appropriately validated.

The successful formulation of predictions from simple empirical models inspired by theory is eventually assimilated as providing "understanding," defined by Pickett et al. (1994) as an objectively determined, empirical match between some set of confirmable, observable phenomena in the natural world and a conceptual construct. Indeed, many parameters in mechanistic models are but empirically determined quantities that are "understood" in the sense that they have been repeatedly tested and are consistent and linked with the consolidated body of principles or laws in ecology (Figure 24.1).

The preceding discussion clearly indicates that the goals of prediction and understanding are self-supportive and that, as a consequence, they advance in parallel (Figure 24.1). Yet, this parallel advancement is dynamic, involving lags and delays, and may be, therefore, somewhat out of phase. Indeed, empirical predictions derived from simple models that are inconsistent with current theory or mechanisms may be suspect but may also hint at flaws in the current understanding or theory. A mismatch between empirical observations and the desired understanding of underlying mechanisms has been a recurrent stumbling block in science, as illustrated by the long resistance to accept the theories of evolution by natural selection and continental drift simply because they lacked a clear mechanism at the time they were formulated (Greene 2001). Hence, ecologists must be prepared to accept that prediction and understanding may be at times discordant and that the process of model development for prediction is dynamic. Indeed, increased understanding results in model reevaluation and improved predictive capacity. For these reasons, the following sections assess the limits to the use of ecological models for prediction and understanding separately.

Limits to Prediction

Uncertainty in the observational data available to develop the models leads to uncertainty about the model predictions, which imposes a limit on the precision of the predictions (Amthor et al. 2001; Chapter 8). Similarly, uncertainties about the data available to test and validate models lead directly to uncertainty about the accuracy of model predictions (Amthor et al. 2001; Chapter 8). A key related difficulty is that "biological principles which have to form the base of model-building are too fragmentary to embark on straightforward model-building along the same lines as in the physical sciences" (de Wit 1970, 20). Ecological modelers must continuously make compromises to overcome knowledge gaps, and modeling may be reduced to expressions of intuition or educated guesses in some cases. Indeed, de Wit (1970, 23) concluded that success "is only possible when we have the common sense to recognize that we know only bits and pieces of nature around us and restrict ourselves to quantitative and dynamic analyses of the simplest ecological systems." This is not encouraging when the goal is to understand and predict complex entities such as ecosystems.

Models often develop from a need to formulate predictions based on processes that elude direct observation, such as processes that are very slow (e.g., seagrass meadow formation; Duarte 1995), operate at very small spatial scales (e.g., sub-millimeter plankton patchiness; Duarte and Vaqué 1992), or result from episodic or extreme phenomena that are difficult to observe directly (e.g. floods, extreme droughts; Turner et al. 1998, Changnon 2000), as well as future events (e.g., global change; National Assessment Synthesis Team 2000). Lack of adequate observations precludes the assessment of predictive power in these circumstances, so that the model's reliability is entirely dependent on the confidence in its mechanistic basis. The extensive evaluation of model outputs is particularly useful to help assess the models and examine their behavior in detail. For example, the ability of a model to hindcast a historical event such as a drought or a fire improves the model's acceptance for prediction.

Models are, to a variable extent, specific, and their predictive power is restricted to a particular domain (e.g., independent variable range in regression models, particular vegetation type or species), beyond which the reliability of the predictions requires additional testing. Because model components are typically multivariate, their corresponding domains are complex, multivariate spaces that cannot be reliably probed even if multiple validation tests are conducted. The model domain also encompasses the assumptions built into the model (e.g., homogeneity of the variances, steady states, equilibrium conditions, etc.), which must be met by the subject if they are to be applied with any confidence. Any extension of models beyond the domain over which they were originally developed or tested entails uncertainty in the achievable predictive power. Most models predicting NPP were developed for specific biomes, although they have been since applied to other vegetation types or climatic regimes. This extension involves a reexamination of the underlying theory in the model, including its appropriateness to these new biomes, as well as the parameterization of the model parameters for these new biomes. General models do exist, such as the so-called Miami model (equations 12-1 and 12-2 in Leith 1975)—a simple statistical model relating NPP to annual mean temperature and annual precipitation based on field data/estimates from multiple biomes. Such general models, however, have been found to poorly represent the dynamic changes occurring within a particular site, which seem to be due, at least in the case of the Miami model, to time lags in responses of vegetation structure to the changing conditions (Lauenroth and Sala 1992).

Model outputs are not always deterministic, for even simple models can display complex behavior derived from both deterministic and stochastic components of the model. Such results lead to undefined predictions, where very different outcomes are equally likely. These sources of uncertainty are, however, poorly addressed by current sensitivity analysis in ecological modeling, which, by and large, fails to address the consequences of simultaneous variability in the driving parameters and variables, and tests for alternative expressions of ecological processes.

Limits to Understanding

The history of science provides evidence that a model's capacity to accurately predict cannot be used to infer the veracity of the processes underlying a mechanistic model. This implies no more than the acknowledgement that ecological modeling does not escape the generic limitation of science's incapacity to fully prove hypotheses. Yet, the successful application of a model to a variety of different subjects, encompassing the broadest possible ranges in the key traits or variables, would increase the confidence about the robustness of the model's principles. Even if the results of an explanatory ecological model correspond to observations of the system being modeled, "there is room for doubt regarding the correctness of the model" (de Wit 1993, 9). Whenever discrepancies between model output and reality occur, the model may be adjusted ("tuned") to obtain better agreement, and since there are typically many equations and parameters, this is easy to do. It is, however, a "disastrous" way of working because the model then degenerates from an explanatory model into a descriptive model (de Wit 1970). The word "degenerates" does not mean that descriptive models are inferior, but simply that they no longer explain the system. The attribution of explanation to such descriptive models "is the reason why many models made in ecological studies . . . have done more harm than good" (de Wit 1993, 21). The limitation of our fragmented ecological knowledge can and will, therefore, undermine explanatory modeling.

Mechanism-driven models cannot be formulated in cases where there is little or no understanding of the problems or where the empirical base is thin and critical data are missing. In these instances, the contribution of modeling must derive from the development of simple empirical or conceptual models that promote the emergence of sufficient understanding as to render the development of mechanism-driven models possible (Chapter 25).

Increasing model complexity may increase the extent to which models are believed to reproduce nature, but at the same time, they become more open to unexpected behavior. This situation has both potential advantages and disadvantages. One advantage is that a model's prediction of unexpected behavior, if it can be tested against observations, can offer either strong corroboration or rejection of the model. Another advantage is that complex models are likely to produce a variety of outputs (intermediate-level output) other than merely that of the particular variables of interest. These outputs may provide predictions that can be tested against independent data, thus partially checking the model. Unexpected behavior, including erratic predictions, may derive from either deterministic or stochastic processes in the model. The uncertainties in the numerous mechanisms may add or multiply, and, if there is no way of observing intermediate variables in order to correct for this process, lead to a large, and perhaps unknown, amount of uncertainty in the variables of interest and a model that is difficult to interpret or understand (Figure 24.1). This results in the paradox that the more mechanism-rich a model becomes (the more knowledge it incorporates), the more uncertain it becomes (Chapter 2). Much of the art of modeling lies in constructing models that avoid this paradox by producing a variety of outputs that can be tested against as much independent observational data as possible.

Conclusions and Recommendations

Consequences for Model Use

Efforts to promote the use of existing models must be enhanced, because only widespread use of a model can help assess the model's domain and its predictive power under sufficiently contrasting situations. Failures and successes derived from using a model open opportunities for improved model design and reformulation. Therefore, model validation should rely heavily on the widest possible use of the models by the scientific community. Consequently, modelers—and the funding agencies that fund model development—must take all possible steps to make the models widely available as stand-alone products. The users, however, must be fully aware of the model's assumptions, limitations, behavior, and domain of construction, which must be supplied as extensive meta-data accompanying the model. The latter is the responsibility of the model developers (Aber 1997). Obviously, this lack of communication on behalf of model developers has led to the misinterpretation or misuse of their models by able model users (Chapter 13).

Consequences for Model Design

Limited data availability implies that site- or subject-specific models may be possible only for a limited number of subjects (ecosystems, communities, or species). As a consequence, there is little hope that the large-scale problems of ecosystem alteration, habitat loss, and biodiversity erosion may be addressed merely through a mosaic of subject-specific models. Addressing these questions on a large scale will require the development of ecological models aimed at achieving generality at the expense of detail. This does not reduce the usefulness of site- and subject-specific models when data are available for them. Tactical models that address key specific goals, such as the conservation of high-profile endangered species or of critical ecosystem functions, can be designed to be flexible enough to utilize a variety of types of available data (Chapter 5); however, for the most part, more generic models will be necessary.

We contend, therefore, that model design should proceed hierarchically, beginning with simple empirical rules to formulate a parsimonious, generic simple model that eventually results in case-specific applications. A modular construction approach should clearly specify the reduction in the model domain and generality at each development stage, so that the transferability of the models is explicit at each level. Effort allocation in ecological models needs to be readjusted to provide the needed attention to the development of general models, even if these are perceived to be imprecise or provide less understanding.

Model construction must engage experimental ecologists and modelers throughout the entire process from model conception to validation, ensuring, thereby, a more adequate match between data acquisition and model requirements. The full potential of models to achieve synthesis should also be explored further than is currently done, and large research programs will benefit from such exercises, which should be facilitated by program managers. An improved feedback between statistical and mechanistic models must be established. The development of ecosystem models will also greatly benefit from increased interdisciplinary connections (Figure 24.2), which would allow the rapid implementations of new approaches to modeling complex systems—the stone against which many disciplines (physics, health science, etc.) are presently stumbling—

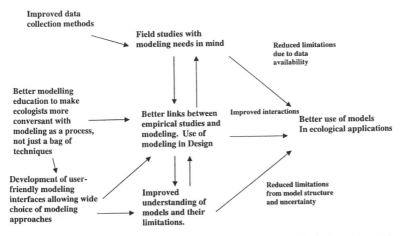

Figure 24.2. Network of suggested pathways to alleviate present limitations in model development.

and to analyze model output and behavior, as well as to better represent human intervention in ecological systems.

In summary, the present demand for predictions in ecology implies that models must be at the forefront of ecology. The limitations to model development and achievements outlined here are largely derived from poor coordination within the scientific community. A commitment to the participation in model construction and validation by the entire community is, therefore, required (Chapter 27). This effort must extend beyond the boundaries of ecology to benefit from inputs and progress in other disciplines (Figure 24.2). Ecologists must also be sufficiently educated as to be able to understand models and model requirements and to formulate, at least, simple models (Chapter 23; Figure 24.2). Progress in model development is also likely to benefit from a better coordination, as opposed to competition, between the development of simple statistical models and their integration into mechanistic models. These actions require a substantial effort from all ecologists. Benefits will become apparent in the form of better synthesis, through modeling, and a better service to society through the use of the models for the effective preservation of biological diversity and ecosystem function.

References

Aber, J.D. 1997. Why don't we believe in models? *Bulletin of the Ecological Society of America* 78: 232–233.

Aber, J., and J. Melillo. 2001. *Terrestrial Ecosystems.* 2d Ed. Burlington, MA: Harcourt/Academic Press.

Amthor J.S., J.M. Chen, J.S. Clein, S.E. Frolking, M.L. Goulden, R.F. Grant, J.S. Kimball, A.W. King, A.D. McGuire, N.T. Nikolov, C.S. Potter, S. Wang, and S.C. Wofsy. 2001. Boreal forest CO_2 exchange and evapotranspiration predicted by nine ecosystem process models: Intermodel comparisons and relationships to field measurements. *Journal of Geophysical Research–Atmospheres* 106 (D24): 32,623–33,648.

Bridgman, P.W. 1927. *The Logic of Modern Physics.* New York: MacMillan.

Cartwright, D.E. 2001. *Tides: A Scientific History.* New York: Cambridge University Press.

Coops, N.C., and R.H. Waring. 2001. The use of multi-scale remote sensing imagery to derive regional estimates of forest growth capacity with 3-PGS. *Remote Sensing of Environment* 75: 324–334.

Changnon, S.A. 2000. Flood prediction: Immersed in the quagmire of national flood mitigation strategy. Pp. 85–106 in D. Sarewitz, R.A. Pielke, and R. Byerly Jr., editors. *Prediction: Science, Decision Making, and the Future of Nature.* Washington, DC: Island Press.

de Wit, C.T. 1970. Dynamic concepts in biology. Pp. 17–23 in I. Setlik, editor. *Prediction and Measurement of Photosynthetic Productivity.* Wageningen, the Netherlands: Center for Agricultural Publishing and Documentation.

————. 1993. Philosophy and terminology. Pp. 3–9 in P.A. Leffelaar, editor. *On Systems Analysis and Simulation of Ecological Processes.* Dordrecht: Kluwer.

Dillon, P.J., and F.R. Rigler. 1975. A simple method for predicting the capacity of a lake for development based on trophic state. *Canadian Journal of Fisheries and Aquatic Science* 31: 1518–1531.

Duarte, C.M. 1995. Submerged aquatic vegetation in relation to different nutrient regimes. *Ophelia* 41: 87–112.

Duarte, C.M., and D. Vaqué. 1992. Scale dependence of bacterioplankton patchiness. *Marine Ecology Progress Series* 84: 95–100.

Greene, M. 2001. Mechanism: A tool, not a tyrant. *Nature* 410: 8.

Lauenroth, W.K., and O.E. Sala. 1992. Long-term forage production of North American shortgrass steppe. *Ecological Applications* 2: 397– 403.

Law, B.E., R.H. Waring, P.M. Anthoni, and J.D. Aber. 2000. Measurements of gross and net ecosystem productivity and water vapor exchange of a *Pinus ponderosa* ecosystem, and an evaluation of two generalized models. *Global Change Biology* 6: 155–168.

Lehman, J.T. 1986. The goal of understanding in limnology. *Limnology and Oceanography* 31: 1160–1166.

Leith, H. 1975. Modeling the primary productivity of the world. Pp. 237–263 in H. Leith and R.H. Whittaker, editors. *Primary Productivity of the Biosphere.* New York: Springer-Verlag.

Loomis, R.S., R. Rabbinge, and E. Ng. 1979. Explanatory models in crop physiology. *Annual Review of Plant Physiology* 30: 339-367.

National Assessment Synthesis Team. 2000. *Climate Change Impacts on the United States: The Potential Consequences of Climate Variability and Change, Overview.* U.S. Global Change Research Program. New York: Cambridge University Press.

Pace, M.L. 2001. Prediction and the aquatic sciences. *Canadian Journal of Fisheries and Aquatic Science* 58: 63–72.

Penning de Vries, F.W.T. 1983. Modeling of growth and production. Pp. 117–150 in O.L. Lange, P.S. Nobel, C.B. Osmond, and H. Ziegler, editors. *Plant Physiological Ecology.* Vol. 4. Berlin: Springer-Verlag

Peters, R.H. 1986. The role of prediction in limnology. *Limnology and Oceanography* 31: 1143–1159.

———. 1991. *A Critique for Ecology.* New York: Cambridge University Press.

Pickett S.T.A., J. Kolasa, and C.G. Jones. 1994. *Ecological Understanding.* San Diego: Academic Press.

Rastetter, E.D., G.I. Ågren, and G.R. Shaver. 1997. Responses of N-limited ecosystems to increased CO_2: A balanced nutrition, coupled element-cycles model. *Ecological Applications* 7: 444–460.

Running, S.W. 1994. Testing FOREST-BGC ecosystem process simulations across a climatic gradient in Oregon. *Ecological Applications* 4: 238–247.

Running, S.W., P.E. Thornton, R.R. Nemani, and J.M. Glassy. 2000. Global terrestrial gross and net primary productivity from the earth observing system. Pp. 44–57 in O. Sala, R. Jackson, and H. Mooney, editors. *Methods in Ecosystem Science.* New York: Springer-Verlag.

Schimel, D., J. Melillo, S.W. Running, et al. 2000. Contribution of increasing CO_2 and climate to carbon storage by ecosystems in the United States. *Science* 287: 2004–2006.

Smith V.H. 1998. Cultural eutrophication of inland, estuarine and coastal waters. Pp. 7–49 in M.L. Pace and P.M. Groffman, editors. *Successes, Limitations, and Frontiers in Ecosystem Science.* New York: Springer-Verlag.

Turner, M.G., W.L. Baker, C.J. Peterson, and R.K. Peet. 1998. Factors influencing succession: Lessons from large, infrequent natural disturbances. *Ecosystems* 1: 511–523.

VEMAP Members.1995. Vegetation/ecosystem modeling and analysis project: Comparing biogeography and biogeochemistry models in a continental-scale study of terrestrial ecosystem responses to climate change and CO_2 doubling. *Global Biogeochemical Cycling* 9(4): 407–437.

Vollenweider, R.A. 1976. Advances in defining critical loading levels for phosphorus in lake eutrophication. *Memorie dell'Istituto Italiano di Idrobiologia* 33: 53–83.

Waring, R., and S.W. Running. 1998. *Forest Ecosystems: Analysis at Multiple Scales.* San Diego: Academic Press.

Part V

Concluding Comments

25

The Need for Fast-and-Frugal Models

Stephen R. Carpenter

Fast-and-frugal ecosystem models are relatively simple, inexpensive to build and analyze, easily confronted with data, and disposable. Such models accelerate learning, especially by those who participate in the model building but also by the ecosystem science community as a whole. In my opinion, fast-and-frugal models are underused in ecosystem ecology, could be employed more widely, and could significantly accelerate progress.

What Is a Model?

Understanding requires abstraction, or consideration of one aspect of the subject while ignoring others (Heisenberg 1974). A familiar case is an object in a gravitational field, which can be represented by an infinitesimal point having all the mass of the object, regardless of shape, size, or density. The best abstractions capture multiple insights and meanings in relatively simple form (Root-Bernstein and Root-Bernstein 1999).

A model translates an abstraction into something we can manipulate (Root-Bernstein and Root-Bernstein 1999; Starfield et al. 1994), and through the process of manipulation, we learn about the abstraction, the model, and relevant aspects of the world. Diverse media can be used to create models, including poems, stories, verbal descriptions, sculptures, pictures, or mathematics. Pictures—such as conceptual diagrams, flow charts, and graphics—are especially important models in ecosystem science. Mathematical models are also central. Ecosystem data are mostly quantitative, and mathematical models enable us to confront abstractions with quantitative data.

The abstract notion of competition for limiting resources governed by conservation of mass is considered by Pastor (Chapter 15). Using mathematics, Pastor translated this abstraction into models of plant communities that can be calibrated and compared with data. Manipulation of models allowed Pastor to evaluate the abstraction in comparison with alternative ideas. Thus, he reached certain conclusions about the ecosystems for which the abstraction was useful and informative. In addition, Pastor's paper showed that the evaluation of the abstraction is inseparable from the evaluation of the models. Thus, the invention of models that artfully capture abstractions is an important craft in its own right.

In order to be useful, a model must be a simplification of the ecosystem it portrays. In statistical terms, the model must have fewer degrees of freedom than the ecosystem. Therefore, the model is certainly wrong in the sense that it cannot perfectly represent ecosystem behavior. Nevertheless, the model may be valuable if it allows us to understand or predict important features of the ecosystem. It is trivial and useless to discover that a model is wrong. It is important and useful to discover surprising, systematic discrepancies that lead to revised abstractions and to new models that are simple, effective, and better. The faster this process occurs, the faster we learn.

Cycles of Learning

Many writers have noted the cyclic or iterative nature of scientific learning. Box (1980) separates learning cycles into phases of estimation and criticism. Estimation is the calculation of model parameters from data under the assumption that the model accurately represents the ecosystem. The estimated parameters can be used to make predictions of other data, which are different from those used to estimate the parameters. Criticism is the confrontation of this new data with models. It can lead to revisions of the model or to drastic changes in the research program, such as consideration of completely new abstractions. Algorithms can be written for the cycle of estimation, criticism, and revision (Box 1980; Chapter 12). In contrast, the creation of novel, innovative models cannot be engineered, although certain modes of thinking seem likely to foster the creation of novel models in ecosystem science (Fisher 1997; Root-Bernstein and Root-Bernstein 1999; Holling and Allen 2002).

I distinguish two types of trajectories for learning about ecosystems, named r and K after the parameters of the logistic equation for population growth (Figure 25.1). To understand these trajectories, at least three variables are helpful. The rate of change in models is a quantitative index of the scientific learning rate (Brock and Durlauf 1999). It could be measured as the inverse of the time elapsed between adoption of a model and its replacement by a new model. Model complexity could be measured by the number of distinct structures or processes represented in the model. In statistical curve fitting, this is just the number of parameters. In ecosystem models, complexity might be indexed by the number of types of state variables or by the number of types of processes. Note that a model could have low complexity if it consisted of many replicates of the same relatively simple structure (Chapter 5). The degree of satisfaction with a model is related to the extent to which the model meets the needs of scientists or other users and could be measured by the proportion of scientists (and/or managers) in the field who actively use the model.

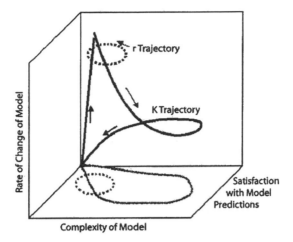

Figure 25.1. Contrasting trajectories (r, dashed line, and K, solid line) for model development in ecosystem science. Arrows show movement over time in relation to three variables: rate of model change, model complexity, and degree of satisfaction with model predictions. Black lines show three-dimensional trajectories, and gray lines show shadow of the trajectories on the bottom plane.

In the r-trajectory, models are replaced rapidly. Model complexity is low and so, therefore, is the cost of discarding a model. The satisfaction index is also generally low. Only a few scientists, possibly close collaborators, may use the model. Also, frequent confrontations with data lead to rapid mortality of models. Occasionally, however, a model may yield important insights: it may reveal surprising ecosystem patterns or connections or capture a powerful idea in an especially useful or evocative way. These nuggets are the payoff from working with fast-and-frugal models. Pace (Chapter 4) discusses an r-trajectory, or fast-and-frugal, style of modeling, and DeAngelis (1992) and Scheffer (1998) illustrate the richness that can emerge from repeated cycles on the r-trajectory. Many other examples could be selected from a diverse literature.

In the K-trajectory, a model is conserved and elaborated. Complexity grows, and with it grows the cost of discarding the model. The satisfaction index can increase as more and more scientists adopt the model. Government agencies or other clients of science may adopt the model and expand the community of users, and the model may even become an "industry" for scientists who depend on it for funding. Such dependencies buffer the impact on the model of confrontations with data. Because sustaining the model is an overriding concern, ad hoc fixups may be preferred to wholesale changes. The former tend to complexify the model, perhaps increasing incidence of problems (such as bugs in the software) and decreasing the transparency of the model to other users, and eventually these issues may erode satisfaction with the model. At this point, the model is vulnerable to breakdown (movement to zero satisfaction followed by creation of a new model), which may be triggered by many events or factors, such as retirement of a key individual, appearance of a superior rival model, or a shift in research objectives away from the processes represented in the model.

The K trajectory is already well represented in ecosystem ecology and will become more common as the demand for ecosystem forecasts grows. Ecosystem ecology has become competent at forecasting—enough so that our tools are in demand by mission agencies of many governments and by diverse international organizations (Clark et al. 2001). In many respects, this is a desirable and natural consequence of the growth of our discipline. Many of the papers and commentaries at this Cary Conference have addressed models on the K trajectory, which are represented by computer programs that have become familiar names in the discipline. Such computer models require considerable specialized expertise, evolve through an ongoing process of new computational features and assimilation of new data, and in some cases have become the "gold standard" for subdisciplines of ecology.

This success comes at a cost. There is a growing need for quality assurance and quality control (Chapter 11). While there are good reasons to promulgate standards, they are a form of overhead. Development and enforcement of standards divert human resources and funding from primary scientific research to model maintenance. In addition, the K-trajectory tends to create a cadre of specialists focused on the issues of a particular model, thus separating the modelers from other groups of ecosystem ecologists. Many commentators at the Cary Conference were concerned about this division, and some discussion groups considered mechanisms for increasing the communication between modelers and other ecosystem ecologists (Chapter 23). These changes in the community are perhaps inevitable and must be accommodated as ecology assumes greater importance in societal problem solving. Nonetheless, many important ecosystem problems remain better addressed through the r-trajectory.

Fast-and-Frugal Models

The r-trajectory selects for fast-and-frugal models, that is, models that can be developed rapidly and cheaply and discarded readily. Fast-and-frugal models have several advantages. With them, learning is rapid. Models can be analyzed quickly, and therefore researchers can quickly determine which features of the model are consistent with data and useful for understanding. Simple models that prove useful often become valuable teaching cases. Fast-and-frugal models are well suited to work of small interdisciplinary teams in which most ecosystem science is performed (Pace and Groffman 1998). Simple models are more likely to be transparent to diverse users; thus, they engage more people in modeling. Experience shows that those who actively participate in modeling learn much more than those who merely read about it (Carpenter 1992).

Are fast-and-frugal models underused in ecosystem ecology? At the Cary Conference, fast-and-frugal models were less common than K-trajectory models. This may or may not be an accurate sampling of the discipline. Because r-trajectory models suffer high mortality, fewer of them may be published. Increasing pressure on ecosystem ecology to provide predictions needed by other disciplines or for environmental problem solving is likely to increase investment in K-trajectory models. Quite possibly this could lead to a decline in the amount of time that ecosystem ecologists spend with fast-and-frugal models.

What can be done to increase the frequency of fast-and-frugal modeling in ecosystem ecology? Many of the reforms advocated at the Cary Conference will move the field in this direction (Chapters 22 and 23). It is especially important that interactive, skilled modelers participate actively in interdisciplinary research, an enterprise requiring dedicated individuals and a willingness by all parties to actively exchange ideas rather than partition jobs along disciplinary lines. Several papers in Pace and Groffman (1998) discuss the formation of actively interdisciplinary teams in ecosystem science.

Most importantly, funding panels and agencies should recognize the importance of fast and frugal modeling by allocating salaries for modelers who interact productively in interdisciplinary projects. The rate of learning driven by the r-trajectory more than justifies this expenditure. Interactive, productive, collaborative modelers are as important to the success of ecosystem science as are biogeochemists and autoanalyzers.

Acknowledgments. I thank the Resilience Network for discussions of modeling and creativity in ecology and the National Science Foundation and A.W. Mellon Foundation for research support.

References

Box, G.E.P. 1980. Sampling and Bayes' inference in scientific modeling and robustness. *Journal of the Royal Statistical Society, Series A* 143: 383–430.

Brock, W.A., and S.N. Durlauf. 1999. A formal model of theory choice in science. *Economic Theory* 14: 113–130.

Carpenter, S.R. 1992. Modeling in the Lake Mendota program: An overview. Pp. 377–380 in J.F. Kitchell, editor. *Food Web Management: A Case Study of Lake Mendota.* New York: Springer-Verlag.

Clark, J.S., S.R. Carpenter, M. Barber, S. Collins, A. Dobson, J. Foley, D. Lodge, M. Pascual, R. Pielke Jr., W. Pizer, C. Pringle, W. Reid, K. Rose, O. Sala, W. Schlesinger, D. Wall, and D Wear. 2001. Ecological forecasting: an emerging imperative. *Science* 293: 657–660.

DeAngelis, D.L. 1992. *Dynamics of Nutrient Cycling and Food Webs.* London: Chapman and Hall.

Fisher, S.G. 1997. Creativity, idea generation, and the functional morphology of streams. *Journal of the North American Benthological Society* 16: 305–318.

Heisenberg, W. 1974. *Across the Frontiers*. Translated by Peter Heath. New York: Harper and Row.

Holling, C.S., and C.R. Allen. 2002. Adaptive inference for distinguishing credible from incredible patterns in nature. *Ecosystems* 5(4): 319–328.

Pace, M.L., and P.M. Groffman, editors. 1998. *Successes, Limitations, and Frontiers in Ecosystem Science*. New York: Springer-Verlag.

Root-Bernstein, R., and M. Root-Bernstein. 1999. *Sparks of Genius*. New York: Houghton-Mifflin.

Scheffer, M. 1998. *Ecology of Shallow Lakes*. London: Chapman and Hall.

Starfield, A.M., K.A. Smith, and A.L. Bleloch. 1994. *How to Model It*. Edina, MN: Burgess International.

26

On the Benefits and Limitations of Prediction

Ann P. Kinzig

As I sit down to these comments, I am reminded of a story, perhaps apocryphal, of a press phalanx awaiting the conclusion of a closed-door congressional committee meeting that was running well past its anticipated adjournment. Finally, a spokesperson emerged. "Members of the press, we have reached the point where everything has been said, but not everyone has said it." And so the meeting continued.

Much has already been said, and said well, in this volume concerning the role of quantitative modeling in ecosystem ecology. My colleagues have reviewed the literature and offered new insights into how models can be used to advance the field by guiding observations, promoting synthesis, and fueling prediction. They note the benefit of models for spanning the temporal and spatial scales not amenable to experiment and for allowing a multiplicity of "experiments" (simulations) that might prove logistically and economically impossible to conduct on the ground. There are chapters in the book devoted to the use of models in education, management, and policy and chapters devoted to the advances in validation and verification that are needed to better understand and convey the limits and applicability of model output.

While there are thus many topics to choose from and little I can add to any of them, I will concentrate in these few pages on the role of modeling in prediction. This focus seems particularly timely given the forecasting initiatives emerging from government agencies and the broader ecological community (Clark et al. 2001). These initiatives rest not just on the traditional push for prediction in science—prediction in order to test and verify (or reject) hypotheses —but also on society's need to better understand possible future ecological trajectories in order to make better decisions today. This means that those using models for forecasting need not only grapple with all the usual challenges of representing complex systems but also must tackle the challenges of conveying model results to policymakers and other members of the public.

I will therefore focus on three points in the remainder of this commentary, including

- the requirements of, and limits to, prediction in complex and human-dominated systems,

- the use of predictive models in policy, and

- the potential for predictive models to foster a more inclusive conversation among scientists and citizens.

Oreskes (Chapter 2) began Cary Conference IX by noting that the natural sciences that deal with complex systems have no notable record of success in prediction: "Science has been successful at prediction only when dealing with short-duration repetitive systems in which the controlling variables are known and measurable." Forecasting ecological dynamics far into the future may be even more problematic. Even in the absence of humans, ecological systems are highly complex, with dynamics and feedbacks spanning multiple spatial and temporal scales (Levin 1999). In addition, though, all ecological systems are now, to some extent, structured and impacted by human actions (Vitousek et al. 1997). These actions can continue to influence ecosystem processes and structures for centuries or even millennia after cessation (Turner et al. 1990, Redman 1999). Thus, both near-term and distant ecological trajectories will depend on the vagaries of human behaviors and actions, which are notoriously resistant to prediction.

I would therefore assert that for most of the interesting cases of long-term ecological dynamics, the best we will be able to do is to simulate possible ecological dynamics *given* a priori social, political, and economic scenarios. In some cases, we might be able to incorporate some simple feedback loops—to understand, for instance, how ecological and economic systems might co-evolve. I would argue that neither is a forecast. The scenarios of human activity will likely be without the probabilities or quantitative uncertainties that are implicit to a forecast, and thus the ecological predictions based on these scenarios cannot have unconditional probabilities and quantitative uncertainties assigned to them.

This is not an argument against engaging in ecological prediction, however much it might sound like one. The advances in scientific understanding that have accompanied the development of models for prediction have, in many cases, been immense, and there is no reason to believe these intellectual payoffs will not continue for integrated ecological models. We will have a better understanding of our own field if we engage in prediction and scenario building. This is reason and justification enough for engaging in these activities. Given the poor record of prediction in the sciences dealing with complex systems, however, we should err on the side of humility with regard to our own abilities and on the side of awe with respect to the intricacies of the earth's human and environmental systems.

There is a second reason to promote advances in modeling and scenario building stemming from society's need to make better-informed natural-resource decisions today. This need may motivate some (though it need not motivate all) of the scientists engaged in simulating future ecological trajectories. When the results of predictive models or scenarios are introduced into the decision-making process, however, we must be very clear about what we are providing, and what the challenges and limitations of that provision have been. Further, the words we attach to those accomplishments (e.g., "forecast," "prediction") must resonate with the decision makers in the same way they resonate with the scientific community. This is not merely a matter of semantics but rather a matter of communicating across ideological divides, where perceptions of prediction and the role of science in decision making can differ fundamentally. This topic is covered in more detail, and more eloquently, in the chapter by Pielke (Chapter 7).

Some at Cary Conference IX argued that, in some cases, our lack of predictive success equated to a lack of understanding that should prevent us from introducing our scientific conclusions into the policy process. I must disagree. To assume that no scientific information—rather than a thoughtful articulation of what is (however paltry) and is not (however immense) known—would better serve the policy process would seem to be equivalent to suggesting that withholding information could prevent scientifically ill-advised decisions from being made. Decisions will be made anyway, and they may as well be informed by the best available scientific information, even if that information is highly uncertain.

Models are a particularly useful way to convey the necessary messages about scientific uncertainty. These can either be the fast-and-frugal models discussed by Carpenter (Chapter 25) or more detailed models. Merely demonstrating how simple changes in mechanistic assumptions or quantitative variations in input parameters can lead to differing outcomes can help policy makers grasp the relative resilience of some systems and the potential for surprises or catastrophic change in others. It is difficult to imagine how we might otherwise convey the complicated and conditional behaviors of systems governed by thresholds and other nonlinear responses.

There is much more to be said on the matter of conveying uncertainty to policy makers and the public (and much of it covered in this volume—see particularly Chapters 7, 8, and 9). This hurdle must be cleared if scientific information is to be used more effectively in the policy process. I believe it can be cleared: policy makers and the public make decisions in the face of uncertainty everyday, and they can do the same with scientific uncertainty if we improve our ability to understand, frame, and convey that uncertainty. This is not to suggest, however, that decisions will or even should be governed solely by the proffered scientific information. None of today's most compelling ecological problems can be solved purely based on science. Scientific information can help determine the realm of physically and biologically plausible outcomes, but those outcomes will differ not only in terms of ecological conditions, but also in terms of resource distributions, property rights, equity, justice, and aesthetics, among other things. Scientists are no more expert than other citizens when navigating among these considerations. Potential outcomes therefore can really be differentiated only by society's values and not by scientific assessment, thus making policy makers and the public more generally the final arbiters of environmental trajectories.

But it is here, finally, that I think models may have one of their greatest values. Models can be complicated and tricky to build, but the public has displayed an astonishing capacity to grasp models, formulate simple (fast-and-frugal) models, and use them to explore local or regional ecological trajectories (see, for instance, Margerum 1999 and Ruth and Lindholm 1996). In formulating the models (and the engagement of stakeholders as well as "trained modelers" in this step is crucial), the public is forced to examine and acknowledge the complex interconnectedness of system components and to contemplate how changes in one area have historically engendered changes in other areas. In observing future trajectories, the public increases its awareness of how today's policies can reverberate through time. In trying to engineer desirable outcomes, the players become increasingly aware of how insistence on maximizing one resource or service can compromise other valuable resources and services, better illuminating the often difficult choices that must be made to reconcile competing desires, uses, and needs. And in playing out multiple scenarios, the public begins to glean that even well-meaning policies and management actions can lead to surprises or catastrophes and that the best approaches for coping with environmental problems require flexibility for responding to changing conditions and values.

We need a scientifically educated citizenry to best grapple with today's—and tomorrow's—environmental problems. We also need scientists who are aware of the multitude of perspectives and values that legitimately compete in the decision-making arena. Simple models provide a productive and proven means of facilitating that needed communication and education. They can also serve as the springboard for more complicated models that can be used in scenario building of greater sophistication and that can be used to test the limits of what we do and don't know about ecological processes. We haven't exhausted the limits or even tapped the potential of the use of models in ecosystem ecology.

References

Clark, J., S. Carpenter, M. Barber, S. Collins, A. Dobson, J. Foley, D. Lodge, M. Pascual, R. Pielke Jr., W. Prizer, C. Pringle, W. Reid, K. Rose, O. Sala, W. Schlesinger, D. Wall, and D. Wear. 2001. Ecological forecasts: An emerging imperative. *Science* 293: 657–660.

Levin, S.A. 1999. *Fragile Dominion: Complexity and the Commons.* Reading: Perseus Books.

Margerum, R.D. 1999. Integrated environmental management: The foundations for successful practice. *Environmental Management* 24(2): 151–166.

Redman, C.L. 1999. *Human Impact on Ancient Environments.* Tucson: University of Arizona Press.

Ruth, M., and J. Lindholm. 1996. Dynamic modeling of multispecies fisheries for consensus building and management. *Environmental Conservation* 23(4): 332–342.

Turner, B.L., W.C. Clark, R.W. Kates, J.F. Richards, J.T. Mathews, and W.B. Meyer, eds. 1990. *The Earth as Transformed by Human Action.* Cambridge: Cambridge University Press.

Vitousek, P.M., H.A. Mooney, J. Lubchenco, and J.M. Melillo. 1997. Human domination of Earth's ecosystems. *Science* 277: 494–499.

A Community-Wide Investment in Modeling

Dean L. Urban

In keeping to the symmetry of these three-part commentaries, I will focus on three general points: the role of models in ecosystems science, the modeling community, and the need for education about modeling. I close with a call for more support for the huge amount of work implied by the recommendations made here.

Ecosystem Science *Is* Models

Models play several roles in ecosystem science: they are the basis upon which we organize our ideas about ecosystems, they allow us to explore these ideas formally through analysis, and they are the means by which we communicate these ideas among ourselves and, increasingly, to a client public.

In a very real sense, what we do as ecosystem scientists is a legacy from a previous era of ecosystems studies, the International Biological Programme (IBP), which defined the important questions to be asked of ecosystems and codified the tools used to address these questions. This era was largely defined by a community-wide investment in models, and I would like to emphasize two points on this theme of investment. First, the investment was communal and general, meaning that it was made by funding agencies as well as by nonmodelers (to which I will return momentarily). Second, the investment was in complex models developed to organize our thoughts and data; many of these models, by the strictest definitions we have discussed at this meeting, did not work: they did not provide predictions that matched independent data. Rather, the models often served as integrating frameworks, and in this role they were so successful that it would be almost impossible to divorce the current state of ecosystem science from its beginnings in model-based inquiry. Ecosystems science *is* models.

So, we should bear in mind the various roles that models can serve, from very informal organizing frameworks to rigorous analytic tools to decision support and communication devices for management and policy. All of these functions are valuable—even crucial—to a healthy scientific discipline.

The Modeling Community

The distinction between "modelers" and "nonmodelers" as occupying two distinct camps is immensely troubling. If we recognize modeling as an integral component of ecosystem science, then *everyone* in the field must be a modeler in some sense. By this I do not mean that everyone needs to learn a programming language and cut some code, but rather that everyone needs to be conversant about models—to be savvy about modeling (Chapters 22 and 23). To recognize two camps is to suggest that "modeling is best left to modelers," and this attitude is clearly poisonous to our larger mission.

As an analogy, we might consider the role of statistics in our field. Few of us would claim to *be* statisticians, but we all are invested collectively in statistics. This investment is evidenced in several important ways. We accept (even demand) statistics where applications properly call for them as decision support tools. None of us would accept an author's inferences about data without some sort of statistical justification. For familiar applications we share a standard, community-wide expectation for which statistics are used and how they are reported. For example, we all know when a particular test is called for, and we know what to report (e.g., the value of the test statistic, degrees of freedom, and P value). And when novel applications call for novel statistical tests, we are willing to entertain innovations—so long as they are presented in a compelling fashion that clearly motivates the test and demonstrates that it does what it was designed to do (and does so better than an existing, more familiar method).

We need to engender this same sense of community-wide investment in and expectations of models in ecosystem science. And whose job is it to engender this community-level set of standards and expectations? Ours alone.

Modeling School and Continuing Education

Our challenge is to train ourselves, our students, and our clients to be savvy about models. How do we proceed? Clearly, a multifaceted solution will be required for this multifaceted challenge. We have already heard some excellent suggestions from the working groups on education (Chapters 22 and 23), but it will be important to recognize that different solutions will be appropriate for these three audiences.

First, it might be worth noting that active modelers—the folks actually cutting code—have a pressing need for continuing education about modeling. In ecosystem science especially, some models have been in use for so long that they are essentially historical artifacts. Recall that the heyday of modeling during the IBP era was a time when computer simulation was quite young. Advancements in computer data structures, algorithms, and even languages have come at a remarkable pace in the past few years, and none of us can be so complacent as to ignore these changes. This is not to argue that existing models should be recoded simply because they can be, but rather to point out that there are technologies available today that might make our lives much easier. For example, in the early 1980s I wrote an individual-based bird metapopulation model. The application is clearly a candidate for agent-based simulation using object-oriented code, yet I wrote it in FORTRAN because that was what scientists used in those days (C++ was developed years later). In this case, the power of object-oriented programming easily justifies its use for similar applications today. Similar advancements in event-based processing and data assimilation might be very useful in certain ecosystems applications. Likewise, advancements in data structures and processing algorithms could facilitate simulation models designed to work at very large spatial scales, such as next-generation dynamic global vegetation models. So long as ecosystem models are designed mostly by professional ecosystem scientists who also are good at computing— rather than by professional computer scientists who also are interested in ecosystems—there will be a strong need for continuing education of modelers in our field.

I emphasize continuing education first because it is quite likely that the next generation of ecosystems modelers will be trained by the current working generation (their academic advisors and colleagues), much as this generation was. Yet, in many ways, the training of the next generation is the more daunting task. We have heard a number of recommendations that might facilitate this task (Chapters 22 and 23). These include:

- centers of excellence that could serve as training facilities, offering short courses or semester-long classes on modeling. Certainly, existing programs such as NSF's Integrated Graduate Education and Research Training (IGERT) and Research Experience for Undergraduates (REU) programs could be extended to include modeling. Indeed, opening an IGERT program to visiting students from other institutions would seem an especially easy and effective extension.

- a "visiting scholar" program that allows graduate students to attend other schools specifically to learn modeling techniques also merits some consideration. A student could attend another school for a semester or two, taking either modeling courses or simply learning one-on-one with a mentor.

- self-instructional tutorials for modelers, including packaged programs, demonstration data sets, and exercises geared to teach fundamentals such as sensitivity and uncertainty analysis, error propagation, and model testing protocols. While perhaps not as attractive as on-site learning at a center or modeling institution, we could certainly take more advantage of the growing capabilities for electronic "distance learning" technology using Web-based tools.

- increased accessibility and use of standardized "toolkits" for modeling. Part of an effort to come to consensus on community-level standards for modeling might include an effort by modelers to take advantage of widely accessible, generic modeling components and approaches. This would facilitate model development as well as the communication of modeling results. This standardization might be expressed at a variety of levels. At one level, equivalent to the "formalization" stage of model development, one might simply adopt a common approach to a particular problem ("I used the Farquhar photosynthesis model" or "I used Bonan's implementation of the Priestley-Taylor estimate of PET"). Doing so implies a community-level ratification of these standard approaches, much as we have ratified the t test and other familiar statistics to a point where these need not be explained in detail to an audience of peers. At a far more detailed level, we might embrace modeling conventions such as the Standard Template Library (STL) in the C++ programming language, which is a set of generic functional prototypes that can be adapted to various uses. This level of standardization is probably unrealistic, but some intermediate compromise might be worth pursuing.

These suggestions focus on teaching the techniques of actually developing and using models. Yet I have been careful to stress that a community-level investment in modeling requires that these techniques, and modeling in general, be conveyed to students who are not likely to actually use models personally. This is a group we would like to train to be savvy consumers of models while we also emphasize the fundamental role of modeling in ecosystem science in general. To this end, a somewhat different approach to education is warranted. While "modeling techniques" is a specialized subject best addressed in a specialized class, modeling itself cannot be quarantined to these special-interest classes. We must strive to infuse models and modeling into introductory courses and at all levels of instruction (Chapter 22). How this happens will depend, of course, on the details of particular case and subjects. In my own teaching, I find it easy to use models as a vehicle to present alternative conceptual notions in a very precise way. For example, alternative ideas about how animals might disperse among patchy habitats are readily communicated in modeling terms. Our students must be presented with the clear message that models are a fundamental part of our science.

Finally, we must attend the education of the nonscientists who are the ultimate consumers of our models, the managers and policy makers whose actions are guided by our research models. Again, this clientele need not be beleaguered with the technical details of how models are implemented and exercised, but they do need to be taught why the model does what it does, where the crucial uncertainties lie, which results are robust and confident, and how to use this information. In this arena, we have perhaps done ourselves some disservice by failing to do this properly in the past. Thus, as John Aber has noted, we often tend to mistrust models in general. Adopting a community-wide set of standards for the publication and presentation of models should go far to redress this (Chapter 11).

Thanks For Your Support . . .

Clearly, we have a great deal of work to do. A huge amount of effort is implied by the recommendations above. At present, there is little support and indeed, little incentive for modelers to invest the energy to pursue meaningful solutions to some of these issues. For example, there is a huge difference between coding a model so that I can use it, and coding a model so that others can use it easily. The difference lies in error-trapping the code, documenting its use, and perhaps providing a user-friendly interface. Yet, there is no incentive for a modeler to invest the extra effort to do this, since we are graded on our research productivity (publications) rather than our providing useful tools to our peers. Similarly, we rarely see publications with complete sensitivity analyses of complicated simulation models, partly because few editors would entertain such a manuscript for publication. Some of our most prestigious journals will publish model applications but prefer that we publish the "gory details" elsewhere. The solution to this peer-level problem is also peer-level, based again on a community-level consensus that these activities are worthy of support and publication.

   Finally, it seems to have grown increasingly difficult to secure funding in support of "pure modeling" research (i.e., research on the tools and techniques of modeling), despite a growing consensus that such research is desperately needed. An important product from this Cary Conference should be the clear message to our funding agencies, that we strongly support a sustained investment in modeling in all its stages—from initial conceptualization to implementation to analysis and testing—and at all levels—from student training to methodological research to dissemination of modeling expertise to client consumers.

Index